AF324126

Mechanics of Functionally Graded Material Structures

Mechanics of Functionally Graded Material Structures

Isaac Elishakoff
Florida Atlantic University, USA

Demetris Pentaras
Cyprus University of Technology, Cyprus

Cristina Gentilini
University of Bologna, Italy

World Scientific

NEW JERSEY · LONDON · SINGAPORE · BEIJING · SHANGHAI · HONG KONG · TAIPEI · CHENNAI · TOKYO

Published by

World Scientific Publishing Co. Pte. Ltd.

5 Toh Tuck Link, Singapore 596224

USA office: 27 Warren Street, Suite 401-402, Hackensack, NJ 07601

UK office: 57 Shelton Street, Covent Garden, London WC2H 9HE

British Library Cataloguing-in-Publication Data
A catalogue record for this book is available from the British Library.

MECHANICS OF FUNCTIONALLY GRADED MATERIAL STRUCTURES

ISBN 978-981-4656-58-0

Printed by FuIsland Offset Printing (S) Pte Ltd Singapore

Dedicated to blessed memories of my sister
Lamara (Dvorah) and brother Moshe.
I.E.

Dedicated to my newborn son Kyriacos.
D.P.

Dedicated to my beloved family.
C.G.

Preface

This book deals with the static and dynamic modelling of structures with variable parameters. A variable parameter is a parameter that can assume different values at different points of the structure. In engineering practice, there are designed variabilities and inherent variabilities.

The former is designed in order to obtain desired responses from the structural system. For example, in the new class of materials, called *functionally graded materials*, the mechanical characteristics are variable through the thickness or any other direction in order the structural element made with them to be able to fulfill the designed response. According to Reimanis (2004), "Functionally graded materials (FGMs) are materials that comprise a spatial gradation in structure and/or composition, tailored for a specific performance or function. FGMs are not technically a separate class of materials but rather represent an engineering approach to modify the structural and/or chemical arrangement of materials or elements. This approach is most beneficial when a component has diverse and seemingly contradictory property requirements, such as the cecessity for high hardness and high toughness in wear-resistant coatings."

Jha, Kant and Singh (2013) write: "Alhtough the concept of FGMs, and our ability to fabricate them, appears to be an advanced engineering invention, the concept is not new. These sorts of materials have been occurring in nature...Bones have functional grading. Even our skin is also graded to provide certain toughness, tactile and elastic qualities as a function of skin depth and location on the body." The naturally available materials include bamboo (Nogata and Takahashi, 1996; Ghavami *et al.*, 2003; Ramachandrarao, 2005; Silva *et al.*, 2006, 2008; Tan *et al.*, 2011) and seashells (Li, X.D., 2007).

Apparently the first monograph on the functionally graded materials was pioneered by Suresh and Mortensen (1998) which got over twelve hundred citations since its appearance, by the time the current book was under composition. One should also mention the definitive books by Miyamoto (1999), Shen Jha, Galgali and Misra (2004), (2009b) and Chung (2010) in addition to several volumes of conference proceedings published around the world.

Several disstertaions were also published in recent years in the book form. The interested reader can consult with works by Ebrahimi (2010), Marur (2011) and Reynolds (2011), for example. It is also important to cite reviews by Markworth, Ramesh and Parks (1995), Aboudi, Pindera and Arnold (1999), Paulino (2002), Birman and Byrd (2007), Hilton, Lee and El Fouly (2008), Jha, Kant and Singh (2013), Birman, Keil and Hosder (2013), and Birman (2014a, 2014b). It should be noted that the first papers on inhomogeneity in thickness direction were written many years prior to the uncovering of the FGMs. The interested readers can consult papers by Stavsky (1965), Zaslavsky (1970), Tvergaard (1976), and Eisenberger (1992), as well as the books by Maugin (1993) and Epstein and Elzanowski (2007) to cite just few. The inhomogeieties in other directions were treated in early papers by Jaiani (1975, 1976, 1977, 1998a, 1998b). Modern setting of FGM plates and shells belongs to Reddy (2000), Paulino (2002a, 2013), Batra, Hilton (2003, 2005), Aboudi, Pindera and Arnold (1996), and their respective coworkers, along with numerous contributors around the world as manifested in about 800 referenced papers.

The inherent variabilites are related to the fact that all the structural systems are affected by uncertainties including the he defin lack of complete information. Such uncertainties result from the imprecision of the manufacturing tools, instruments that measure data, and from our incomplete understanding of the laws governing the behavior of natural phenomena. For example, the crack location and depth in a structural element can be modelled through parameters affected by uncertainties. These kinds of problems represent two different aspects of the engineering practice, and, as a consequence, they have to be faced with two different approaches. In order to deal with the first problem, a deterministic method is used, where all the parameters are known exactly and the structural response is determined with the variation of the design parameters. The second problem requires the use of probabilistic models to explicitly incorporate imprecision into the mathematical models that are to be utilized in the analysis and design. Pertinent reviews of structures with uncertaqinties were given in the reviews by Ibrahim (1987) and Manohar and Ibrahim (1999). It was most natural to develop finite element method in stochastic setting. This task was accomplished by Nakagiri and Hisada (1985), Ghanem and Spanos (1991), Kleiber and Hien (1992), Haldar and Mahadevan (2000) and Elishakoff and Ren (2003). Reviews of stochastic finite element method were written by Elishakoff, Ren and Shinozuka (1996), Elishakoff and Ren (1999), Matthies (2008), Sudret and Der Kiureghian (2009), and Stefanou (2009), among others. The second possibility of inherently inhomogeneous materials deals with deterministically variable modulus of elasticity or mass density or any other structural parameters. Naturally there are many works dealing with such systems. There might be an intersection between these two types of variabilities, since once the new system is created via the functional grading, the material characteristics may turn to be varying along some or all coordinates probabilistically. In this book, we limit ourselves with structures made of FGM, and totally concentrate on structures with *fully determined* rather than uncertain parameters.

The book is organized as follows. The first part is devoted to the three-dimensional analysis of rectangular plates made of functionally graded materials (FGMs). The concept of FGMs has been introduced in the '90 apparently by Koizumi (1993, 1997), with the aim of overcoming the drawbacks of composite materials. Traditional composite materials offer numerous superior properties to metallic materials, such as high specific strength and high specific stiffness. This has resulted in the extensive use of laminated composite materials in aircraft, spacecraft and space structures. For example, a layer of a ceramic material bonded to the surface of a metallic structure acts as a thermal barrier in high temperature applications. However, the sudden change in material properties across the interface between discrete materials can results in large interlaminar stresses leading to delamination. Furthermore, large plastic deformations at the interfaces may trigger the initiation and propagation of cracks in the material. One way to overcome these adverse effects is to use functionally graded materials in which material properties, say, modulus of elasticity, mass density, Poisson's ratio, vary continuously. This is achieved either by gradually changing the volume fraction of the constituent materials, usually in the thickness direction only, with the objective of optimizing the performance of the material for a specific application. Since properties of FGMs vary along a direction, they fall under the category of inhomogeneous materials.

FGMs have found application in various branches of engineering, for example, in aerospace structures, power generation industries, machine parts, etc. In recent years, these new classes of materials have gained considerable attention that motivates the importance of a deep understanding of their static and dynamic behavior. The rectangular plate is the most common structural element for such applications. Thus, rectangular plates are under study. In particular, since FG plates may have substantial thickness a three-dimensional analysis is conducted. Bi-dimensional theories may result inadequate for an appropriate design of such structural components. Special attention is devoted to the effects of the thickness, of the material property variation and boundary conditions on the static and dynamic behavior of such plates.

Whereas in the first part of the present book we deal with grading in thickness direction as proposed by the pioneers of the very concept of FGMs, we devote its second part to the functional grading in either axial direction for beams and columns, or in radial direction for circular plates. We extend methodologies presented for the first time in the monograph by Elishakoff (2005) and some papers cited therein. We concentrate on closed-form solutions shedding light on effects of grading on the involved eigenvalues.

It is anticipated that the future technologies will allow performing functional grading in arbitrary direction in the structure in order to tailor it to desired behavior. It is hoped, therefore, this book to become a useful reference for engineers and scientists involved in developing new materials, in conjunction with desired structural performance.

Table of Contents

Part I

Three-Dimensional Analysis of Rectangular Plates Made of Functionally Graded Materials

Introduction

The demand for improved structural efficiency in many engineering sectors, such as aerospace, fast computers, biomedical industry, environmental sensors, has resulted in development of a new class of materials, called functionally graded materials (FGMs) (Suresh and Mortensen, 1998). The continuous change in the material properties, for example in the modulus of elasticity, from one surface of the material to the other surface, distinguishes FGM from conventional composite materials. A common structural element for such applications is the rectangular plate, for which recent studies on statics, buckling and free vibrations have been performed.

Thus, this part deals with the analysis of functionally graded rectangular plates and is organized as follows.

In Chapter 1, the most common theories for the analysis of elastic plates are briefly described. As it is well known, bi-dimensional theories, as Kirchhoff plate theory (Leissa, 1973) and higher order plate theories (Mindlin, 1951; Reddy, 1984), neglect transverse normal deformations, and generally assume that a plane stress state of deformation prevails in the plate. These assumptions may be appropriate for plates with thickness-side ratio equal to 0.2 or less. Since FG plates may have substantial thickness, a three-dimensional analysis results necessary to appropriately describe the behavior of such structural components. The basic relations of the three-dimensional elasticity theory are given in order to provide a background for the developments discussed later in this part. Chapter 2 is devoted to a concise description of the manufacturing processes and the models employed to describe the effective material properties of FGMs. In Chapter 3, the governing equations for the dynamic analysis of FG plates are derived by means of Ritz method. The effects of mechanical property variation, thickness and boundary conditions on the dynamic behavior of FG plates are discussed in detail. In Chapter 4, some considerations on static analysis of FG plates end the first part of the book.

Chapter 1

Elastic Plates

In structural engineering, functionally graded materials (FGMs) are mostly employed as constituent materials of rectangular plates. Thus, this chapter is devoted to a brief overview about the most common elastic plate theories, starting from the classical (Leissa, 1973), and the shear deformation theories (Mindlin, 1951; Reddy, 1984). It is well known that these structural models are based on assumptions concerning the kinematics of deformation or the stress through the thickness of the plate.

In general, these assumptions allow the reduction of a three-dimensional problem to a bi-dimensional one. Thus, these theories are suitable to describe the behavior of thin and/or moderately thick plates. Since functionally graded (FG) plates may have substantial thickness, bi-dimensional theories may result inadequate.

Based on these remarks, it appears important to face the study of FG plates in the three-dimensional setting, where the plate is modelled as a three-dimensional solid.

Summarizing, the analyses of plates have been based on one of the following approaches:

1. Classical plate theory (CPT) (Leissa, 1973).
2. Shear deformation plate theory:

   - First-order shear deformation theory (FSDT) (Mindlin, 1951).
   - Third-order shear deformation theory (TSDT) (Reddy, 1984).

3. Three-dimensional elasticity theory.

Those theories as briefly described in the following sections.

Basic definitions

A rectangular plate of length a, width b and uniform thickness h is considered, as represented in Fig. 1.1. The volume and the upper surface of the plate are indicated with V and S, respectively. The plate geometry and dimensions are defined with respect to a Cartesian coordinate system $(O; x, y, z)$, the origin of which is the center of the plate and the axes are parallel to the edges of the plate. A point of the plate is indicated by the vector $\mathbf{x}$ of its Cartesian coordinates. The generic configuration of the plate is described by the displacement components $u(x, t)$, $v(x, t)$ and $w(x, t)$ along the x, y and z axes, where t denotes time.

It is well known that once the displacement field is, defined it is possible to evaluate the strains associated to each theory. This is done in the hypothesis of small strains in order to study a linear elastic problem. The strain–displacement relations can be written in matrix form as follows:

$$\boldsymbol{\varepsilon}(\mathrm{x},\ t) = \mathbf{D}\mathbf{u}(\mathrm{x},\ t), \tag{1.1}$$

where $\boldsymbol{\varepsilon}$ is the strain vector, $\mathbf{D}$ is the compatibility operator and $\mathbf{u}$ is the displacement vector.

1.1 Bi-dimensional theories

1.1.1 The CPT

A plate is a solid body, bounded by two parallel flat surfaces, whose lateral dimensions (the width and length of a rectangular plate or the diameter of a

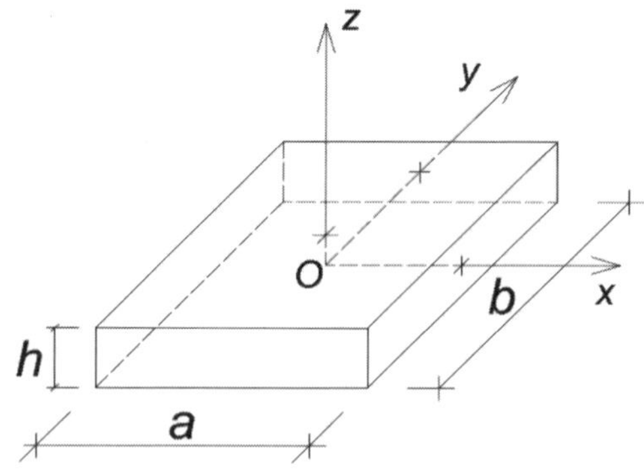

Figure 1.1. Coordinate system and geometry of a rectangular plate

circular plate) are large compared to the distance (the thickness of the plate) between the flat surfaces. The theory of elastic plates is an approximation of three-dimensional elasticity theory to two dimensions and it can be used to determine the state of stress and deformation in plates. The approximations are simplifications that allow to us consider the deformation of the midplane of the plate.

Plates can be classified into two groups: thin plates and thick plates. A plate is said to be thin when the ratio of the thickness, h, to the smaller span length, a, is less than 1/20 ($h/a = 0.05$). The small-deflection theory of bending of thin plates is known as the CPT and is based on assumptions similar to those used in the Euler–Bernoulli theory (or classical beam theory). The fundamental assumptions of the Kirchhoff–Love or CPT are:

1. The displacements of the midsurface are small compared with the thickness of the plate and, therefore, the slope of the deflected surface is very small and the square of the slope is negligible compared to unity.
2. The midplane of the plate remains unstrained, that is, the midplane of the plate remains as the neutral plane, after bending.
3. Kirchhoff–Love assumption: Plane sections initially normal to the midsurface remain plane and normal to the midsurface after bending.

   Analogous to beams, this assumption implies that the transverse shear strains are negligible. The deflection of the plate is thus associated principally with bending strains. Consequently, the transverse normal strain resulting from transverse loading can be neglected.
4. The transverse normal stress is small compared with the other stress components of the plate and, therefore, can be neglected.

These assumptions give results that do not differ significantly from those obtained using the three-dimensional theory of elasticity for a vast majority of thin plate problems.

Under these assumptions, the displacement field of the classical theory can be written in the form:

$$u(x, y, z, t) = u_0(x, y, t) - z\frac{\partial w}{\partial x}, \tag{1.2}$$

$$v(x, y, z, t) = v_0(x, y, t) - z\frac{\partial w}{\partial y}, \tag{1.3}$$

$$w(x, y, z, t) = w_0(x, y, t), \tag{1.4}$$

where u_0, v_0 and w_0 are the displacement components along the x, y and z-axes belonging to the midplane, see Fig. 1.2(a). According to the classical theory, the displacement vector $\mathbf{u}$, the strain vector $\boldsymbol{\varepsilon}$ and the compatibility operator $\mathbf{D}$, introduced in the strain–displacement relation, Eq. (1.1), are in the form:

$$\mathbf{u}(\mathrm{x}, t) = [u_0 \quad v_0 \quad w_0]^{\mathrm{T}},$$

$$\boldsymbol{\varepsilon}(\mathrm{x}, t) = [\varepsilon_{x0} \quad \varepsilon_{y0} \quad \gamma_{xy0} \quad X_x \quad X_y \quad X_{xy}]^{\mathrm{T}},$$

$$\mathbf{D} = \begin{bmatrix} \partial/\partial x & 0 & 0 \\ 0 & \partial/\partial y & 0 \\ \partial/\partial y & \partial/\partial x & 0 \\ 0 & 0 & -\partial^2/\partial x^2 \\ 0 & 0 & -\partial^2/\partial y^2 \\ 0 & 0 & -2\partial^2/\partial x \partial y \end{bmatrix},$$

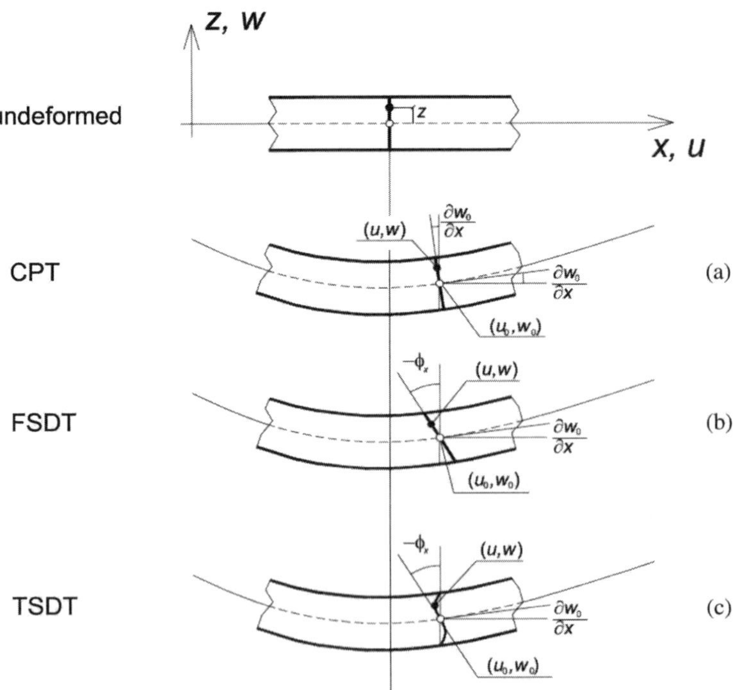

Figure 1.2. Deformation of a transverse normal according to the classical, first-order and third-order plate theories

respectively. Components ε_{x0}, ε_{y0}, γ_{xy0} are the strains due to the in-plane displacements and X_x, X_y and X_{xy} are the curvatures due to the first-order terms.

1.1.2 Shear deformation theories: First-order deformation theory

In the first-order shear deformation plate theory, the Kirchhoff hypothesis is relaxed by removing the condition that the transverse normals remain perpendicular to the midsurface after deformation. This amounts to including transverse shear strains in the theory. The inextensibility of transverse normal requires that w not be a function of the thickness coordinate z. Under the same assumptions and restrictions as in the CPT, the displacement field of the first-order theory is of the form:

$$u(x, y, z, t) = u_0(x, y, t) + z\phi_x(x, y, t) \tag{1.5}$$

$$v(x, y, z, t) = v_0(x, y, t) + z\phi_y(x, y, t) \tag{1.6}$$

$$w(x, y, z, t) = w_0(x, y, t) \tag{1.7}$$

where u_0, v_0 and w_0 denote the displacements of a point on the plane $z = 0$.
Note that

$$\frac{\partial u}{\partial z} = \phi_x, \quad \frac{\partial v}{\partial z} = \phi_y$$

which indicate that ϕ_x and ϕ_y are the rotations of a transverse normal about the y and x axes, respectively. The quantities u_0, v_0, w_0, ϕ_x, and ϕ_y are called generalized displacements, see Fig. 1.2(b). For thin plates, the rotation functions ϕ_x and ϕ_y should approach the respective slopes of the transverse deflection:

$$\phi_x = -\frac{\partial w_0}{\partial x}, \quad \phi_y = -\frac{\partial w_0}{\partial y}.$$

According to the first-order theory, the displacement vector $\mathbf{u}$, the strain vector $\boldsymbol{\varepsilon}$ and the compatibility operator $\mathbf{D}$, introduced in the strain–displacement relation, Eq. (1.1), are in the form:

$$\mathbf{u}(x, t) = [u_0 \quad v_0 \quad w_0 \quad \phi_x \quad \phi_y]^{\mathrm{T}},$$

$$\boldsymbol{\varepsilon}(x, t) = [\varepsilon_{x0} \quad \varepsilon_{y0} \quad \gamma_{xy0} \quad X_x \quad X_y \quad \gamma_{xz0} \quad \gamma_{yz0}]^{\mathrm{T}},$$

$$
\mathbf{D} =
\begin{bmatrix}
\partial/\partial x & 0 & 0 & 0 & 0 \\
0 & \partial/\partial y & 0 & 0 & 0 \\
\partial/\partial y & \partial/\partial x & 0 & 0 & 0 \\
0 & 0 & 0 & \partial/\partial x & 0 \\
0 & 0 & 0 & 0 & \partial/\partial y \\
0 & 0 & 0 & \partial/\partial y & \partial/\partial x \\
0 & 0 & \partial/\partial x & 1 & 0 \\
0 & 0 & \partial/\partial y & 0 & 1
\end{bmatrix},
$$

respectively. Components γ_{yz0} and γ_{xz0} are the shear strains due to the first-order terms.

1.1.3   Shear deformation theories: Third-order deformation theory

The third-order plate theory is based on the same assumptions as the classical and first-order plate theories, except that the assumption on the straightness and normality of a transverse normal after deformation is relaxed by expanding the displacements u, v, w as follows:

$$
u(x, y, z, t) = u_0(x, y, t) + z\phi_x(x, y, t) - cz^3\left(\phi_x + \frac{\partial w}{\partial x}\right), \quad (1.8)
$$

$$
v(x, y, z, t) = v_0(x, y, t) + z\phi_y(x, y, t) - cz^3\left(\phi_y + \frac{\partial w}{\partial y}\right), \quad (1.9)
$$

$$
w(x, y, z, t) = w_0(x, y, t), \quad (1.10)
$$

where $c = 4/(3h^2)$ that is determined by requiring that the transverse shear strain vanishes on the top and the bottom surfaces of the plate. The particular choice for the displacement field, Eqs. (1.8)–(1.10), allow for distortion (of the third-order) of normals to the mid plane of the undeformed plate, see Fig. 1.2(c). This refined theory predicts displacements and natural frequencies accurately for plates with thickness to width ratio up to 0.2. Although a cubic variation of the in-plane displacement through the thickness is taken into account, the displacement field in Eqs. (1.8)–(1.10) contains the same number of independent variables as in the first-order deformation theory, Eqs. (1.5)–(1.7).

According to the third-order theory, the displacement vector $\mathbf{u}$, the strain vector $\boldsymbol{\varepsilon}$ and the compatibility operator $\mathbf{D}$, introduced in the strain–displacement relation, Eq. (1.1), are in the form:

$$\mathbf{u}(x, t) = [u_0 \quad v_0 \quad w_0 \quad \phi_x \quad \phi_y]^{\mathrm{T}},$$

$$\boldsymbol{\varepsilon} = [\varepsilon_{x0} \; \varepsilon_{y0} \; \gamma_{xy0} \; X_x \; X_y \; X_{xy} \; \gamma_{xz0} \; \gamma_{yz0} \; X_{xs} \; X_{ys} \; X_{xys} \; \gamma_{xzs} \; \gamma_{yzs}]^{\mathrm{T}},$$

$$\mathbf{D} = \begin{bmatrix} \partial/\partial x & 0 & 0 & 0 & 0 \\ 0 & \partial/\partial y & 0 & 0 & 0 \\ \partial/\partial y & \partial/\partial x & 0 & 0 & 0 \\ 0 & 0 & 0 & \partial/\partial x & 0 \\ 0 & 0 & 0 & 0 & \partial/\partial y \\ 0 & 0 & 0 & \partial/\partial y & \partial/\partial x \\ 0 & 0 & \partial/\partial x & 1 & 0 \\ 0 & 0 & \partial/\partial y & 0 & 1 \\ 0 & 0 & c\partial^2/\partial x^2 & c\partial/\partial x & 0 \\ 0 & 0 & c\partial^2/\partial y^2 & 0 & c\partial/\partial y \\ 0 & 0 & 2c\partial^2/\partial x\partial y & c\partial/\partial y & c\partial/\partial x \\ 0 & 0 & 3c\partial/\partial x & 3c & 0 \\ 0 & 0 & 3c\partial/\partial y & 0 & 3c \end{bmatrix},$$

respectively. Components X_{xs}, X_{ys}, X_{xys}, γ_{xzs} and γ_{yzs} are the curvatures and shear strains due to the higher-order terms.

1.2 Three-dimensional theory

In the three-dimensional setting, the generic configuration of the plate is described by the displacement vector $\mathbf{u}(\mathrm{x}, t)$, whose independent components are arranged in the following form:

$$\mathbf{u}(\mathrm{x}, t) = [u(\mathrm{x}, t) \quad v(\mathrm{x}, t) \quad w(\mathrm{x}, t)]^{\mathrm{T}}. \tag{1.11}$$

No hypotheses are adopted on the displacement field, Eq. (1.11).

The strain vector $\boldsymbol{\varepsilon}$ in the strain–displacements relation, Eq. (1.1), is:

$$\boldsymbol{\varepsilon} = [\varepsilon_{xx} \quad \varepsilon_{yy} \quad \varepsilon_{zz} \quad \gamma_{xy} \quad \gamma_{xz} \quad \gamma_{yz}]^{\mathrm{T}},$$

and the compatibility operator $\mathbf{D}$ is defined as:

$$\mathbf{D} = \begin{bmatrix} \partial/\partial x & 0 & 0 \\ 0 & \partial/\partial y & 0 \\ 0 & 0 & \partial/\partial z \\ \partial/\partial y & \partial/\partial x & 0 \\ \partial/\partial z & 0 & \partial/\partial x \\ 0 & \partial/\partial z & \partial/\partial y \end{bmatrix}.$$

The strain–displacement relation in components are the well-known relations:

$$\varepsilon_{xx} = \frac{\partial u}{\partial x}, \tag{1.13}$$

$$\varepsilon_{yy} = \frac{\partial v}{\partial y}, \tag{1.14}$$

$$\varepsilon_{zz} = \frac{\partial w}{\partial z}, \tag{1.15}$$

$$\gamma_{xy} = \frac{\partial u}{\partial y} + \frac{\partial v}{\partial x}, \tag{1.16}$$

$$\gamma_{xz} = \frac{\partial u}{\partial z} + \frac{\partial w}{\partial x}, \tag{1.17}$$

$$\gamma_{yz} = \frac{\partial v}{\partial z} + \frac{\partial w}{\partial y}. \tag{1.18}$$

Moreover, for a linear elastic body the constitutive relations are given by:

$$\sigma_{xx} = \lambda(\varepsilon_{xx} + \varepsilon_{yy} + \varepsilon_{zz}) + 2\mu\varepsilon_{xx}, \tag{1.19}$$

$$\sigma_{yy} = \lambda(\varepsilon_{xx} + \varepsilon_{yy} + \varepsilon_{zz}) + 2\mu\varepsilon_{yy}, \tag{1.20}$$

$$\sigma_{zz} = \lambda(\varepsilon_{xx} + \varepsilon_{yy} + \varepsilon_{zz}) + 2\mu\varepsilon_{zz}, \tag{1.21}$$

$$\tau_{xy} = \mu\gamma_{xy}, \tag{1.22}$$

$$\tau_{xz} = \mu \gamma_{xz}, \tag{1.23}$$

$$\tau_{yz} = \mu \gamma_{yz}, \tag{1.24}$$

where λ and μ are the Lamè coefficients:

$$\lambda = \frac{\upsilon E}{(1+\upsilon)(1-2\upsilon)}, \quad \mu = \frac{E}{2(1+\upsilon)}. \tag{1.25}$$

Equations (1.19)–(1.24) can be written in matrix form:

$$\boldsymbol{\sigma} = \mathbf{H}\boldsymbol{\varepsilon},$$

where $\boldsymbol{\sigma}$ is the vector of the stress components:

$$\boldsymbol{\sigma} = [\sigma_{xx} \quad \sigma_{yy} \quad \sigma_{zz} \quad \tau_{xy} \quad \tau_{xz} \quad \tau_{yz}]^{\mathrm{T}}$$

and $\mathbf{H}$ is the matrix of elastic stiffness moduli. For isotropic materials, matrix $\mathbf{H}$ has the following form:

$$\mathbf{H} = \frac{E(1-\upsilon)}{(1+\upsilon)(1-2\upsilon)}$$
$$\times \begin{bmatrix} 1 & \dfrac{\upsilon}{1-\upsilon} & \dfrac{\upsilon}{1-\upsilon} & 0 & 0 & 0 \\[2ex] \dfrac{\upsilon}{1-\upsilon} & 1 & \dfrac{\upsilon}{1-\upsilon} & 0 & 0 & 0 \\[2ex] \dfrac{\upsilon}{1-\upsilon} & \dfrac{\upsilon}{1-\upsilon} & 1 & 0 & 0 & 0 \\[2ex] 0 & 0 & 0 & \dfrac{1-2\upsilon}{2(1-\upsilon)} & 0 & 0 \\[2ex] 0 & 0 & 0 & 0 & \dfrac{1-2\upsilon}{2(1-\upsilon)} & 0 \\[2ex] 0 & 0 & 0 & 0 & 0 & \dfrac{1-2\upsilon}{2(1-\upsilon)} \end{bmatrix}. \tag{1.26}$$

Since FGM are considered, matrix $\mathbf{H}$ is a function of the z-coordinate, if the properties are assumed to vary through the thickness of the plate.

1.3 Some remarks

In this chapter, some aspects of the most common elastic plate theories have been analyzed. It should be remarked that bi-dimensional theories are able to predict displacements, stresses and natural frequencies from thin ($h/a = 0.01$) to moderately thick plates ($h/a = 0.2$). Keeping this in mind, the dynamic and static studies on FG plates are conducted in the three-dimensional elasticity context.

Chapter 2

Introduction to Functionally Graded Materials

The superior properties of advanced composite materials, such as specific high strength and high stiffness, have led to their widespread use in high performance aircrafts, spacecrafts, automobile parts and space structures. In conventional laminated composite structures, homogeneous elastic laminae are bonded together to obtain enhanced mechanical and material properties.

Composite materials are man-made and, therefore, the constituents of composite materials can be selected and combined so as to produce a useful material that has desired properties, such as high strength, high stiffness, greater corrosion resistance, greater fatigue life, low weight and so on.

The anisotropic constitution of laminated composite structures often results in stress concentrations near material and geometric discontinuities, Fig. 2.1, that can lead to damage in the form of delamination, matrix cracking and adhesive bond separation.

Functionally graded materials (FGMs) are a class of composites that have a continuous variation of material properties from one surface to another and thus alleviate the stress concentrations found in laminated composites.

The gradation in properties of the material reduces thermal stresses, residual stresses and stress concentration factors. The gradual variation results in a very efficient material tailored to suit the needs of the structure and therefore is called a FGM. They are typically manufactured from isotropic components such as metals and ceramics, since they are employed also as thermal barrier structures in environments with severe thermal gradients (e.g. thermoelectric devices for energy conversion, semiconductor industry). In such applications, the ceramic provides heat and corrosion resistance; meanwhile the metal provides the strength and toughness.

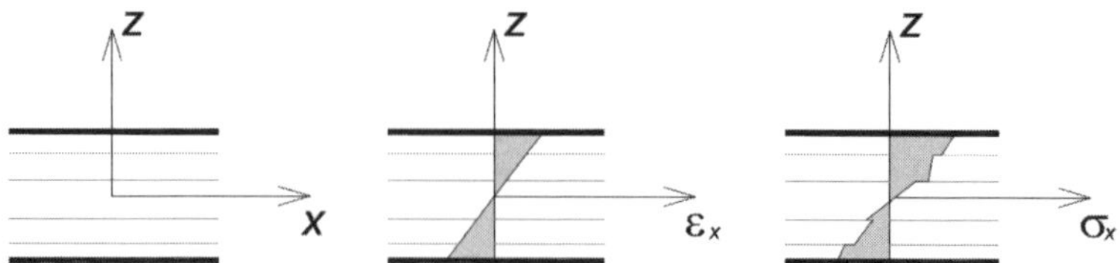

Figure 2.1. Variation of strains and stresses through the thickness of a laminate plate

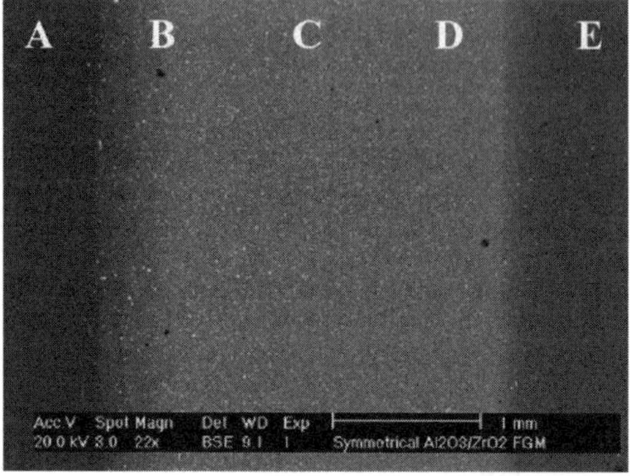

Figure 2.2. FGM: Microscopic view

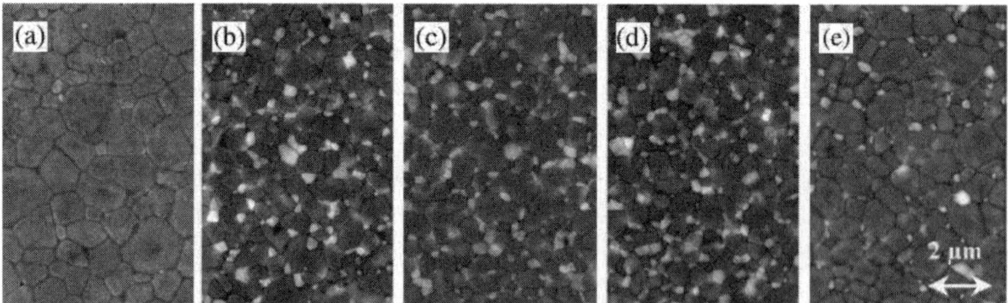

Figure 2.3. FGM: Microstructure

FGMs are therefore composite materials with a microscopically inhomogeneous character. Continuous changes in their microstructure distinguish FGMs from conventional composite materials, see Figs. 2.2 and 2.3.

In this chapter, some features of FGMs are presented. In particular, manufacturing systems of FGMs are briefly introduced and attention is devoted to the description of micromechanics models employed to describe the effective properties of FGMs.

2.1 Fabrication methods

The synthesis of FGMs has been successfully demonstrated through a variety of methods, including thermal spray, powder metallurgy, physical and chemical vapor deposition and self-propagating high-temperature synthesis (SHS) or combustion synthesis. High-temperature synthesis is particularly well suited to fabricating FGMs, because of the rapidity of the combustion reaction. It consists of a simultaneous combustion synthesis reaction and compaction of the virgin material powders under a hydrostatic pressure to fabricate dense FGMs in a single processing operation (Zhan *et al.*, 2000).

Functionally graded structures can be manufactured by a high-speed centrifugal casting techniques (Fukui, 1991), in which layers are formed in the radial direction due to different mass densities of the constituents. The centrifugal force enables the ceramic powder in a metal to create a gradient distribution for a ceramic/metal FGM. In other fabrication processes, the mixture of the component phases are sprayed by means of the plasma torches onto the material surface to pile up a graded material. The FGM is heat-treated afterward.

A technique for preparing model FGMs using polyester resin and cenosphere is also employed (Parameswaran and Shukla, 2000). An inhomogeneous distribution of the cenospheres in the polyester matrix is achieved by employing a buoyancy assisted casting process. Cenospheres are, obtained from the fly ash of thermal power plants, hollow spheres made of aluminum silicates.

2.2 Modelling of the effective material properties

Consider a functionally graded composite material fabricated by mixing two distinct material phases, for example, a metal and a ceramic. Often, precise information about the size, the shape and the distribution of particles may not be available and the effective elastic moduli of the graded composite must be evaluated only based on the volume fraction distribution and the approximate shape of the dispersed phase. Several micromechanics models have been developed over the years to infer the effective properties of macroscopically homogeneous composite materials.

The most common homogenization techniques for modelling the effective material properties are the rule of mixtures (Markworth and Saunders, 1995), the Mori–Tanaka method (Mori and Tanaka, 1973; Tanaka, 1997) and Hill's self-consistent approach (Hill, 1965) that are summarized as follows.

These models are available to estimate the overall properties of composites from the knowledge of the material composition and constituent properties.

The inclusions and the matrix are assumed to be made of isotropic materials and the macroscopic response of the composite is modelled as isotropic.

2.2.1 The rule of mixtures

According to the rule of mixtures, an arbitrary material property, denoted as P, of the FGM is assumed to vary smoothly along a direction, as a function of the volume fractions and properties of the constituent materials. Since FG plates are considered, the varying direction is the thickness direction, see Fig. 2.4.

In this context, P can represent, for example, the modulus of elasticity, the mass density and/or Poisson's ratio. This property can be expressed as a linear combination.

$$P(z) = P_1 V_1 + P_2 V_2 \tag{2.1}$$

Figure 2.4. A functionally graded plate

where z is the varying direction, P_1, V_1 and P_2, V_2 are the material properties and volume fractions of the constituent materials 1 and 2, respectively. The volume fractions of all the constituent materials should add up to unity

$$V_1 + V_2 = 1 \tag{2.2}$$

The volume fraction V_1 is assumed to have the following power-law distribution, Markworth and Saunders (1995):

$$V_1 = \left(\frac{z}{h} + \frac{1}{2}\right)^N \tag{2.3}$$

and as a consequence:

$$V_2 = 1 - \left(\frac{z}{h} + \frac{1}{2}\right)^N$$

where h is the thickness of the plate. Figure 2.5 represents the variation through the thickness of the volume fraction V_1, and consequently the variation of the material properties. Parameter N is the volume fraction exponent which takes positive real values and dictates the material variation profile through the thickness. The constituent materials 1 and 2 can be, for example, ceramic and metal, respectively. According to this distribution, the bottom surface, $z = -h/2$, of the functionally graded plate is pure metal and the top surface, $z = h/2$, is pure ceramic. This power-law assumption reflects a simple rule of mixtures used to obtain the effective properties of the FGM. The material 2 content in the plate increases as the value of N increases. The value of $N = 0$ represents a homogeneous material 1 plate.

The model (2.1) provides exact values for the mass density ρ and fairly good values for other mechanical properties. A more precise determination of the macroscopic material properties requires a better understanding of the microstructure.

The Mori–Tanaka and the self-consistent methods are accurate micromechanics models that are employed to describe FGMs properties.

2.2.2 The Mori–Tanaka model

The Mori–Tanaka (1973) model (Tanaka, 1997) is used for estimating the effective moduli of the material. It accounts approximately for the interaction among neighboring inclusions and is generally applicable to regions of

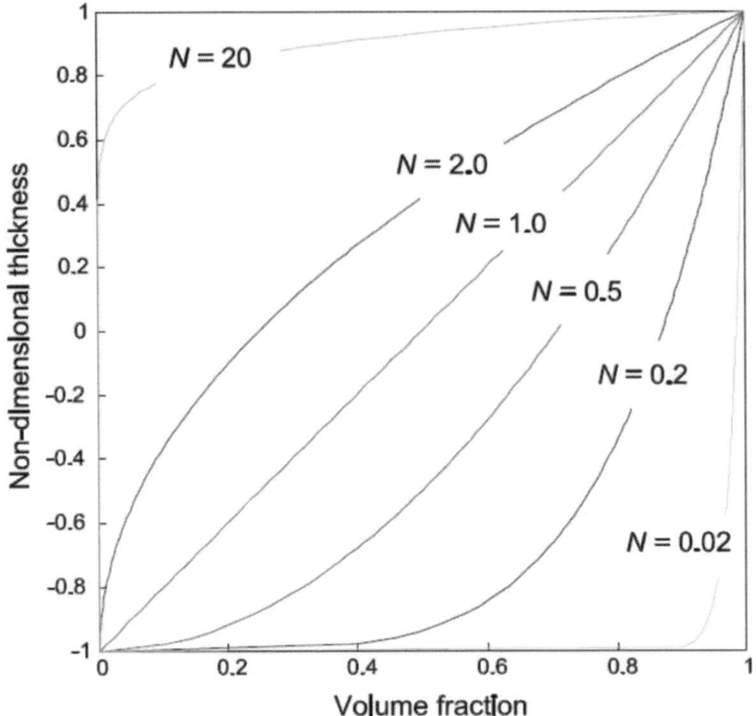

Figure 2.5. Variation of the volume fraction in the thickness direction

the graded microstructure that have a well-defined continuous matrix and a discontinuous particulate phase as depicted in Fig. 2.6(a). It is assumed that the matrix phase, denoted by the subscript 1, is reinforced by spherical particles of a particulate phase, denoted by the subscript 2. In this notation, K_1, μ_1 and V_1 denote the bulk modulus, the shear modulus and the volume fraction of the matrix phase respectively; K_2, μ_2 and V_2 denote the corresponding material properties and the volume fraction of the particulate phase. It is known that Lamè constant λ is related to the bulk and shear moduli by the following relation:

$$\lambda = K - \frac{2\mu}{3}$$

The effective mass density, ρ, is given exactly by the rule of mixtures, Eq. (2.1):

$$\rho = \rho_1 V_1 + \rho_2 V_2.$$

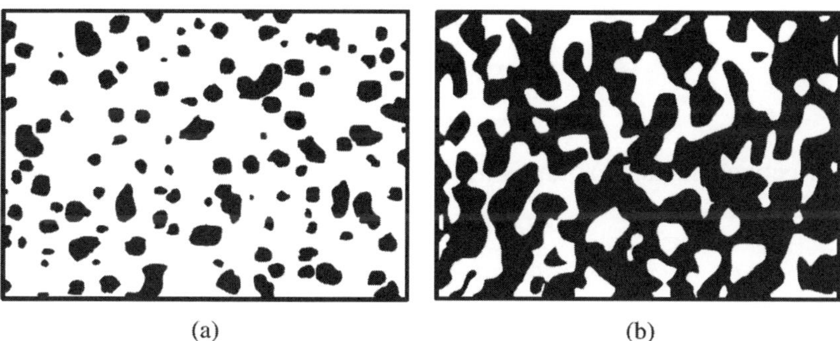

Figure 2.6. Two phase material with (a) a particulate microstructure and (b) a skeletal microstructure

According to the Mori–Tanaka method for a random distribution of isotropic particles in an isotropic matrix, the locally effective bulk modulus K and shear modulus μ are given by:

$$\frac{K - K_1}{K_2 - K_1} = \frac{V_2}{1 + (V_1 - V_2)\frac{K_2 - K_1}{3K_1 + 4\mu_1}}$$

$$\frac{\mu - \mu_1}{\mu_2 - \mu_1} = \frac{V_2}{1 + (V_1 - V_2)\frac{\mu_2 - \mu_1}{\mu_1 + f_1}}$$

where:

$$f_1 = \mu_1 \frac{9K_1 + 8\mu_1}{6(K_1 + 2\mu_1)}.$$

Through the thickness variation of V_2 is assumed to be given by the following power-law function:

$$V_2 = V_2^- + \left(V_2^+ - V_2^-\right)\left(\frac{1}{2} + \frac{z}{h}\right)^N$$

where superscripts $+$ and $-$ signify, respectively, values of the quantity on the top and the bottom surfaces of the structural element, and the parameter N describes the variation of Phase 2.

For example, $N = 0$ and $N = \infty$ correspond to uniform distributions of Phase 2 with volume fractions V_2^+ and V_2^-, respectively. Recalling that the bulk modulus K and the shear modulus μ are related to Young's modulus E and Poisson's ratio ν by the following relations

$$K = \frac{E}{3(1 - 2\nu)}, \quad \mu = \frac{E}{2(1 + \nu)}$$

the effective values of Young's modulus and Poisson's ratio are found:

$$E = \frac{9K\mu}{3K + \mu}, \quad \nu = \frac{3K - 2\mu}{2(3K + \mu)}$$

In Fig. 2.7, the through the thickness distributions of the modulus of elasticity, Poisson's ratio, shear modulus and bulk modulus are represented.

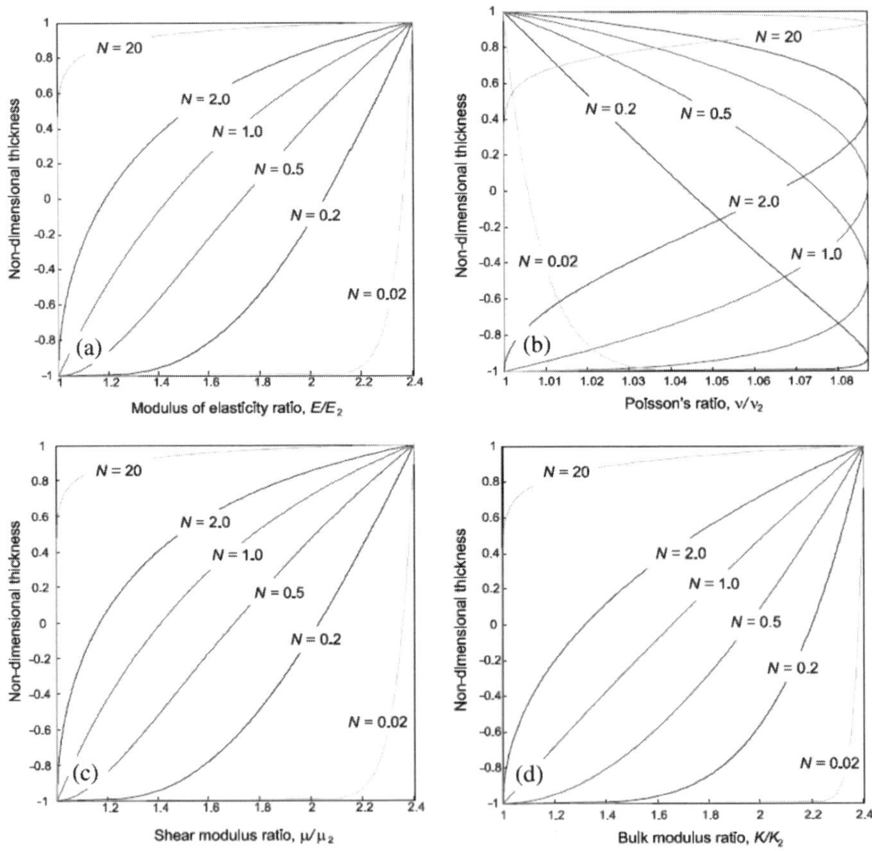

Figure 2.7. Through the thickness distribution of the (a) modulus of elasticity, (b) Poisson's ratio, (c) shear modulus and (d) bulk modulus

2.2.3 Self-consistent model

The self-consistent method (Hill, 1965) assumes that each reinforcement inclusion is embedded in a continuum material whose effective properties are those of the composite. This method does not distinguish between matrix and reinforcement phases and the same overall moduli are predicted in another composite in which the roles of the phases are interchanged. This makes it particularly suitable for determining the effective moduli in those regions which have an interconnected skeletal microstructure as depicted in Fig. 2.6(b). The locally effective elastic moduli, K and μ, by the self-consistent method are given by:

$$\frac{\delta}{K} = \frac{V_1}{K - K_2} \frac{V_2}{K - K_1}, \tag{2.4}$$

$$\frac{\eta}{\mu} = \frac{V_1}{\mu - \mu_2} + \frac{V_2}{\mu - \mu_1} \tag{2.5}$$

where:

$$\delta = 3 - 5\eta = \frac{K}{K + \frac{4\mu}{3}} \tag{2.6}$$

These are implicit expressions for the unknowns K and μ. After substitution for δ, Eq. (2.4) can be solved for K in terms of μ in the form:

$$\frac{1}{K + \frac{4\mu}{3}} = \frac{V_1}{K_1 + \frac{4\mu}{3}} + \frac{V_2}{K_2 + \frac{4\mu}{3}}$$

that gives:

$$K = \frac{1}{\frac{V_1}{K_1 + \frac{4\mu}{3}}} + \frac{1}{\frac{V_2}{K_2 + \frac{4\mu}{3}}} - \frac{4\mu}{3} \tag{2.7}$$

Parameter μ is obtained by solving the following equation:

$$(1 - \eta)\,\mu^2 + [\eta\,(\mu_1 + \mu_2) - (V_1\mu_1 + V_2\mu_2)]\,\mu - \eta\mu_1\mu_2 = 0 \tag{2.8}$$

where parameter η can be found from Eq. (2.6). Since Eq. (2.8) has to be solved to find the shear modulus μ, in general it is easier to use the Mori–Tanaka method than the self-consistent scheme. It should be noted that the differences in the material property distribution between the Mori–Tanaka and the self-consistent scheme are negligible.

 Introduction to FGMs

2.3 Some remarks

In this chapter, some fundamental features of FGMs have been considered. In particular, manufacturing processes and micromechanics models for the description of the material properties of a FGM were discussed. In the forthcoming chapter, the basic concept of the rule of mixture model will be adopted in the description of FGM properties.

Chapter 3

Dynamic Analysis of Plates Made
of Functionally Graded Materials

This chapter deals with the dynamic analysis of rectangular plates made of functionally graded materials (FGMs). There are several papers devoted to vibrations of FG structures in the literature. Among these, Loy *et al.* (1999a) and Pradhan *et al.* (2000) examined the free vibration of FG cylindrical shells by using the Rayleigh–Ritz method. The first-order deformation theory is employed by Praveen and Reddy (1998) to analyze the nonlinear static and dynamic response of heated FG ceramic-metal plates subjected to dynamic lateral loads by the finite element method. Reddy (2000) developed both theoretical and finite element formulations for thick FG plates according to the higher order deformation theory, and studied the nonlinear dynamic response of FG plates subjected to sudden applied uniform pressure. Yang and Shen (2002) dealt with free vibration and transient response due to impulsive lateral patch load for initially stressed FG thin plates by means of the classical plate theory together with a differential quadrature method. After, the same authors developed their studies on free and forced vibration of FG plates in Yang and Shen (2002) employing Reddy's higher-order theory. Higher order theory together with a meshless Petrov–Galerkin procedure was employed for the static and dynamic analysis of FG plates in Qian *et al.* (2004).

A question is raised as whether such approximate models are sufficiently accurate for the study of plates and shells made of FGMs. To address this issue, it is necessary to develop exact solutions within the framework of the three-dimensional theory to provide a basis of comparison for the two-dimensional theories and numerical methods. Also because the thickness of structures made of FGMs may be comparable to the side length, it was felt that three-dimensional analyses were called for dynamic problems. Thus, three-dimensional studies on FG structures were conducted by Reddy and Cheng (2003) that proposed an asymptotic approach formulated in terms of transfer matrix. The method is applied to a simply supported FG plate made of monel and zirconia. A three-dimensional exact solution is presented for free and forced vibrations of FG plates in Vel and Batra (2004) without considering damping.

The authors provided with the closed-form solution for all-round simply supported rectangular plates by employing the power series method. Results were presented for two constituent metal–ceramic plates whose effective material properties at a point are estimated by either the Mori–Tanaka or the self-consistent schemes.

Based on the above remarks, the present study utilizes an approximate, yet versatile Ritz method to derive the governing eigenvalue equation for FG rectangular plates in the three-dimensional setting. The in-plane displacements and deflections are approximated by sets of one-dimensional polynomial series.

These functions are chosen to be Chebyshev polynomials multiplied by boundary functions, in order the geometric conditions at the edges are identically satisfied. Chebyshev polynomials employed as admissible functions have been proven to be highly efficient, stable and accurate in many numerical applications, see Zhou *et al.* (2002) for isotropic case.

In particular, the approach by Siu and Bert (1974), Leissa (1978; 1984) and Elishakoff *et al.* (1988) that was originally developed for isotropic plates, is extended to the problem of forced damped vibrations of FG simply supported and clamped rectangular plates in the three-dimensional setting. The classical method for analyzing the forced vibrations of structural elements is to express the displacements as superposition of the free vibration modes. This is only possible for those relatively few problems where the exact eigenfunction solutions exist. According to Leissa, "mode shapes usually are not known with sufficient accuracy to give meaningful results for stresses." On the other hand, the Ritz method is widely used to obtain approximate solutions for free undamped vibration problems. The present analysis demonstrates how the same method can be generalized to analyze forced vibrations of a rectangular FG plate with damping.

This is done directly without requiring the free vibration eigenfunctions. The Rayleigh–Ritz method is extended to include the dissipation energy and the work done by the external excitation.

Numerical results are presented for a two-phase ceramic/metal graded material with a power-law variation of the volume fraction of the constituents through the thickness. Material properties, namely, modulus of elasticity, mass density and damping coefficient, are assumed to vary across the thickness.

Natural frequencies are also determined, after convergence studies. The dynamic response as well as exact natural frequencies of homogeneous plate are used to assess the accuracy of the present formulation. A parametric study with respect to varying volume fraction of the constituents, thickness ratio and boundary conditions is conducted. Preliminary results for free vibrations have been presented in Gentilini (2005) and for forced vibrations in Elishakoff *et al.* (2005).

3.1 Statement of the problem

3.1.1 Basic definitions

Some basic definitions given in Chapters 1 and 2 are recalled for sake of completeness. A FG rectangular plate of length a, width b and uniform thickness h, as represented in Fig. 3.1, is considered.

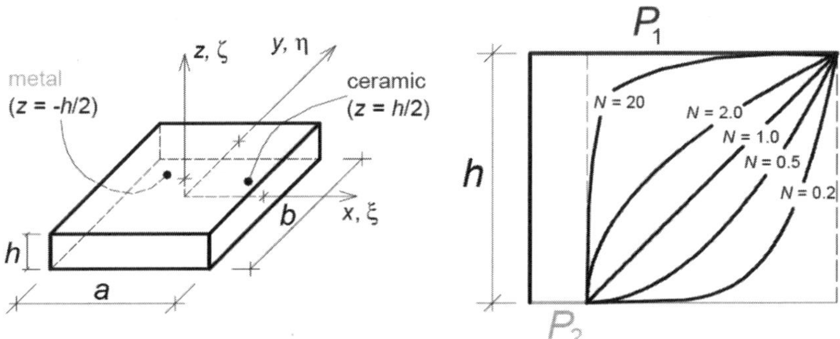

Figure 3.1. Coordinate system and dimensions of a FG rectangular plate

For simplicity, the following non-dimensional coordinate system $(O; \xi, \eta, \zeta)$ is also introduced:

$$\xi = \frac{2x}{a}, \quad \eta = \frac{2y}{b}, \quad \zeta = \frac{2z}{h} \tag{3.1}$$

It is assumed that the FGM is made of a mixture of a ceramic (material 1) and a metallic (material 2) component. The mechanical properties of the plate are assumed to vary across the thickness only. The rule of mixtures is used for modelling the effective moduli of the material, Section 2.2.1. Thus, the local effective material property P at a given point of the plate is assumed to have the following power-law distribution:

$$P(z) = (P_1 - P_2)V_1 + P_2 \tag{3.2}$$

$$V_1 = \left(\frac{z}{h} + \frac{1}{2}\right)^N \tag{3.3}$$

where P_1 and P_2 are the corresponding properties of the ceramic and metal, N is the volume fraction exponent and V_1 is the ceramic volume fraction. The value of N equal to zero represents a fully ceramic plate, when N is approaching infinity the case of the fully metallic plate is obtained. Metal–ceramic FGMs combine synergically the positive properties of metals and ceramics. The ceramic in a FGM offers thermal barrier effects and protects the metal from corrosion and oxidation, while the FGM is toughened and strengthened by the metallic composition.

The material properties along the thickness of the plate, such as Young's modulus E and mass density ρ can be determined according to Eqs. (3.2) and (3.3), yielding:

$$E(z) = (E_1 - E_2)V_1 + E_2 = \sqrt{E_1 E_2} E_{FGM}(z) \tag{3.4}$$

$$\rho(z) = (\rho_1 - \rho_2)V_1 + \rho_2 = \sqrt{\rho_1 \rho_2} \rho_{FGM}(z) \tag{3.5}$$

where $E_{FGM}(Z)$ and $\rho_{FGM}(z)$ are given as follows:

$$E_{FGM}(z) = \left(\sqrt{\frac{E_1}{E_2}} - \sqrt{\frac{E_2}{E_1}}\right)V_1 + \sqrt{\frac{E_2}{E_1}} \tag{3.6}$$

$$\rho_{FGM}(z) = \left(\sqrt{\frac{\rho_1}{\rho_2}} - \sqrt{\frac{\rho_2}{\rho_1}}\right)V_1 + \sqrt{\frac{\rho_2}{\rho_1}} \tag{3.7}$$

In Reddy (2000), the same variation for $E(z)$ and $\rho(z)$ is adopted.

In the present study, it is assumed that the damping is internal; this damping is postulated to be representable as an equivalent viscous damping with c_1, c_2 denoting the associated damping coefficients to the ceramic and metallic constituents, respectively. In particular, the local effective damping c at a given point of the plate is assumed to vary according to the power law distribution Eqs. (3.2) and (3.3):

$$c(z) = (c_1 - c_2)V_1 + c_2 \qquad (3.8)$$

The hypothesis of proportional damping is adopted hereinafter:

$$\frac{c_1}{c_2} = \frac{\rho_1}{\rho_2} \qquad (3.9)$$

This assumption appears not to be restrictive; it corresponds to one of the possible realizations of the proportional damping in a multi-degree of freedom vibrational system. From this assumption, the following simplified power-law distribution for the damping coefficient is obtained:

$$c(z) = \sqrt{c_1 c_2}\rho_{FGM}(z) \qquad (3.10)$$

in which $\rho_{FGM}(z)$ is indicated in Eq. (3.7).

3.2 Three-dimensional analysis

In the following sections, the energy terms needed for the dynamic analysis of FG plates are introduced.

3.2.1 Strain energy

The strain energy of a three-dimensional solid is given by

$$\Phi(x, y, z, t) = \frac{1}{2} \int_V \left\{ (\lambda + 2\mu)\left(\varepsilon_{xx}^2 + \varepsilon_{yy}^2 + \varepsilon_{zz}^2\right) \right.$$
$$\left. + \lambda\left(\varepsilon_{xx} + \varepsilon_{yy} + \varepsilon_{zz}\right) + \mu\left(\gamma_{xy}^2 + \gamma_{xz}^2 + \gamma_{yz}^2\right) \right\} dV \qquad (3.11)$$

or in matrix form:

$$\Phi(x, t) = \frac{1}{2} \int_V \varepsilon^{\mathrm{T}} H \varepsilon \, dV$$

where ε is the strain vector and H is matrix of elastic coefficients, given in Eqs. (1.12) and (1.26), respectively.

In the non-dimensional coordinate system, considering the dependence of the Lamè constants from E and v, Eq. (1.25), the strain energy Φ becomes:

$$\Phi(\xi, \eta, \zeta, t) = \frac{abh}{16} \iiint_{-1}^{1} \frac{E}{1+v} \left\{ \frac{1-v}{1-2v} \left(\varepsilon_{xx}^2 + \varepsilon_{yy}^2 + \varepsilon_{zz}^2 \right) \right.$$

$$+ \frac{2v}{1-2v} \left(\varepsilon_{xx}\varepsilon_{yy} + \varepsilon_{xx}\varepsilon_{zz} + \varepsilon_{yy}\varepsilon_{zz} \right)$$

$$\left. + \frac{1}{2} \left(\gamma_{xy}^2 + \gamma_{xz}^2 + \gamma_{yz}^2 \right) \right\} d\xi d\eta d\zeta \tag{3.12}$$

where the modulus of elasticity E is function of the ζ coordinate, Eq. (3.4).

3.2.2 Kinetic energy

The kinetic energy of the plate during the vibratory motion is given by:

$$T(x, y, z, t) = \frac{1}{2} \int_V \rho(z)(\dot{u}^2 + \dot{v}^2 + \dot{w}^2) \, dV$$

where the dot denotes differentiation with respect to time t and $\rho(z)$ is given in Eq. (3.5). In matrix form:

$$T(\boldsymbol{x}, t) = \frac{1}{2} \int_V \rho(z)\dot{\boldsymbol{u}}^{\mathrm{T}} \dot{\boldsymbol{u}} \, dV \tag{3.13}$$

In the non-dimensional coordinate system $(O; \xi, \eta, \zeta)$, the kinetic energy T is:

$$T(\xi, \eta, \zeta, t) = \frac{abh}{16} \iiint_{-1}^{1} \rho(\zeta)(\dot{u}^2 + \dot{v}^2 + \dot{w}^2) d\xi d\eta d\zeta. \tag{3.14}$$

3.2.3 Dissipation functional

A dissipation functional D, introduced by Siu and Bert (1974) and Leissa (1978; 1984), see also Elishakoff *et al.* (1988), is here extended for a three-dimensional solid:

$$D(x, y, z, t) = \frac{1}{2} \int_V c(z)(\dot{u}u + \dot{v}v + \dot{w}w) \, dV \tag{3.15}$$

where $c(z)$ is the viscous damping coefficient having the expression in Eq. (3.10). In matrix form:

$$D(\boldsymbol{x}, t) = \frac{1}{2} \int_V c(z)\dot{\boldsymbol{u}}^{\mathrm{T}}\boldsymbol{u} \, dV$$

In the non-dimensional coordinate system $(O; \xi, \eta, \zeta)$, the dissipation functional D is given by

$$D(\xi, \eta, \zeta, t) = \frac{abh}{16} \iiint_{-1}^{1} c(\zeta)(\dot{u}u + \dot{v}v + \dot{w}w)\, d\xi d\eta d\zeta \qquad (3.16)$$

3.2.4　Work done by the external forces

The FG plate is subjected to a transverse harmonic force applied on the top surface,

$$q(x, y, t) = Q(x, y) \cos(\bar{\omega}t) \qquad (3.17)$$

where $\bar{\omega}$ is the frequency of the exciting force. For sake of simplicity, in the right hand side of Eq. (3.17) the expression $\cos(\bar{\omega}t)$ is replaced by $e^{i\bar{\omega}t}$, where $i = (-1)^{1/2}$.

With this substitution, the modified governing equation yields a complex value for the displacements, the real part of the final expression constitutes the sought solution. The work done by the exciting force is

$$W(x, y, t) = \int_{S} q(x, y, t)w(x, y, h/2, t)\, dS \qquad (3.18)$$

where $w(x, y, h/2, t)$ is the transverse displacement evaluated at the upper surface, $z = h/2$. In the non-dimensional coordinate system $(O; \xi, \eta, \zeta)$, the work done by the applied load is:

$$W(\xi, \eta, t) = \frac{ab}{4} \iint_{-1}^{1} q(\xi, \eta, t)\, w(\xi, \eta, 1, t)\, d\xi d\eta \qquad (3.19)$$

3.3　Ritz method: Displacement representation

Due to the geometric simplicity of the problem, Ritz method is employed. The following expressions for the displacement components in the non-dimensional coordinate system $(O; \xi, \eta, \zeta)$ are assumed:

$$\begin{aligned}
u(\xi, \eta, \zeta, t) &= U(\xi, \eta, \zeta)\, e^{i\bar{\omega}t}, \\
v(\xi, \eta, \zeta, t) &= V(\xi, \eta, \zeta)\, e^{i\bar{\omega}t}, \qquad (3.20) \\
w(\xi, \eta, \zeta, t) &= W(\xi, \eta, \zeta)\, e^{i\bar{\omega}t},
\end{aligned}$$

where $U(\xi, \eta, \zeta)$, $V(\xi, \eta, \zeta)$ and $W(\xi, \eta, \zeta)$ are expressed in terms of a triple series of Chebyshev polynomials in the following form:

$$U(\xi, \eta, \zeta) = F_u(\xi, \eta) \sum_i^I \sum_j^J \sum_k^K c_{ijk}^u P_i(\xi) P_j(\eta) P_k(\zeta) \qquad (3.21)$$

$$V(\xi, \eta, \zeta) = F_v(\xi, \eta) \sum_l^L \sum_m^M \sum_n^N c_{lmn}^v P_l(\xi) P_m(\eta) P_n(\zeta) \qquad (3.22)$$

$$W(\xi, \eta, \zeta) = F_w(\xi, \eta) \sum_p^P \sum_q^Q \sum_r^R c_{pqr}^w P_p(\xi) P_q(\eta) P_r(\zeta) \qquad (3.23)$$

In Eqs. (3.21)–(3.23), the generic function $P_s(x)$ is the one-dimensional s-th Chebyshev polynomial, c_{ijk}^u, c_{lmn}^v, c_{pqr}^w are complex unknown coefficients to be determined and F_u, F_v, F_w are appropriate boundary functions.

In matrix form, the above relations can be written as:

$$\boldsymbol{u}(\boldsymbol{\xi}, t) = \boldsymbol{U}(\boldsymbol{\xi})\, e^{i\bar{\omega}t} \qquad (3.24)$$

where $\boldsymbol{u}(\boldsymbol{\xi}, t)$ is the displacement vector and $\boldsymbol{U}(\boldsymbol{\xi})$ is the displacement amplitude vector.

Vector $\boldsymbol{U}(\boldsymbol{\xi})$ is written in the following form:

$$\boldsymbol{U}(\boldsymbol{\xi}) = \boldsymbol{P}(\boldsymbol{\xi})\boldsymbol{C} \qquad (3.25)$$

According to the above displacement representations, Eqs. (3.21)–(3.23), matrix $\boldsymbol{P}(\boldsymbol{\xi})$ in Eq. (3.25) of the admissible functions takes the form:

$$\boldsymbol{P}(\boldsymbol{\xi}) = \begin{bmatrix} P_{111}^u(\boldsymbol{\xi}) & \cdots & P_{IJK}^u(\boldsymbol{\xi}) & 0 & \cdots \\ 0 & \cdots & 0 & P_{111}^v(\boldsymbol{\xi}) & \cdots \\ 0 & \cdots & 0 & 0 & \cdots \end{bmatrix}$$

$$\begin{bmatrix} 0 & 0 & \cdots & 0 \\ P_{LMN}^v(\boldsymbol{\xi}) & 0 & \cdots & 0 \\ 0 & P_{111}^w(\boldsymbol{\xi}) & \cdots & P_{PQR}^w(\boldsymbol{\xi}) \end{bmatrix} \qquad (3.26)$$

where

$$P_{ijk}^u(\boldsymbol{\xi}) = F_u(\xi, \eta) P_i(\xi) P_j(\eta) P_k(\zeta),$$

$$P_{lmn}^v(\boldsymbol{\xi}) = F_v(\xi, \eta) P_l(\xi) P_m(\eta) P_n(\zeta),$$

$$P_{pqr}^w(\boldsymbol{\xi}) = F_w(\xi, \eta) P_p(\xi) P_q(\eta) P_r(\zeta),$$

and C in Eq. (3.25) is the following column vector of the unknown coefficients:

$$C = \begin{bmatrix} C^u \\ C^v \\ C^w \end{bmatrix} = \begin{bmatrix} C^u_{111} \cdots C^u_{IJK} & C^v_{111} \cdots C^v_{LMN} & C^w_{111} \cdots C^w_{PQR} \end{bmatrix}^T$$

(3.27)

3.3.1 Chebyshev polynomials

In the expressions of the displacement components, Eqs. (3.21)–(3.23), Chebyshev polynomials multiplied by boundary functions have been employed.

Thus, $P_s(x)$ is the one-dimensional s-th Chebyshev polynomial which can be written via a recurrence relation (Snyder, 1966):

$$P_{s+1}(x) = 2xP_s(x) - P_{s-1}(x)$$

with $s = 2, 3, \ldots$ and $x = \xi, \eta, \zeta$. When $s = 1$, $P_1(x)$ is constant equal to 1.

The first five Chebyshev polynomials are given as follows:

$$P_1(x) = 1,$$
$$P_2(x) = x,$$
$$P_3(x) = 2x^2 - 1,$$
$$P_4(x) = 4x^3 - 3x,$$
$$P_5(x) = 1 + 8x^4 - 8x^2$$

and represented in Fig. 3.2 in the interval $[-1, 1]$. It is known that Chebyshev polynomials constitute an orthogonal polynomial sequence with respect to the weighting function

$$\varpi(x) = (1 - x^2)^{-\frac{1}{2}}$$

(3.28)

on the interval $[-1, 1]$, so that the following relation is satisfied:

$$\int_{-1}^{1} P_s(x) P_{s'}(x) \varpi(x) dx = \begin{cases} \pi & s = s' = 1 \\ \pi/2 & s = s' \neq 1 \\ 0 & s \neq s' \end{cases}$$

(3.29)

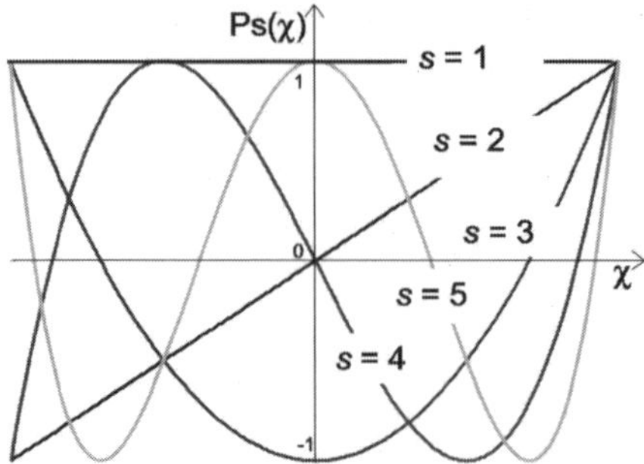

Figure 3.2.   First five Chebyshev polynomials

In numerical approximation, Chebyshev polynomials have shown high accuracy and better stability (Zhou *et al.*, 2002), compared with other polynomial series, as for example simple algebraic polynomials (Leissa and Zhang, 1983; Bhat, 1985), and generated orthogonal polynomials (Liew *et al.* 1993; 1995).

3.3.2 Boundary functions

In this study, all round simply supported and clamped plates are considered. The boundary conditions for the simply supported plate are specified as follows:

$$v = 0, \quad w = 0, \quad \sigma_x = 0, \quad at\ \xi = \pm 1,$$
$$u = 0, \quad w = 0, \quad \sigma_y = 0, \quad at\ \eta = \pm 1.$$

While the boundary conditions for the clamped plate are given by:

$$u = 0, \quad v = 0, \quad w = 0, \quad at\ \xi = \pm 1, \quad at\ \eta = \pm 1.$$

In order to satisfy the above edge conditions according to the Ritz formulation, F_u, F_v and F_w in Eqs. (3.21)–(3.23) can be expressed as a product of two one-dimensional functions in ξ and η direction:

$$F_\delta(\xi, \eta) = f_\delta^1(\xi) f_\delta^2(\eta) \quad \text{with } \delta = u, v, w$$

where $f_\delta^1(\xi)$ and $f_\delta^2(\eta)$ are given in Table 3.1.

It is worth noting that, according to the three-dimensional analysis, the boundary conditions are satisfied in a point-wise manner in the lateral surface of the plate rather than in an integral manner.

Table 3.1. Boundary functions

	$f_u^1(\xi))$	$f_v^1(\xi)$	$f_w^1(\xi)$	$f_u^2(\eta)$	$f_v^2(\eta)$	$f_w^2(\eta)$
Simply supported	1	$1-\xi^2$	$1-\xi^2$	$1-\eta^2$	1	$1-\eta^2$
Clamped	$1-\xi^2$	$1-\xi^2$	$1-\xi^2$	$1-\eta^2$	$1-\eta^2$	$1-\eta^2$

3.4 Solution methodology

Leissa and Young (1984), generalized the functional $T_{max} - \Phi_{max}$ used in free, undamped vibration analysis by minimizing the following functional

$$\Pi_{max} = (\Phi_{max} - W_{max}) - (T_{max} - D_{max}) \tag{3.30}$$

where the terms on the right hand side of Eq. (3.30) are the maximum values of the functionals previously given by Eqs. (3.12), (3.14), (3.16) and (3.19), obtained from Eqs. (3.20) and (3.21)–(3.23) setting $|e^{2i\omega t}| = 1$. The expressions of the functionals in Eq. (3.30) are:

$$T_{max} = \frac{abh}{16}\varpi^2\sqrt{\rho_1\rho_2}\iiint_{-1}^{1}\rho_{fgm}(\zeta)(U^2 + V^2 + W^2)d\xi d\eta d\zeta \tag{3.31}$$

$$D_{max} = \frac{abh}{16}i\bar{\omega}\sqrt{c_1c_2}\iiint_{-1}^{1}\rho_{fgm}(\zeta)(U^2 + V^2 + W^2)d\xi d\eta d\zeta \tag{3.32}$$

$$\Phi_{max} = \frac{bh}{4a}\sqrt{E_1E_2}\iiint_{-1}^{1}\frac{E_{FGM}(\zeta)(1-v)}{(1+v)(1-2v)}\left\{\left(\frac{\partial U}{\partial\xi}\right)^2 + \frac{a^2}{b^2}\left(\frac{\partial V}{\partial\eta}\right)^2\right.$$

$$+ \frac{a^2}{h^2}\left(\frac{\partial W}{\partial\zeta}\right)^2\frac{2v}{1-v}\left[\frac{a}{b}\frac{\partial U}{\partial\xi}\frac{\partial V}{\partial\eta} + \frac{a}{h}\frac{\partial U}{\partial\xi}\frac{\partial W}{\partial\zeta} + \frac{a^2}{hb}\frac{\partial V}{\partial\eta}\frac{\partial W}{\partial\zeta}\right]$$

$$+ \frac{1-2v}{2(1-v)}\left[\frac{a^2}{b^2}\left(\frac{\partial U}{\partial\eta}\right)^2 + \frac{a^2}{h^2}\left(\frac{\partial U}{\partial\zeta}\right)^2 + \left(\frac{\partial V}{\partial\xi}\right)^2\right.$$

$$+ \frac{a^2}{h^2}\left(\frac{\partial V}{\partial\zeta}\right)^2 + \left(\frac{\partial W}{\partial\xi}\right)^2 + \frac{a^2}{b^2}\left(\frac{\partial W}{\partial\eta}\right)^2$$

$$\left.\left.+ \frac{2a}{b}\frac{\partial U}{\partial\eta}\frac{\partial V}{\partial\xi} + \frac{2a}{h}\frac{\partial U}{\partial\zeta}\frac{\partial W}{\partial\xi} + \frac{2a^2}{hb}\frac{\partial V}{\partial\zeta}\frac{\partial W}{\partial\eta}\right]\right\}d\xi d\eta d\zeta \tag{3.33}$$

$$W_{max} = \frac{abh}{4} \iint_{-1}^{1} Q(\xi, \eta,)W(\xi, \eta, 1)d\xi d\eta. \tag{3.34}$$

The Ritz method requires Π_{max} in Eq. (3.30) to attain minimum, so that:

$$\frac{\partial \Pi_{max}}{\partial C_{ijk}^{u}} = 0$$

$$\frac{\partial \Pi_{max}}{\partial C_{lmn}^{v}} = 0 \tag{3.35}$$

$$\frac{\partial \Pi_{max}}{\partial C_{pqr}^{w}} = 0$$

which leads to the following governing equation:

$$RC = \frac{a^2}{2h}Q \tag{3.36}$$

where C is given in Eq. (3.27),

$$Q = \begin{bmatrix} 0 \\ 0 \\ Q^z \end{bmatrix} \tag{3.37}$$

and

$$R = \sqrt{E_1 E_2}K + \left(\frac{\sqrt{c_1 c_2}}{\sqrt{\rho_1 \rho_2}} i\bar{\omega} - \bar{\omega}^2 \right) a^2 \sqrt{\rho_1 \rho_2}M \tag{3.38}$$

with

$$K = \begin{bmatrix} K_{uu} & K_{uv} & K_{uw} \\ K_{uv}^{T} & K_{vv} & K_{vw} \\ K_{uw}^{T} & K_{vw}^{T} & K_{ww} \end{bmatrix} \tag{3.39}$$

$$M = \begin{bmatrix} M_{uu} & 0 & 0 \\ 0 & M_{vv} & 0 \\ 0 & 0 & M_{ww} \end{bmatrix}. \tag{3.40}$$

In Eqs. (3.36)–(3.40), C, Q, K and M are the vectors of the displacement coefficients, the load vector, the stiffness and mass matrices, respectively.

The explicit form of the elements in the stiffness sub-matrices $\boldsymbol{K}_{ij}$ with $i, j = u, v, w$ in Eq. (3.39), are given by:

$$\boldsymbol{K}_{uu} = D^{11}_{uiu\bar{i}} E^{00}_{uju\bar{j}} F^{00}_{k\bar{k}} + \left(\frac{a}{b}\right)^2 D^{00}_{uiu\bar{i}} E^{11}_{uju\bar{j}} T^{00}_{k\bar{k}} + \left(\frac{a}{h}\right)^2 D^{00}_{uiu\bar{i}} E^{00}_{uju\bar{j}} T^{11}_{k\bar{k}},$$

$$\boldsymbol{K}_{uv} = \frac{a}{b}\left(D^{10}_{uivl} E^{01}_{ujvm} S^{00}_{kn} + D^{01}_{uivl} E^{10}_{ujvm} T^{00}_{kn}\right),$$

$$\boldsymbol{K}_{uw} = \frac{a}{h}\left(D^{10}_{uiwp} E^{00}_{ujwq} S^{00}_{kr} + D^{01}_{uiwp} E^{00}_{ujwq} T^{10}_{kr}\right),$$

$$\boldsymbol{K}_{vv} = \left(\frac{a}{b}\right)^2 D^{00}_{vlv\bar{l}} E^{11}_{vmv\bar{m}} F^{00}_{n\bar{n}} + \left(\frac{a}{h}\right)^2 D^{00}_{vlv\bar{l}} E^{00}_{vmv\bar{m}} T^{11}_{n\bar{n}} + D^{11}_{vlv\bar{l}} E^{00}_{vmv\bar{m}} T^{00}_{n\bar{n}},$$

$$\boldsymbol{K}_{uw} = \frac{a^2}{hb}\left(D^{00}_{vlwp} E^{10}_{vmwq} S^{01}_{nr} + D^{00}_{vlwp} E^{01}_{vmwq} T^{10}_{nr}\right),$$

$$\boldsymbol{K}_{ww} = \left(\frac{a}{h}\right)^2 D^{00}_{wpw\bar{p}} E^{00}_{wqw\bar{q}} F^{11}_{r\bar{r}} + \left(\frac{a}{b}\right)^2 D^{00}_{wpw\bar{p}} E^{11}_{wqw\bar{q}} T^{00}_{r\bar{r}}$$
$$+ D^{11}_{wpw\bar{p}} E^{00}_{wqw\bar{q}} T^{00}_{r\bar{r}}.$$

The elements in the mass sub-matrices $\boldsymbol{M}_{ij}$, with $i, j = u, v, w$ in Eq. (3.40), are given by:

$$\boldsymbol{M}_{uu} = \frac{1}{4} D^{00}_{uiu\bar{i}} E^{00}_{uju\bar{j}} M_{k\bar{k}},$$

$$\boldsymbol{M}_{vv} = \frac{1}{4} D^{00}_{vlv\bar{l}} E^{00}_{vmv\bar{m}} M_{n\bar{n}}, \tag{3.41}$$

$$\boldsymbol{M}_{ww} = \frac{1}{4} D^{00}_{wpw\bar{p}} E^{00}_{wqw\bar{q}} M_{r\bar{r}},$$

where

$$D^{rs}_{\alpha\sigma\beta\theta} = \int_{-1}^{1} \frac{d^r[f^1_\alpha(\xi) P_\sigma(\xi)]}{d\xi^r} \frac{d^s[f^1_\beta(\xi) P_\theta(\xi)]}{d\xi^s} d\xi \tag{3.42}$$

$$E^{rs}_{\alpha\sigma\beta\theta} = \int_{-1}^{1} \frac{d^r[f^2_\alpha(\eta) P_\sigma(\eta)]}{d\eta^r} \frac{d^s[f^2_\beta(\eta) P_\theta(\xi)]}{d\eta^s} d\eta \tag{3.43}$$

$$F^{rs}_{\sigma\theta} = \int_{-1}^{1} \frac{E_{fgm}(\zeta)(1-\upsilon)}{(1+\upsilon)(1-2\upsilon)} \frac{d^r P_\sigma(\zeta)}{d\zeta^r} \frac{d^s P_\theta(\zeta)}{d\zeta^s} d\zeta \tag{3.44}$$

$$S_{\sigma\theta}^{rs} = \int_{-1}^{1} \frac{E_{fgm}(\zeta)\upsilon}{(1+\upsilon)(1-2\upsilon)} \frac{d^r P_\sigma(\zeta)}{d\zeta^r} \frac{d^s P_\theta(\zeta)}{d\zeta^s} d\zeta \tag{3.45}$$

$$T_{\sigma\theta}^{rs} = \int_{-1}^{1} \frac{E_{fgm}(\zeta)}{2(1+\upsilon)} \frac{d^r P_\sigma(\zeta)}{d\zeta^r} \frac{d^s P_\theta(\zeta)}{d\zeta^s} d\zeta \tag{3.46}$$

$$M_{\sigma\theta} = \int_{-1}^{1} \rho_{fgm}(\zeta) P_\sigma(\zeta) P_\theta(\zeta) d\zeta \tag{3.47}$$

$$(\sigma = i, j, k, l, m, n, p, q, r; \quad \theta = \bar{i}, \bar{j}, \bar{k}, \bar{l}, \bar{m}, \bar{n}, \bar{p}, \bar{q}, \bar{r})$$

where $r, s = 0, 1$ and the subscripts α and β indicate the corresponding displacement amplitude functions u, v and w. Young's modulus, $E_{FGM}(\zeta)$, and mass density, $\rho_{FGM}(\zeta)$, have the expressions in Eqs. (3.6) and (3.7), respectively.

In free vibration analysis, the damping term, D_{max}, and the work done by the applied load, W_{max}, are set to be zero in the energy functional, Eq. (3.30). Thus, the maximum energy functional Π_{max} of the plate becomes:

$$\Pi_{max} = T_{max} - \Phi_{max} \tag{3.48}$$

where Φ_{max} and T_{max} are the maximum values in time of the strain and kinetic energies given in Eqs. (3.12) and (3.14). By means of these simplifications, the expression of the modified $\mathbf{R}$ matrix, Eq. (3.38), becomes:

$$\mathbf{R} = \sqrt{E_1 E_2}\,\mathbf{K} - \omega^2 a^2 \sqrt{\rho_1 \rho_2}\,\mathbf{M}$$

where ω is the natural frequency. Thus, the following governing eigenvalue equation is obtained:

$$(\mathbf{K} - \widehat{\Omega}^2 \mathbf{M})\mathbf{C} = 0 \tag{3.49}$$

in which

$$\widehat{\Omega} = \omega a \left(\frac{\rho_1 \rho_2}{E_1 E_2}\right)^{\frac{1}{4}} \tag{3.50}$$

$\mathbf{C}$, $\mathbf{K}$ and $\mathbf{M}$ denote the vector of the undetermined coefficients, the stiffness and mass matrices, respectively, that have the same expressions of Eqs. (3.27), (3.39) and (3.40). A non-trivial solution is obtained by setting

Table 3.2. Material properties

	Young's modulus	**Poisson's ratio**	**Mass density**
Zirconia	$E_1 = 168\,\text{GPa}$	$\upsilon_1 = 0.3$	$\rho_1 = 5700\,\text{Kg/m}^3$
Aluminum	$E_2 = 70\,\text{GPa}$	$\upsilon_2 = 0.3$	$\rho_2 = 2707\,\text{Kg/m}^3$

the determinant of the coefficient matrix of Eq. (3.49) equal to zero. Roots of the determinant are the square of the frequency parameter $\widehat{\Omega}$.

3.5 Numerical results

The continuum three-dimensional approach has been applied to study the free and forced vibration of FG all-round simply supported and clamped square ($a/b = 1$) plates. A wide range of thickness to width ratios is considered, from thin ($h/a = 0.05$), moderately thick ($h/a = 0.1$ and 0.2) to very thick plates ($h/a = 0.3 \div 0.6$). The influence of the power-law exponent N, thickness ratio and boundary conditions on the vibration frequencies is examined in detail.

The two constituents of the FG plate are taken to be zirconia and aluminum, with the material properties shown in Table 3.2 (Touloukian, 1967).

The non-dimensional frequency parameter Ω is introduced:

$$\Omega = \frac{b^2}{ah\pi^2}\sqrt{12(1 - \upsilon^2)}\widehat{\Omega},$$

where $\widehat{\Omega}$ is given in Eq. (3.50). In the calculation, the Poisson's ratio υ is considered constant equal to 0.3. The first 12 non-dimensional frequency parameters Ω are calculated. The integrations in Eqs. (3.42)–(3.48) were performed by using numerical integration schemes, as Gaussian quadrature, when analytical solutions are not available.

3.5.1 Free vibrations

Convergence studies

To demonstrate the accuracy of the present approach, convergence tests are carried out for the two kinds of boundary conditions. The objective of these studies is to determine the number of terms in x, y and z direction in the series of the displacement, Eqs. (3.21)–(3.23), that are needed to get the first 12 frequency parameters accurate up to 5 significant figures.

In general, in the representation of the displacement components, Eqs. (3.21)–(3.23), the number of terms in x, y and z direction can be different. In practice, in the following calculations the same number of terms, indicated with N_x, N_y and N_z, in the x, y and z-direction are considered

$$(I = L = P = N_x, J = M = Q = N_y, K = N = R = N_z).$$

Since the convergence properties are not affected by the homogeneity or inhomogeneity of the plates, convergence studies are conducted for FG plates with power-law exponent $N = 0.2$.

The error ε is calculated as

$$\varepsilon = \frac{\Omega_{i+1} - \Omega_i}{\Omega_i} 100$$

where Ω_i corresponds to the frequency parameter found with a certain number of terms in the series and Ω_{i+1} corresponds to the frequency parameter calculated with an increasing number of terms. In Figs. 3.3–3.9, the error ε

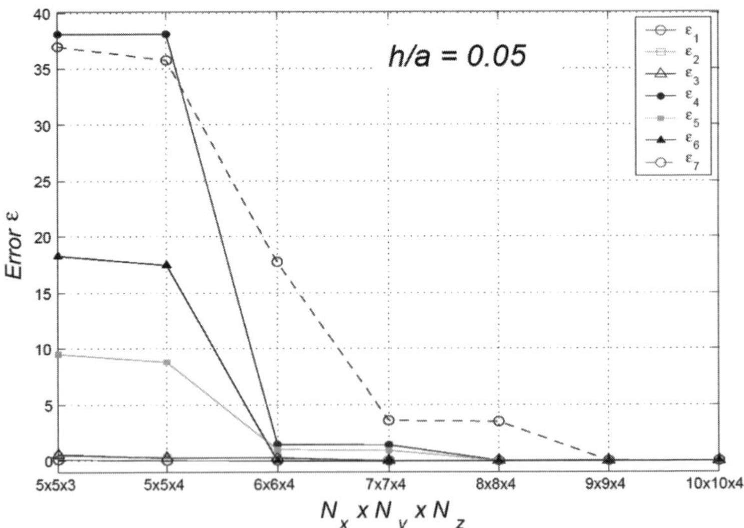

Figure 3.3. Rate of convergence of frequency parameters, Ω, for a FG all round simply supported plate with $h/a = 0.05$ and $N = 0.2$

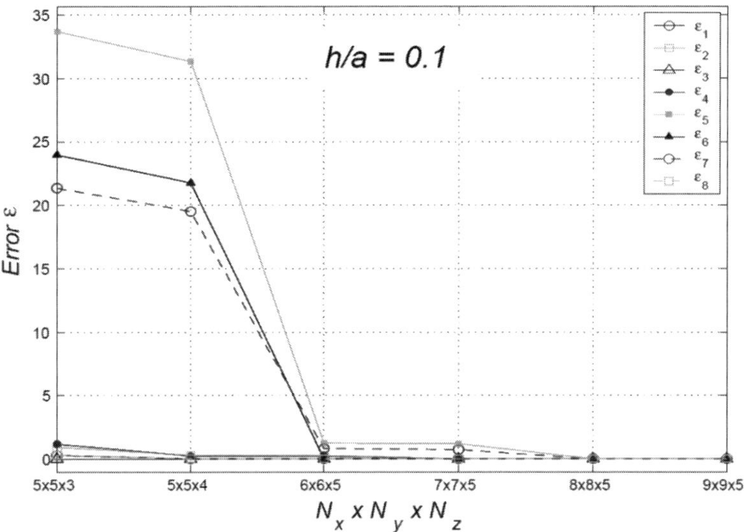

Figure 3.4. Rate of convergence of frequency parameters, Ω, for a FG all round simply supported plate with $h/a = 0.1$ and $N = 0.2$

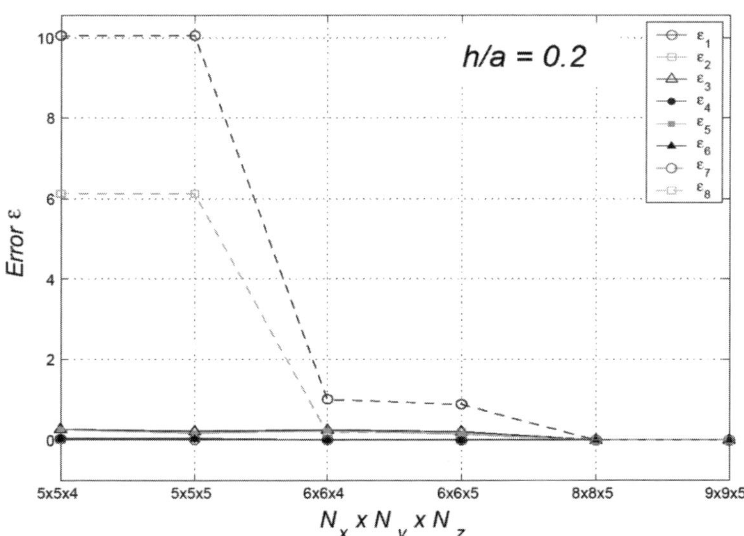

Figure 3.5. Rate of convergence of frequency parameters, Ω, for a FG all round simply supported plate with $h/a = 0.2$ and $N = 0.2$

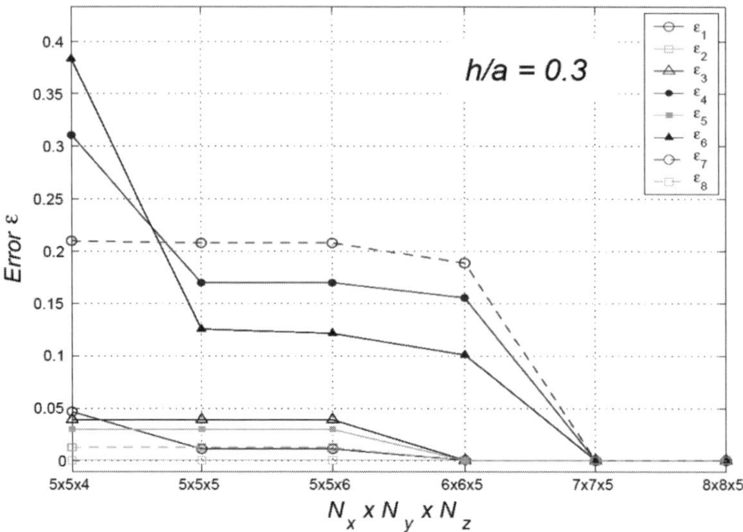

Figure 3.6. Rate of convergence of frequency parameters, Ω, for a FG all round simply supported plate with $h/a = 0.3$ and $N = 0.2$

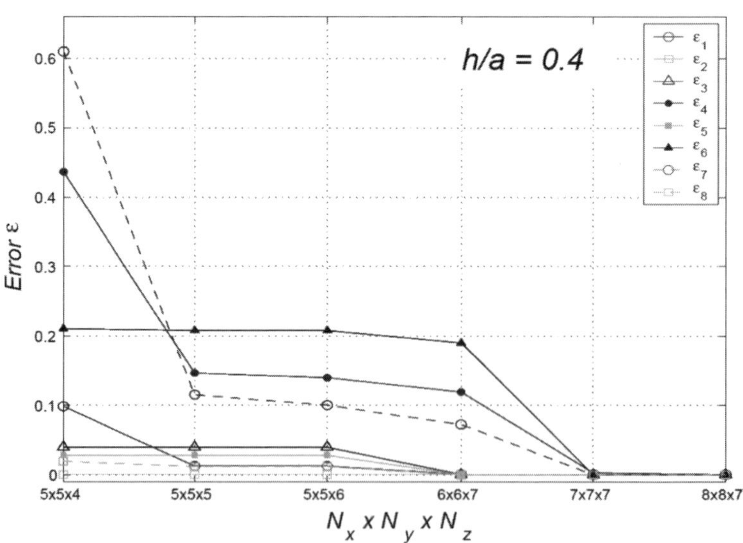

Figure 3.7. Rate of convergence of frequency parameters Ω, for a FG all round simply supported plate with $h/a = 0.4$ and $N = 0.2$

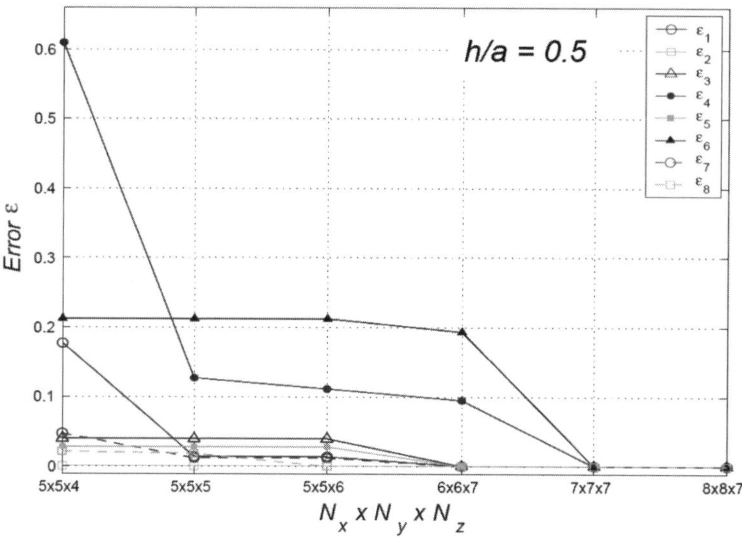

Figure 3.8. Rate of convergence of frequency parameters, Ω, for a FG all round simply supported plate with $h/a = 0.5$ and $N = 0.2$

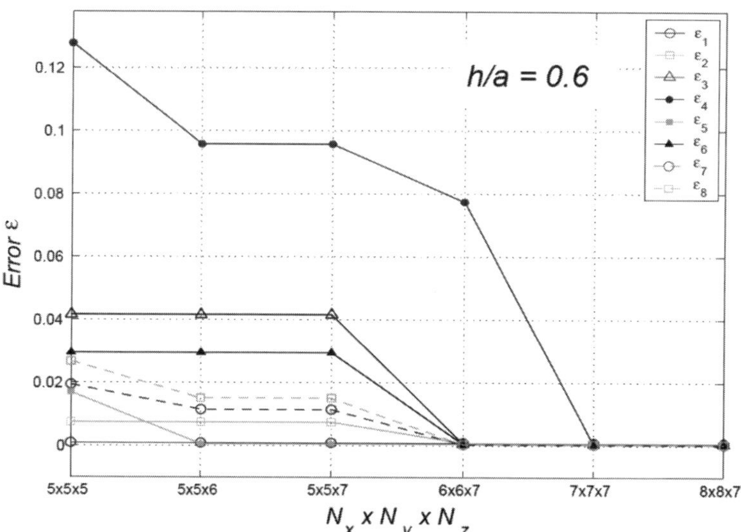

Figure 3.9. Rate of convergence of frequency parameters, Ω, for a FG all round simply supported plate with $h/a = 0.6$ and $N = 0.2$

on the first 12 frequency parameters, Ω, for a FG simply supported plate is shown for thickness ratios from $h/a = 0.05$ to $h/a = 0.6$. Figures 3.10–3.14 show the error ε on the first 12 frequency parameters, Ω, for a FG all round clamped plate for thickness ratios from $h/a = 0.05$ to $h/a = 0.4$. In all the figures, only the errors corresponding to distinct frequency parameters are reported.

A careful scrutiny of the convergence figures and tables reveals that convergence rate is very rapid and the frequency parameters monotonically decrease with the increase in the number of terms of admissible functions.

Furthermore, as the thickness ratio increases, the number of terms needed for the polynomial function in the thickness direction (z-direction) increases.

For a relatively thin simply supported plate ($h/a = 0.05$), 4 terms in the z-direction are sufficient to give convergent results. However, it needs as

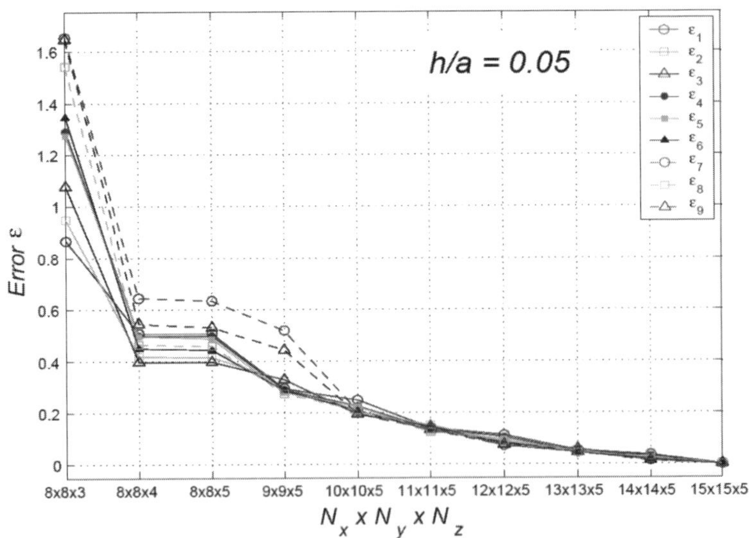

Figure 3.10. Rate of convergence of frequency parameters, Ω, for a FG all round clamped plate with $h/a = 0.05$ and $N = 0.2$

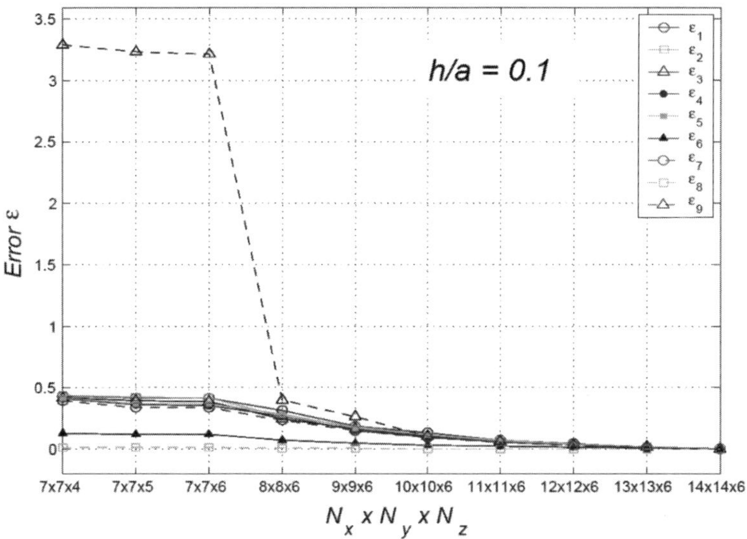

Figure 3.11. Rate of convergence of frequency parameters, Ω, for a FG all round clamped plate with $h/a = 0.1$ and $N = 0.2$

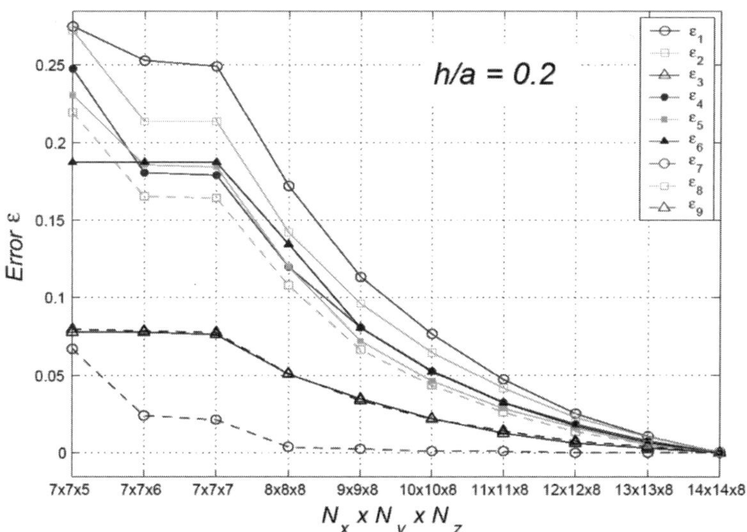

Figure 3.12. Rate of convergence of frequency parameters, Ω, for a FG all round clamped plate with $h/a = 0.2$ and $N = 0.2$

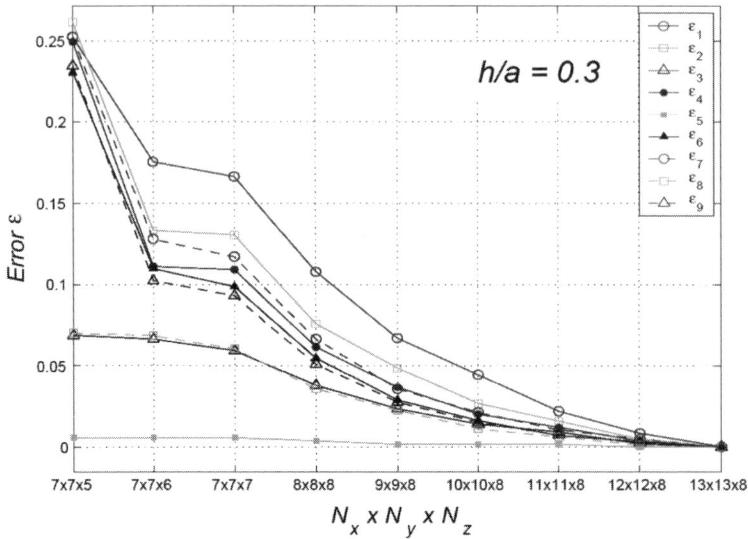

Figure 3.13.   Rate of convergence of frequency parameters, Ω, for a FG all round clamped plate with $h/a = 0.3$ and $N = 0.2$

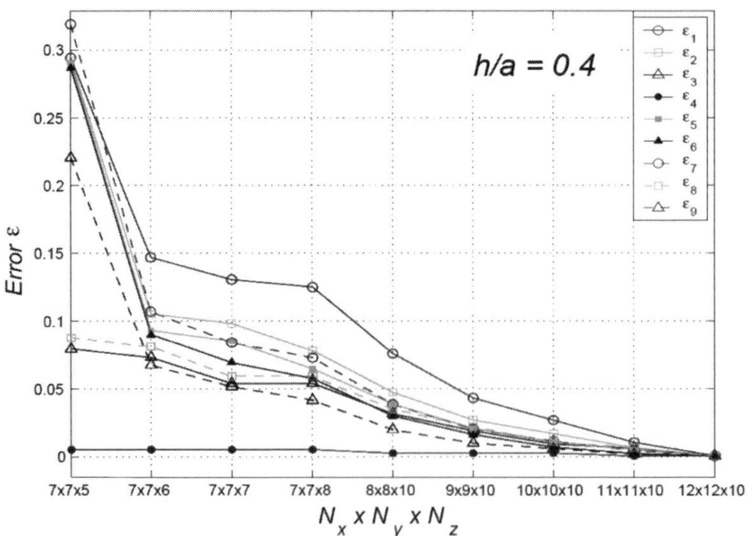

Figure 3.14.   Rate of convergence of frequency parameters, Ω, for a FG all round clamped plate with $h/a = 0.4$ and $N = 0.2$

high as 7 terms for a plate with thickness ratio $h/a = 0.6$ to achieve the same degree of accuracy. On the other hand, the number of terms needed in the admissible functions along the plate surface (x and y directions) reduces from 9×9 to 7×7 terms as the plate thickness dimension becomes significant. In general, few terms are needed to get a good convergence also when the plate is very thick. In fact, for the simply supported plate with $h/a = 0.6$, only 5 terms in the three directions are sufficient to get a good accuracy (i.e., low error ε) for the first frequencies, see Fig. 3.9.

For FG all round clamped plates, as the thickness of the plate increases from $h/a = 0.05$ to $h/a = 0.4$, the number of terms in z-direction increases proportionally from 5 terms to 10 terms and the number of terms in x and y directions decreases from 14×14 to 11×11. Thus, it can be further noticed that the boundary conditions have a great influence on the rate of convergence of the frequency parameters, in fact the number of terms needed for clamped plates is much greater than that needed for a simply supported plate.

From the preceding convergence studies, it can be inferred that the number of terms needed in the series is summarized in Table 3.3 for simply supported plates and in Table 3.4 for clamped plates, for all the thickness ratios considered.

Table 3.3. FG all round simply supported plate ($N = 0.2$): Minimum number of terms required to give the first 12 frequency parameters Ω convergent to 5 significant figures

h/a	No. terms $N_x \times N_y \times N_z$
0.05	$10 \times 10 \times 4$
0.1	$8 \times 8 \times 5$
0.2	$8 \times 8 \times 5$
0.3	$7 \times 7 \times 5$
0.4	$8 \times 8 \times 7$
0.5	$7 \times 7 \times 7$
0.6	$7 \times 7 \times 7$

Table 3.4. FG all round clamped plate ($N = 0.2$): Minimum number of terms required to give the first 12 frequency parameters Ω convergent to 5 significant figures

h/a	No. terms $N_x \times N_y \times N_z$
0.05	$14 \times 14 \times 5$
0.1	$13 \times 13 \times 6$
0.2	$13 \times 13 \times 8$
0.3	$12 \times 12 \times 8$
0.4	$11 \times 11 \times 10$

Comparison studies with homogeneous plates

To further verify the present approach, the first 10 frequency parameters $\bar{\Omega} = \omega a^2 / h \sqrt{\rho_2 E_2}$, computed for a homogeneous aluminum simply supported square plate with thickness side ratios $h/a = 0.1, 0.2$ and 0.5, are compared with results available in the literature.

The frequency parameters are listed in Table B-13 in Appendix B.

It is found that the results from the present formulation are in good agreement with the three-dimensional study by Liew and Teo (1999); Malik and Bert (1998) obtained by using the differential quadrature method, with the exact solution by Srinivas *et al.* (1970) and with the three-dimensional studies by Vel and Batra (2004) and Liew *et al.* (1995). Solutions from the classical plate theory by Leissa (1973) and from the Mindlin plate theory by Liew *et al.* (1995) are also presented for comparison and verification purposes.

In Table B-14 in Appendix B, the first 7 frequency parameters $\bar{\Omega} = \omega a^2 / h \sqrt{\rho_2 E_2}$, computed for a homogeneous aluminum clamped square plate with thickness side ratios $h/a = 0.1$ and $h/a = 0.2$, are listed for verification purposes. These frequencies are compared with results available in literature and calculated by means of different theories (classical theory Leissa, (1973), Mindlin theory (Liew *et al.* 1995) (Dawe and Roufaeil, 1980), three-dimensional solutions (Liew and Teo, 1999)).

After having established the rate of convergence and degree of accuracy of the present formulation, the three-dimensional continuum method is applied to compute the frequency ratios:

$$f_r = \frac{\omega}{\omega_{zir}},$$

between the first 12 natural frequency of FG all round simply supported and clamped plates for different values of the power-law exponent N (0.2, 0.5, 1 and 2) and the first 12 natural frequency of the homogeneous plate made of zirconia. Also the extreme case of the homogeneous plate made of aluminum is considered for comparison.

In Figs. 3.15–3.21, the frequency ratios, f_r, are reported for simply supported plates with thickness side ratios belonging to the interval $h/a = 0.05 \div 0.6$. As it can be seen from Figs. 3.15–3.19, natural frequencies of FG plates have an intermediate value between the natural frequencies of the limit cases of homogeneous plates of zirconia and aluminum, as it is expected. While in Figs. 3.20 and 3.21, there are some frequencies that do not lie in-between the frequencies of pure metal plate and those of pure ceramic plate. In particular, it can be noticed that, there are some frequencies

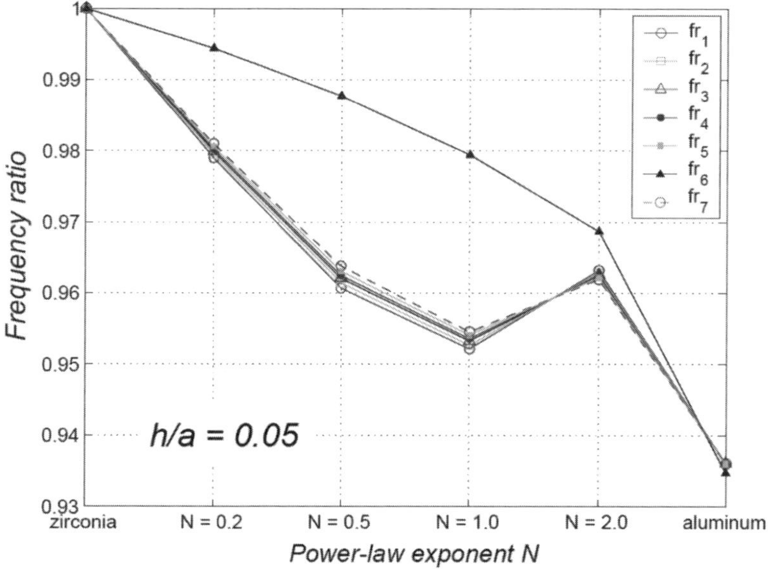

Figure 3.15. Frequency ratio, ω/ω_{zir}, for FG all round simply supported plates with $h/a = 0.05$

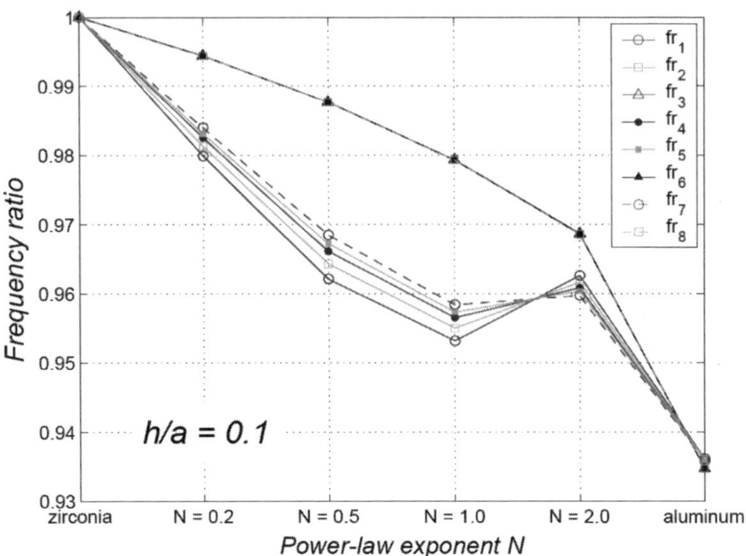

Figure 3.16. Frequency ratio, ω/ω_{zir} for FG all round simply supported plates with $h/a = 0.1$

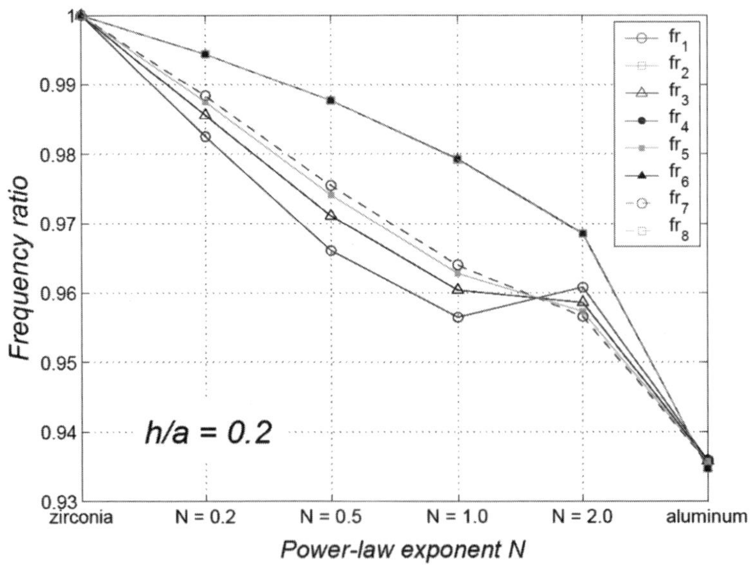

Figure 3.17. Frequency ratio, ω/ω_{zir}, for FG all round simply supported plates with $h/a = 0.2$

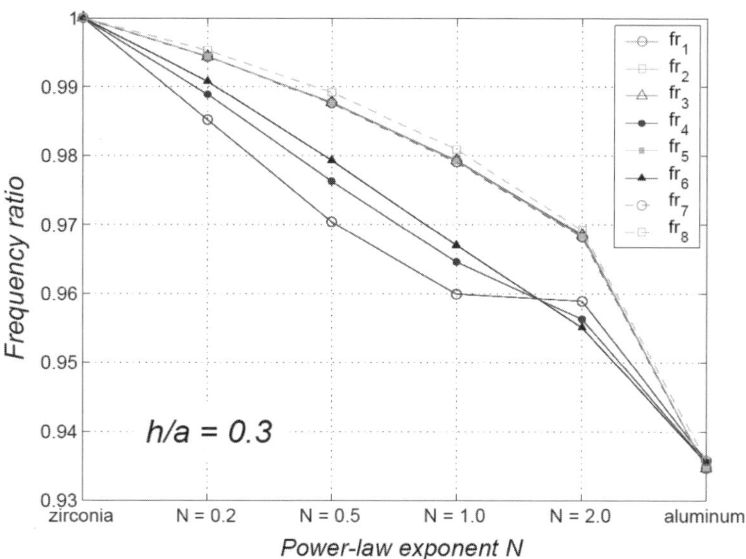

Figure 3.18. Frequency ratio, ω/ω_{zir}, for FG all round simply supported plates with $h/a = 0.3$

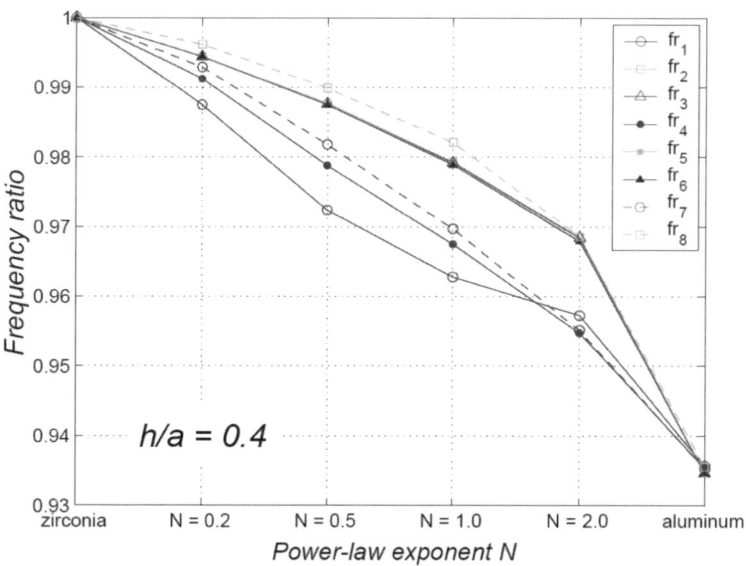

Figure 3.19. Frequency ratio, ω/ω_{zir}, for FG all round simply supported plates with $h/a = 0.4$

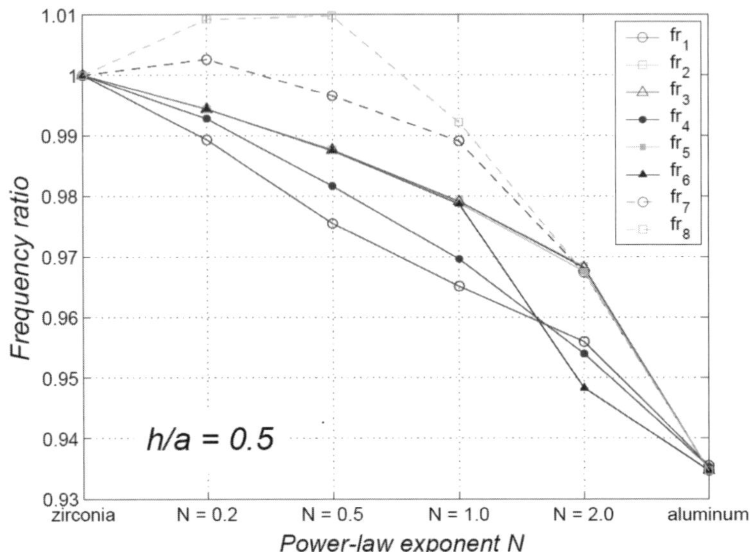

Figure 3.20. Frequency ratio, ω/ω_{zir}, for FG all round simply supported plates with $h/a = 0.5$

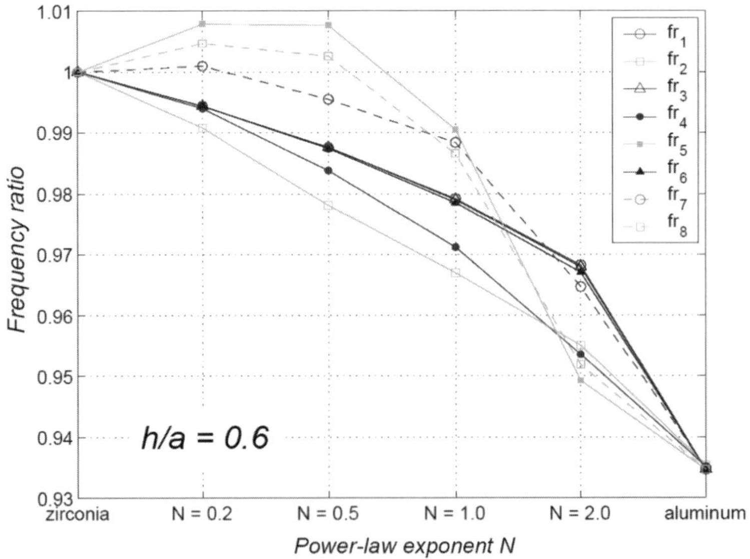

Figure 3.21. Frequency ratio, ω/ω_{zir}, for FG all round simply supported plates with $h/a = 0.6$

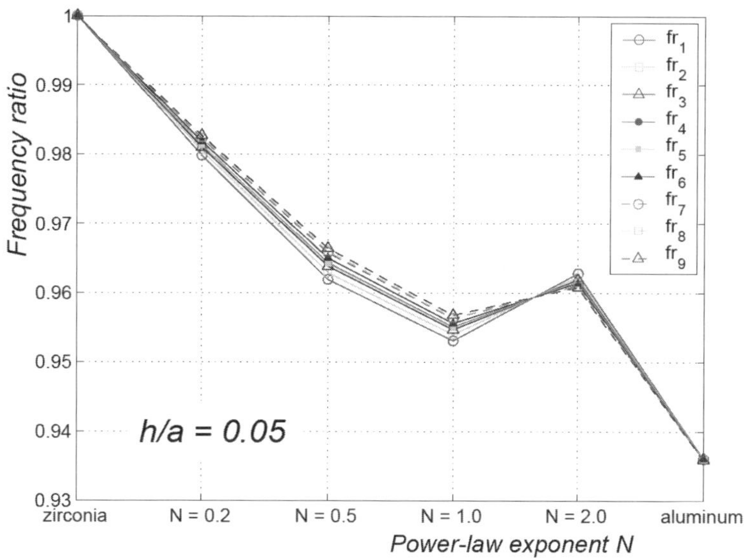

Figure 3.22.   Frequency ratio, $\omega/\omega zir$, for FG all round clamped plates with h/a $= 0.05$

that have a monotonic descending behavior between the two extreme values, while there are the remaining frequencies that exhibit a knee in correspondence of the FG plate with power-law exponent $N = 2$. A further study on the mode shapes of FG simply supported plates, see Figs. 3.22–3.28, shows that the frequencies that have the regular behavior correspond to distortional mode shapes, i.e. in-plane vibration modes, while the others correspond to flexural mode shapes. It is known that as the thickness-side ratio increases, the distortional mode shapes increase in number and migrate to the lower vibration spectrum. In particular, for $h/a = 0.6$ the first frequency corresponds to a distortional mode. On the other hand, it should be noted that when the thickness side ratio h/a is equal to 0.5, the first frequencies corresponding to thickness modes appear, see in Fig. 3.37 the seventh (f_{r7}) and the eighth (f_{r8}) mode shape and in Fig. 3.28, the fifth (f_{r5}), seventh (f_{r7}) and eighth (f_{r8}) mode shape. To these thickness-mode shapes correspond in Figs. 3.20 and 3.21, the not-intermediate behavior of the frequency ratios.

In Figs. 3.29–3.33, the results for the all round clamped plates are depicted.

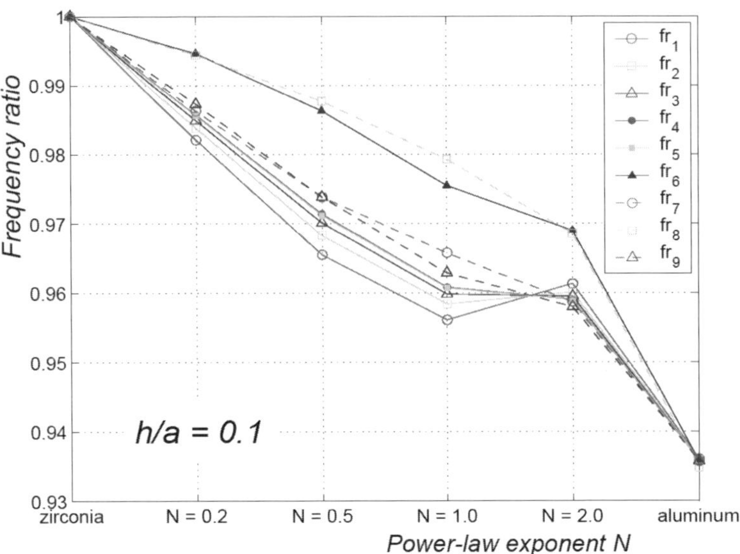

Figure 3.23. Frequency ratio, ω/ωzir, for FG all round clamped plates with h/a $= 0.1$

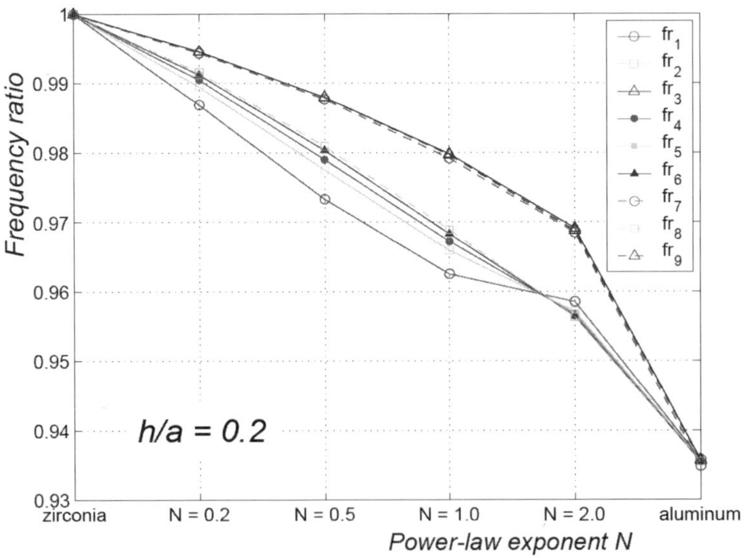

Figure 3.24. Frequency ratio, ω/ωzir, for FG all round clamped plates with h/a $= 0.2$

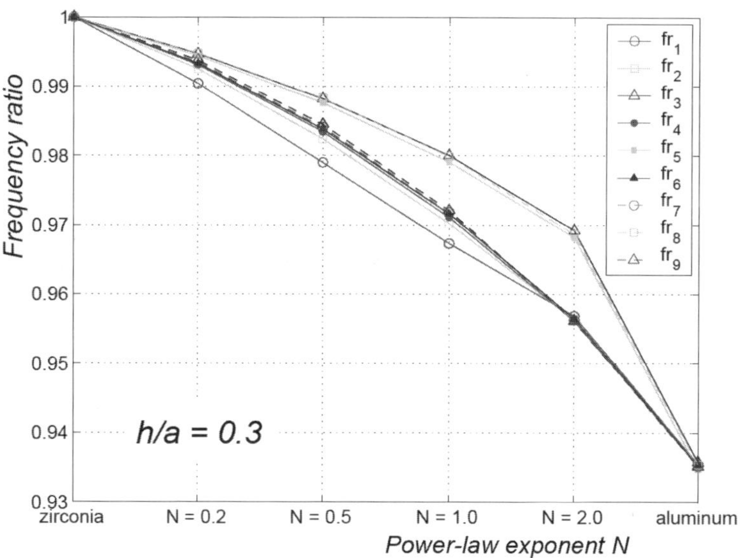

Figure 3.25. Frequency ratio, $\omega/\omega zir$, for FG all round clamped plates with $h/a = 0.3$

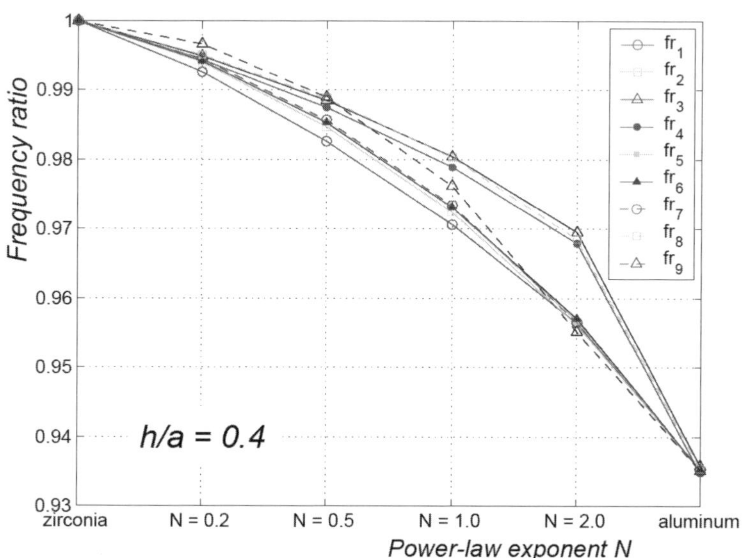

Figure 3.26. Frequency ratio, $\omega/\omega zir$, for FG all round clamped plates with $h/a = 0.4$

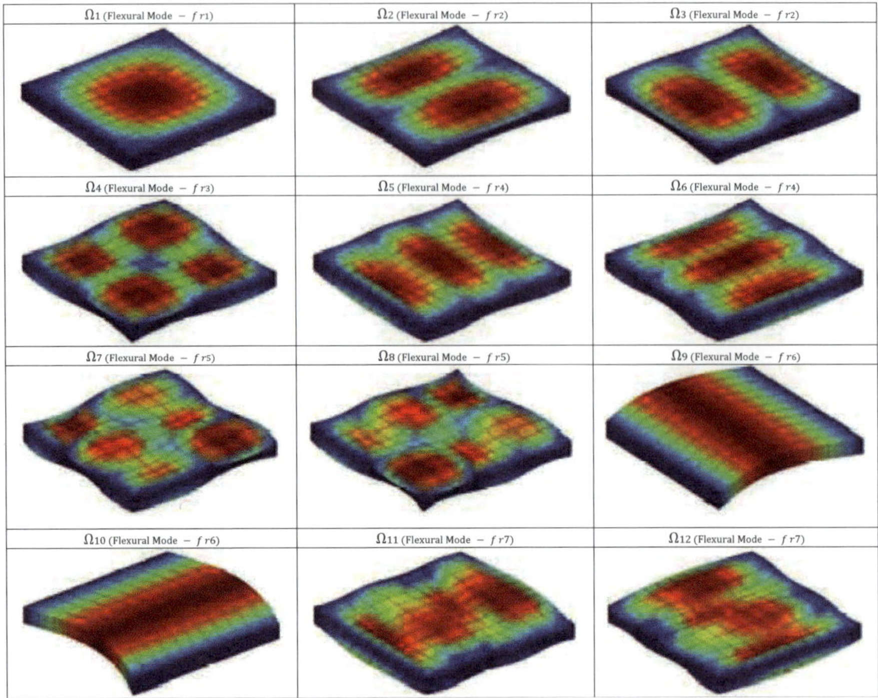

Figure 3.27. Mode shapes for a FG simply supported plate with $h/a = 0.5$ and $N = 0.2$

The behavior is similar to that of all round simply supported plates. In particular, for $h/a = 0.05$ there are no distortional mode shapes that appear only for a plate with $h/a = 0.1$. Since only thickness ratios lower and equal to 0.4 are considered, the thickness modes do not appear.

The presence of surface parallel, thickness modes in the lower vibration spectrum restrict the applicability of the refined plate theory such as Mindlin theory to the analysis of moderately thick plates. Since the first order Mindlin theory only considers flexural modes and in this study, it is found that the thickness modes (which include thickness shear; thickness twist and pure in-plane motion) often precede some of the flexural modes in thick plate vibrations. Although the method has the capability of analyzing accurately very thick plates, which two-dimensional theories cannot, it can also be applied to thin plates, thereby determining conclusively the accuracies of the plate theories. In fact, the in-plane modes of vibration are

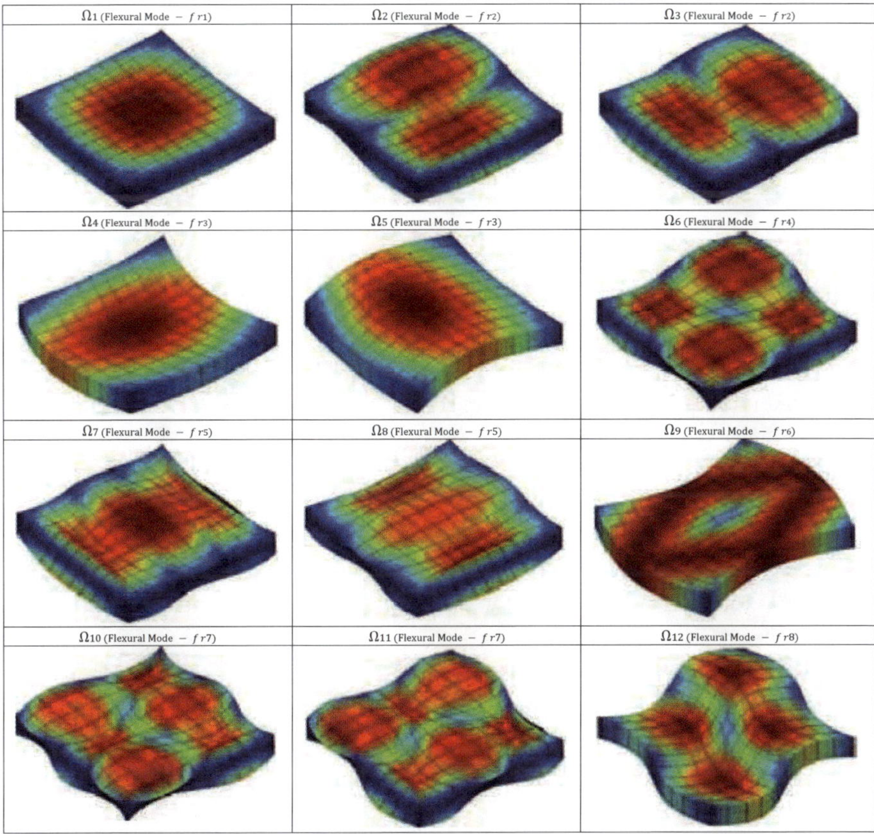

Figure 3.28. Mode shapes for a FG simply supported plate with $h/a = 0.6$ and $N = 0.2$

not predicted by a plate theory, such as the classical one, that neglects the in-plane displacements u and v.

From the tabulated results, it is observed that for plates with prescribed boundary conditions, the non-dimensional frequency parameters, $\bar{\Omega}$, decreases as the thickness side ratio, h/a, increases. Comparing the frequency parameters for plates with different boundary conditions, it is evident that plates with more constraints imposed on the boundaries have a higher value of $\bar{\Omega}$. In fact, the frequency parameters of the all round clamped plate are generally many times higher than the corresponding modes of the all round simply supported plate with the same geometrical

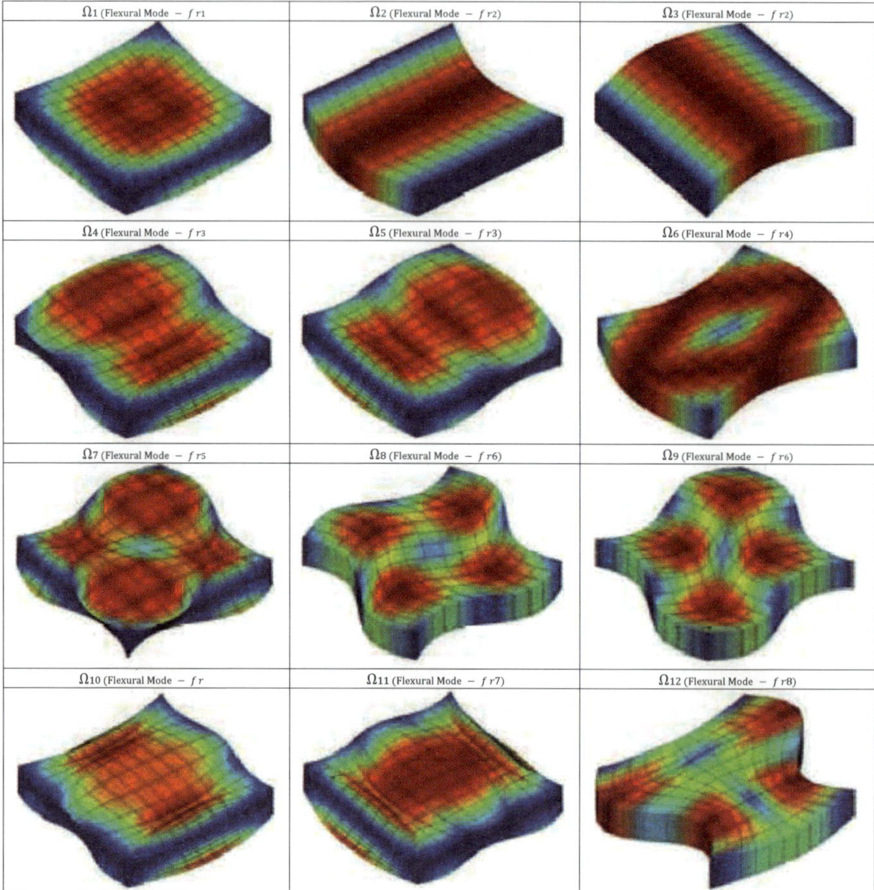

Figure 3.29. Frequency ratio, ω/ω_{zir}, for FG all round clamped plates with $h/a = 0.05$

specifications. It is interesting to note that the frequencies attain a minimum value for the plate made only of metal, due to the fact that aluminum has a much smaller Young's modulus than zirconia. Furthermore, frequencies of the FG plate are close to those of pure zirconia for small value of N (around 0.2). These tables show how it is possible to modify natural frequencies of FG plates by varying the power law exponent of the volume fraction.

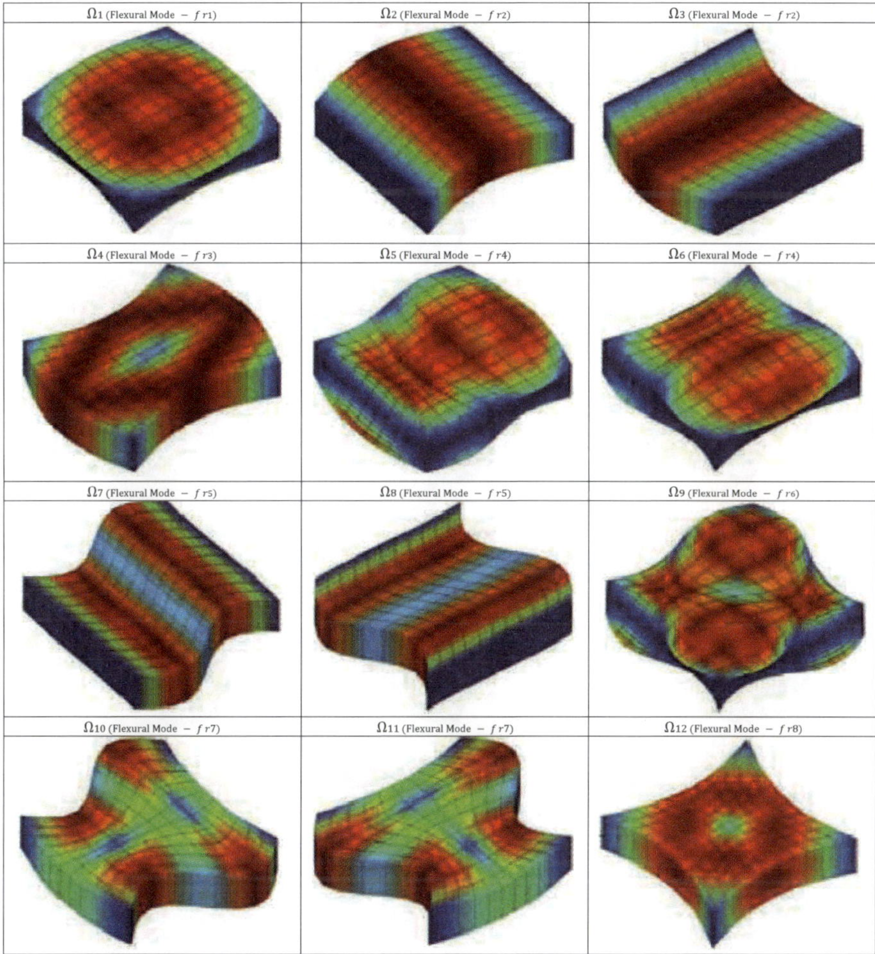

Figure 3.30. Frequency ratio, ω/ω_{zir}, for FG all round clamped plates with $h/a = 0.1$

3.5.2 Forced vibrations

Now the computing of the forced vibration response is considered.

In order to perform verification of this study it is needed to compare the present solution to some benchmark problems. Several papers have been published on forced vibrations of classical plates. Siu and Bert (1974)

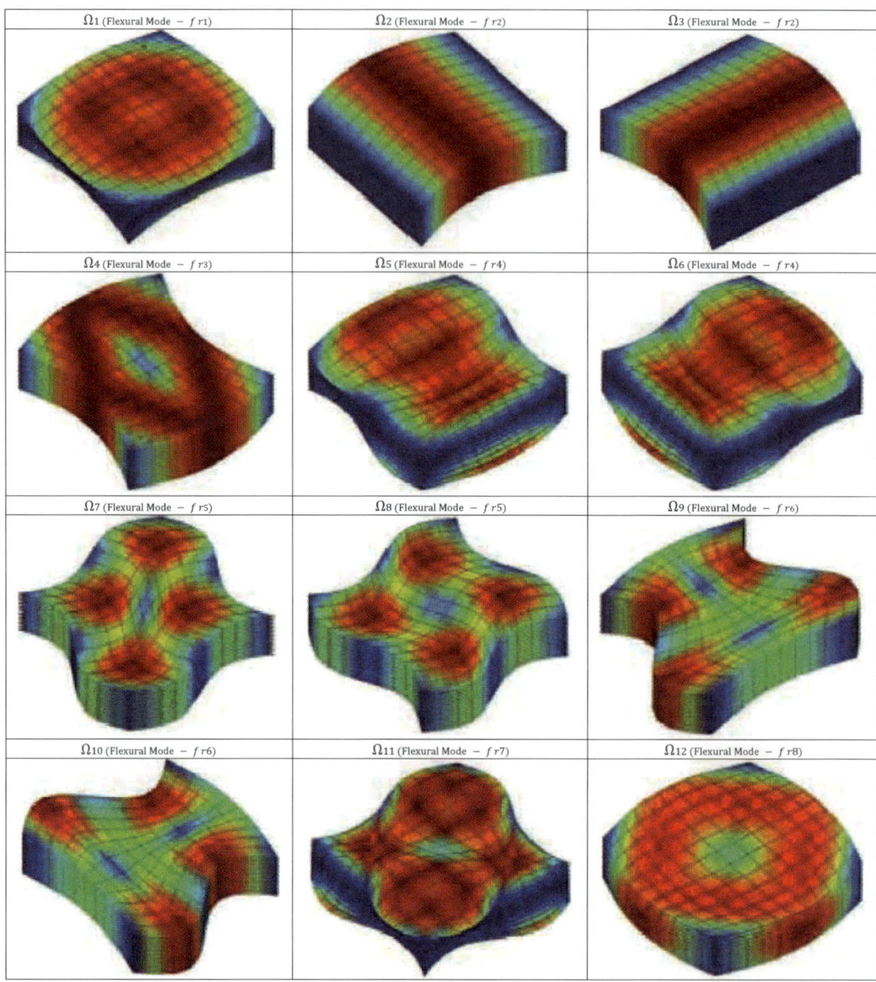

Figure 3.31. Frequency ratio, ω/ω_{zir}, for FG all round clamped plates with $h/a = 0.2$

presented a general forced vibration analysis for laminated isotropic rectangular plates including material damping. Mindlin plate theory and the Rayleigh–Ritz method were used in order to obtain the solution. Forced damped vibrations of circular plates were investigated by Leissa (1978) Rayleigh–Ritz method; Leissa (1978) studied forced vibrations of beams and rectangular plates. Elishakoff *et al.* (1988) extended the approach by

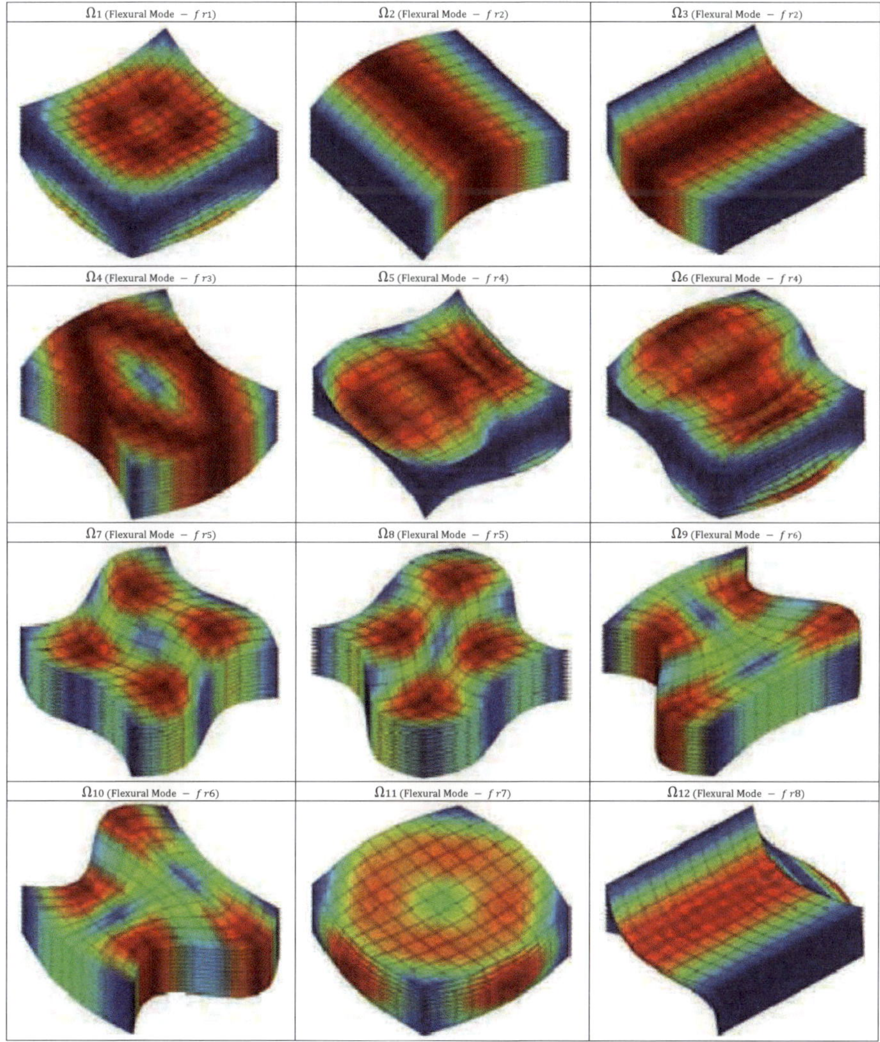

Figure 3.32. Frequency ratio, ω/ω_{zir}, for FG all round clamped plates with $h/a = 0.3$

Leissa by means of Wilson trial functions for the forced vibration problem of a cantilever beam.

In the paper by Laura and Duran (1975), the response was calculated at various frequencies expressed as the ratio of the excitation frequency to the natural frequency; they also calculated static response corresponding to

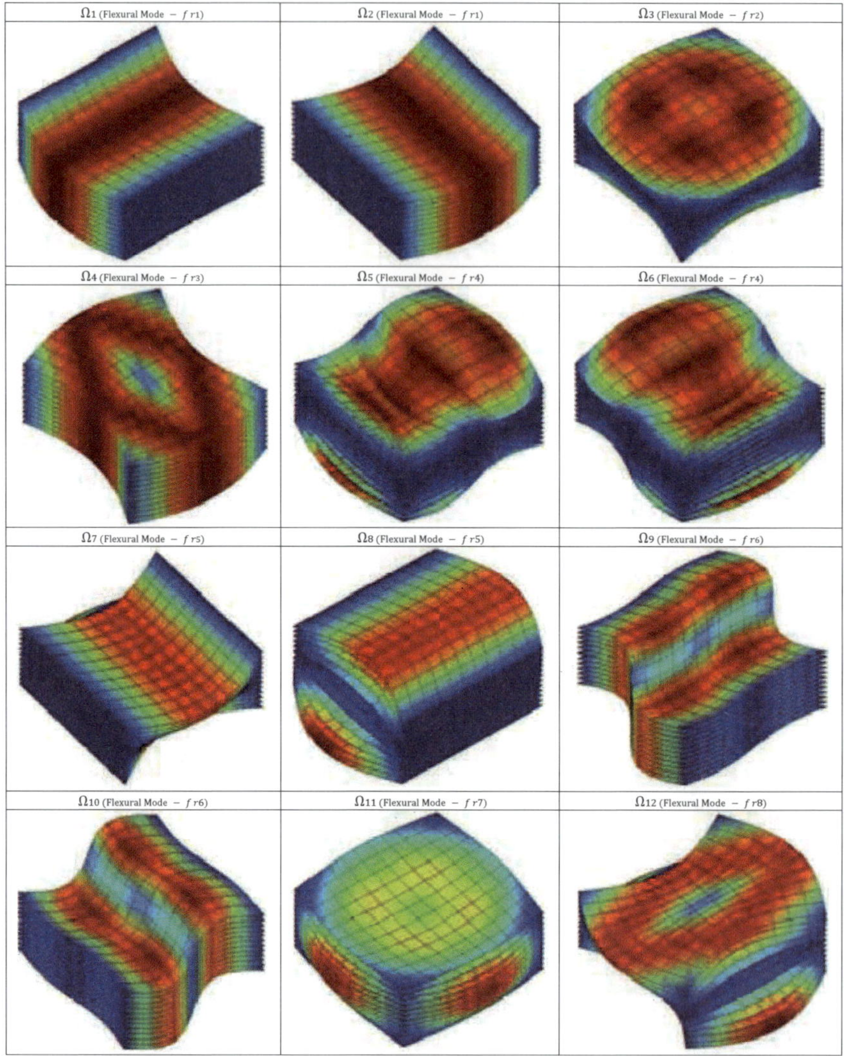

Figure 3.33. Frequency ratio, ω/ω_{zir}, for FG all round clamped plates with $h/a = 0.4$

$\bar{\omega} = 0$, as well as dynamic ones at $\bar{\omega}/\omega_1 = 0.3, 0.5$ and 0.8, where ω_1 is the first natural frequency. An equivalent critical damping c_{c1} is expressed in terms of the first natural frequency ω_1:

$$c_{c1} = 2\sqrt{\rho_1 \rho_2}\,\omega_1. \tag{3.51}$$

The ratio:

$$\delta_1 = \frac{\sqrt{c_1 c_2}}{c_{c1}} \tag{3.52}$$

is introduced, so that the expression for $\sqrt{c_1 c_2}/\sqrt{\rho_1 \rho_2}$ appeared in Eq. (3.38) becomes:

$$\sqrt{\frac{c_1 c_2}{\rho_1 \rho_2}} = 2\delta_1 \omega_1 \tag{3.53}$$

Thus, the matrix $\mathbf{R}$, in Eq. (3.38) can be rewritten in terms of these new quantities:

$$\mathbf{R} = \sqrt{E_1 E_2}\mathbf{K} + \omega_1^2 \gamma(2i\delta_1 - \gamma)\sqrt{\rho_1 \rho_2}a^2\mathbf{M}$$

where $\gamma = \bar{\omega}/\omega_1$. From Eq. (3.36), the vector of the displacements coefficients C can be found. Substituting the elements of C in Eqs. (3.21)–(3.23), a complex value for the amplitude displacements is found. In order to obtain the response the real part of it needs to be evaluated.

In Table 3.5, the non-dimensional transverse displacement amplitude in the center for a homogeneous square aluminum thin plate with thickness-side ratio $h/a = 0.01$ is compared with the results given by Laura and Duran (1975).

In this case the viscous damping is not considered, the static displacement ($\gamma = 0$) and the dynamic responses close to the resonance condition ($\gamma = 0.3, 0.5$ and 8) are calculated. The results for the response at the center of the plate yielded nearly identical comparisons.

Table **3.5.** Comparison of the displacement amplitudes $W(0, 0, 0)$ $E_2 h^3/[12(1 - v)a^4 q_0]$ for the dynamic case for a homogeneous square aluminum plate with thickness-side ratio $h/a = 0.01$

$\gamma = \bar{\omega}/\omega_1$	Laura and Duran (1975)	Classical solution	Present solution
0	0.004062	0.00406	0.004064
0.3	0.0045	0.004473	0.004475
0.5	0.0055	0.005448	0.00545
0.8	0.0115	0.011445	0.01146

For convenience, the arbitrary amplitude load $Q(\xi, \eta)$ in Eq. (3.17) is represented in the general form of a double series:

$$Q(\xi, \eta) = \sum_\alpha \sum_\beta Q_{\alpha\beta} P_\alpha(\xi) P_\beta(\eta) \tag{3.54}$$

where $Q_{\alpha\beta}$ is the generic coefficient and $P_\alpha(\xi)$, $P_\beta(\eta)$ are Chebyshev polynomials.

According to Eq. (3.54), the element $\mathbf{Q}^z$ of the load vector $\mathbf{Q}$ in Eq. (3.37) reads:

$$\mathbf{Q}^z = \sum_\alpha \sum_\beta Q_{\alpha\beta} \int_{-1}^{1} P_\alpha(\xi) P_p(\xi) f_w^1(\xi) d\xi \int_{-1}^{1} P_\beta(\eta) P_q(\eta) f_w^2(\eta) d\eta \, P_r(1).$$

$$\tag{3.55}$$

In view of Eq. (3.36) in order to evaluate the response, the $\mathbf{R}$ matrix in Eq. (3.38), needs to be determined numerically. From Eq. (3.36), a set of equations has to be solved for the real and imaginary parts of the vector of the complex displacement coefficients $\mathbf{C}$. Using the relation, $x + iy = Ce^{i\phi}$ where $C = \sqrt{x^2 + y^2}$ and $\tan\phi = y/x$, the displacement components Eq. (3.20) are expressed as follows

$$\begin{aligned}
u(\xi, \eta, \zeta, t) &= U(\xi, \eta, \zeta)e^{i\bar{\omega}t} = \overline{U}(\xi, \eta, \zeta)e^{i(\bar{\omega}t - \phi_u)} \\
v(\xi, \eta, \zeta, t) &= V(\xi, \eta, \zeta)e^{i\bar{\omega}t} = \overline{V}(\xi, \eta, \zeta)e^{i(\bar{\omega}t - \phi_v)} \\
w(\xi, \eta, \zeta, t) &= W(\xi, \eta, \zeta)e^{i\bar{\omega}t} = \overline{W}(\xi, \eta, \zeta)e^{i(\bar{\omega}t - \phi_w)}
\end{aligned} \tag{3.56}$$

where U, V, W, ϕ_u, ϕ_v and ϕ_w are the amplitude responses and the phase angles in the ξ, η and ζ directions, respectively. The steady-state solution corresponding to the harmonic load in Eq. (3.54) is given by the real part of Eq. (3.56).

After these comparison studies conducted in order to verify the present approach for homogeneous plates, now forced vibrations of FG simply supported square plates are investigated. In the following, a square plate with thickness-side ratio $h/a = 0.2$ and 4 different values of the exponent $N = 0.2, 0.5, 1$ and 2 are considered. The viscous damping is small $\delta 1 = 0.01$.

The ratio

$$R_w = \frac{\mathrm{Re}\left(\overline{W}e^{i(\bar{\omega}t - \phi_w)}\right)}{W_{st}} = \frac{W^R}{W_{st}}$$

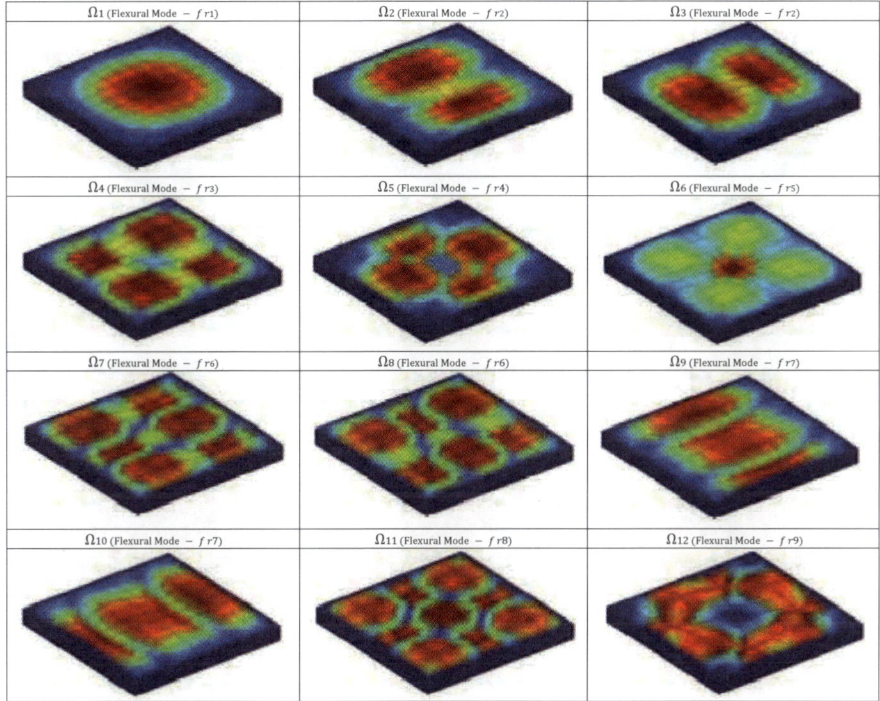

Figure 3.34. Mode shapes for a FG clamped plate with $h/a = 0.05$ and $N = 0.2$

between the real part of the dynamic response amplitude $\overline{W}e^{i(\bar{\omega}t-\phi_w)}$, W^R, and the static displacement W_{st} in the ζ direction at the center of the plate is introduced. The results obtained for R_w calculated at different values of the frequency ratio γ are listed in Table 3.6. For comparison purposes, also the extreme cases of homogeneous plates made of aluminum and zirconia are considered.

As expected, the ratio R_w becomes larger as the frequency ratio γ approaches the unity. The value of R_w is not always in between the two extreme values of the pure metal and pure ceramic plate. For example, at the resonance condition, the FG plate with $N = 2$ presents the greatest value of R_w.

This can be due to the fact that since R_w is given by the ratio between the dynamic response and the static displacement, the pure aluminum plate performs the greatest values of the static and dynamic displacement at the

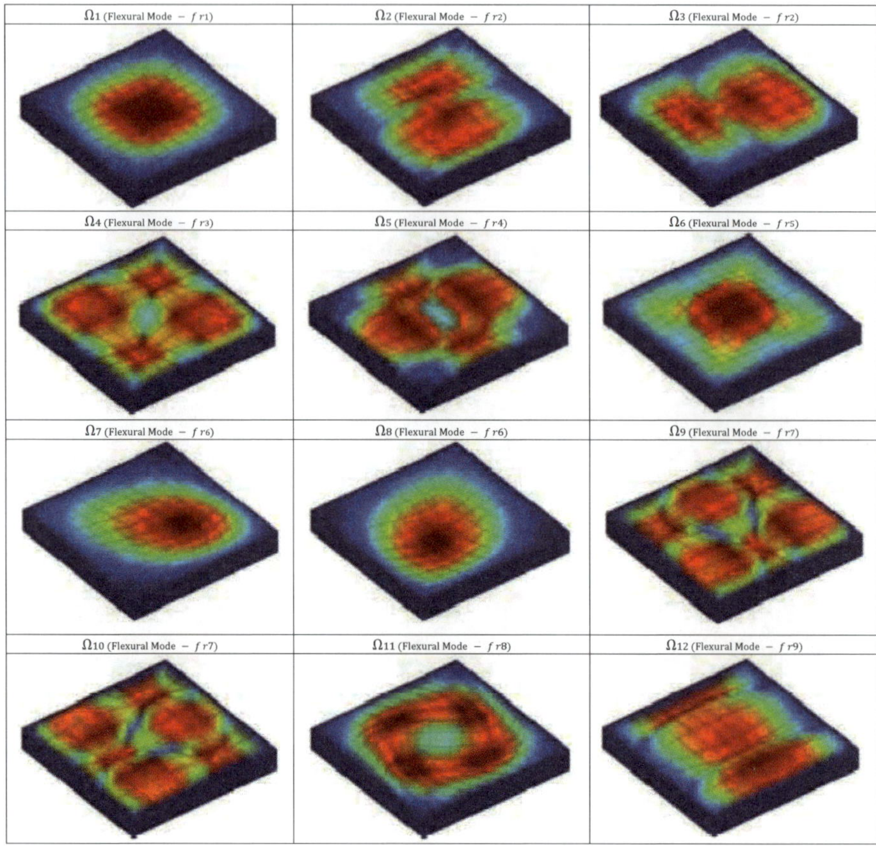

Figure 3.35. Mode shapes for a FG clamped plate with $h/a = 0.1$ and $N = 0.2$

center of the plate, but the ratio results smaller than the one of the FG plate with $N = 2$.

In Fig. 3.39, the results for the non-dimensional deflection in the thickness direction of a FG plate for different values of the exponent N are plotted. The non-dimensional transverse deflection is represented in Figs. 3.39(a)– 3.39(d) for the case of the static load ($\gamma = 0$) and for γ equals 0.8, 0.95 and 1, respectively. It can be seen that the response for the FGM plates presents intermediate values between the pure metal plate and the pure ceramic plate also in the case of the resonance condition.

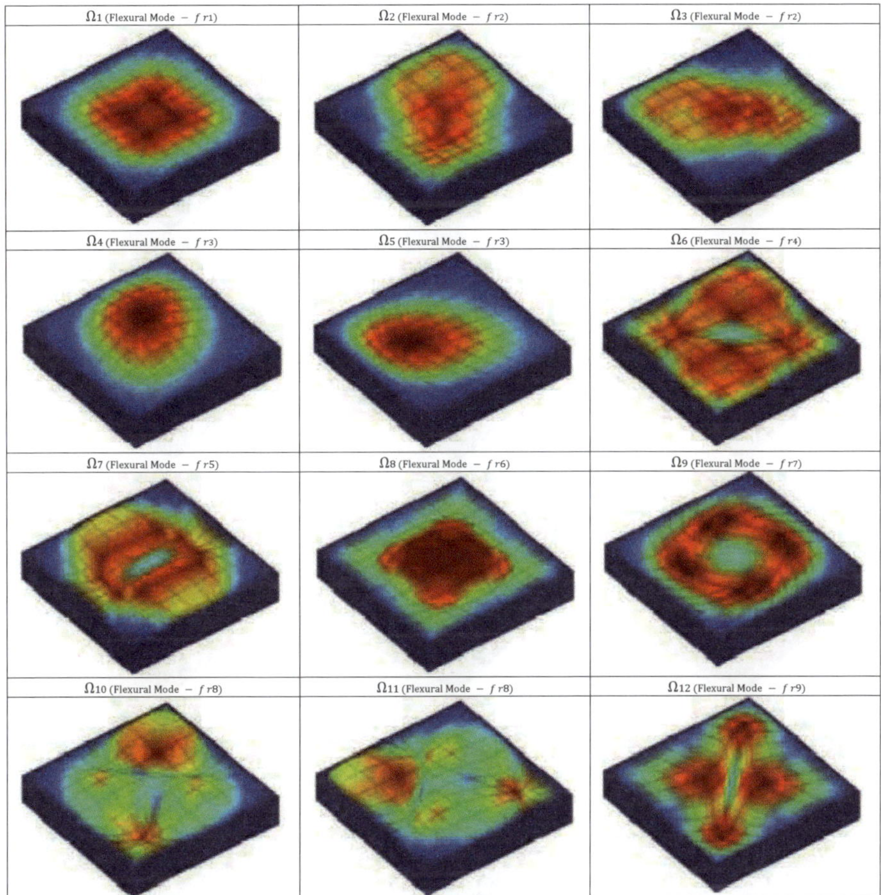

Figure 3.36. Mode shapes for a FG clamped plate with $h/a = 0.2$ and $N = 0.2$

From these figures, it can be pointed out that the gradients in material properties play an important role in determining the dynamic response of FG plates both inside and outside the resonance condition.

Through the thickness variation of the non-dimensional in-plane displacement $100h^2 E_2 U^R/(q_0 a^3)$, evaluated at $(\xi, \eta) = (1/2, 1/2)$, for FG plates are represented for different values of γ and N in Figs. 3.40–3.43, where U^R is the real part of $\overline{U} e^{i(\bar{\omega}t - \phi_u)}$ in Eq. (3.56). The non-dimensional in-plane displacement of a pure aluminum square plate is also represented for comparison purposes.

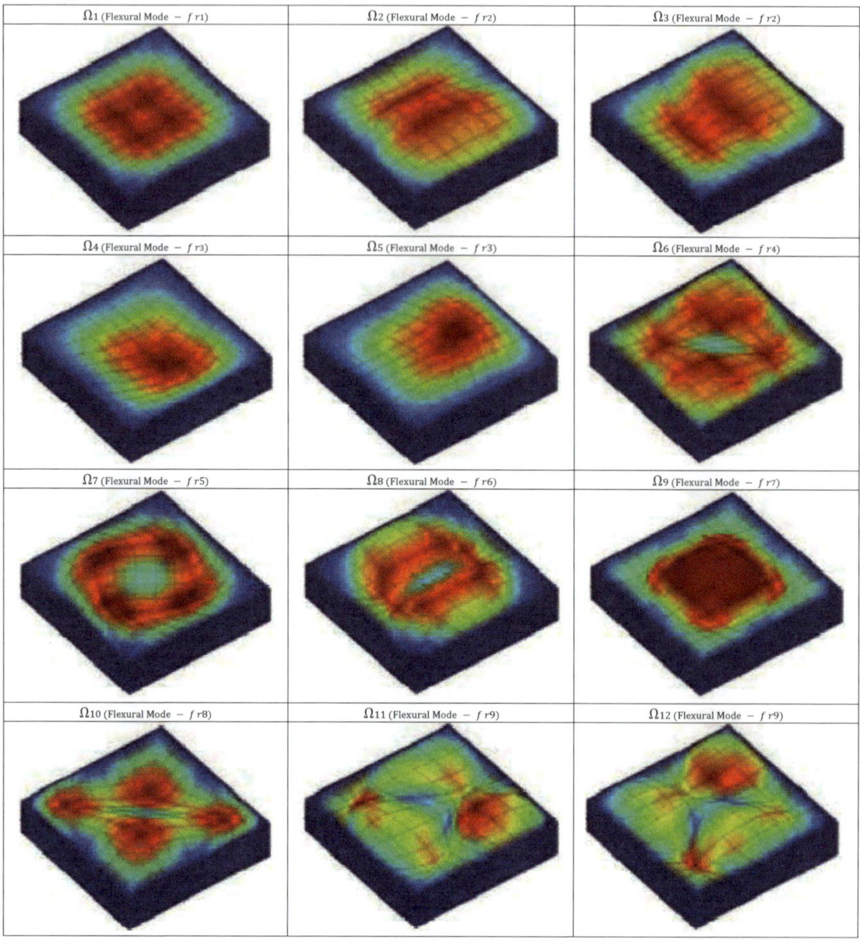

Figure 3.37. Mode shapes for a FG clamped plate with $h/a = 0.3$ and $N = 0.2$

In Fig. 3.40, the behavior of the non-dimensional in-plane displacement for a FG square plate with $N = 0.2$ is depicted. The displacement is evaluated for the static case ($\gamma = 0$), in the resonance condition ($\gamma = 1$) and close to the resonance condition ($\gamma = 0.95$). Similarly to the behavior of the transverse displacement, the in-plane displacement is minimum when $\gamma = 0$ and it becomes larger as the ratio γ approaches to unity.

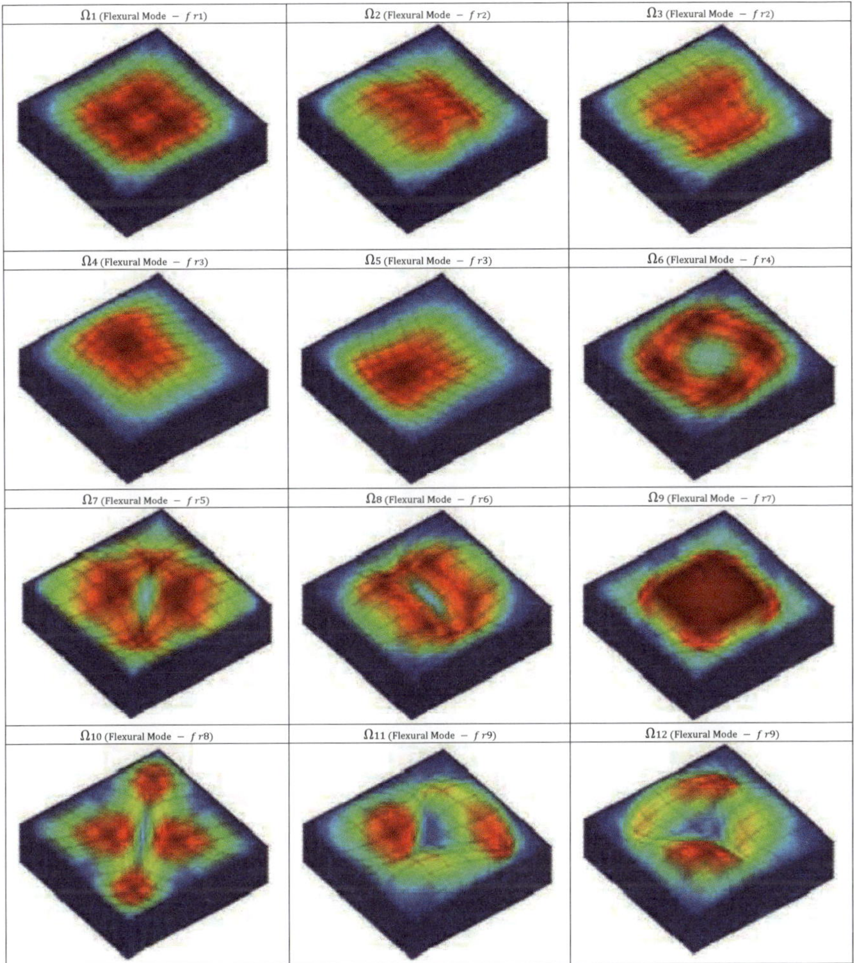

Figure 3.38. Mode shapes for a FG clamped plate with $h/a = 0.4$ and $N = 0.2$

In Figs. 3.41–3.43, the non-dimensional in-plane displacement in the thickness direction for FG plates with $N = 0.5, 1$ and 2 is represented, respectively. It can be seen that the differences between the in-plane displacement for the FG plates and the homogeneous one at the bottom and upper surfaces ($\zeta = \pm 1$) become smaller with the increasing value of the exponent N. In particular, the point where the homogeneous plate and the

Table 3.6. R_w for a FG square plate ($h/a = 0.2$) at different values of the frequency ratio

$\gamma = \bar{\omega}/\omega_1$	**Zirconia**	$N = 0.2$	$N = 0.5$	$N = 1$	$N = 2$	**Aluminum**
0	1	1	1	1	1	1
0.8	2.8415	2.8415	2.8416	2.8423	2.8444	2.8416
0.95	10.4146	10.4143	10.4148	10.4187	10.4298	10.4148
1	51.9287	51.9264	51.9293	51.9509	52.0127	51.9295
1.05	9.9676	9.9670	9.9676	9.9722	9.9855	9.9677

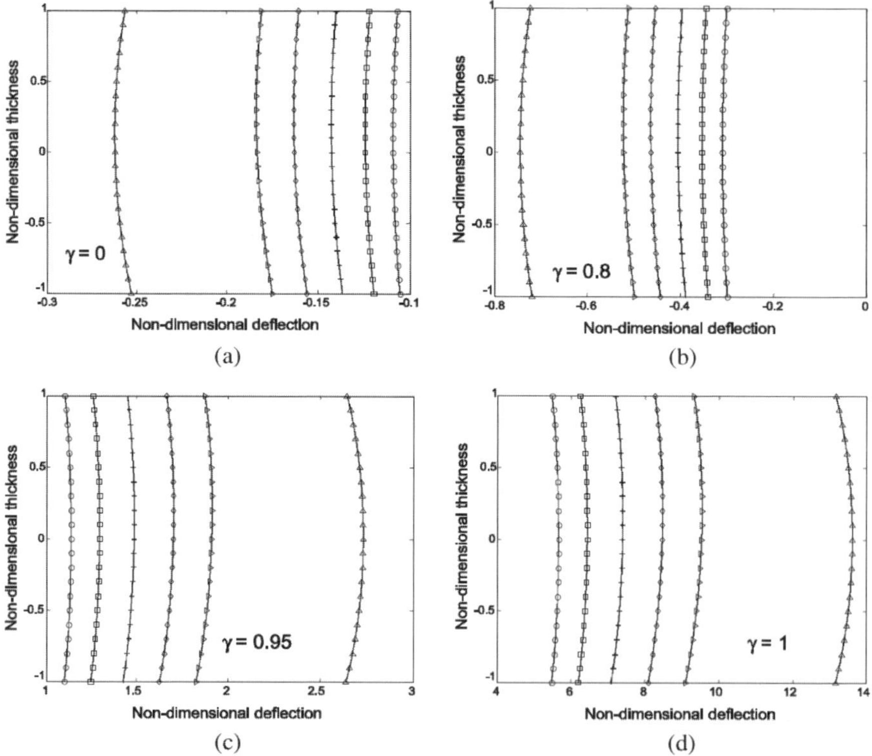

Figure 3.39. Non-dimensional deflection $100h^2 E_2 W^R/(q_0 a^2)$ versus non-dimensional thickness for FG and homogeneous plates (pure aluminum plate, $N = 0.2$, $N = 0.5$, $N = 1$, $N = 2$, pure ceramic plate): (a) $\gamma = 0$ — static case, (b) $\gamma = 0.8$, (c) $\gamma = 0.95$, (d) $\gamma = 1$ — resonance condition

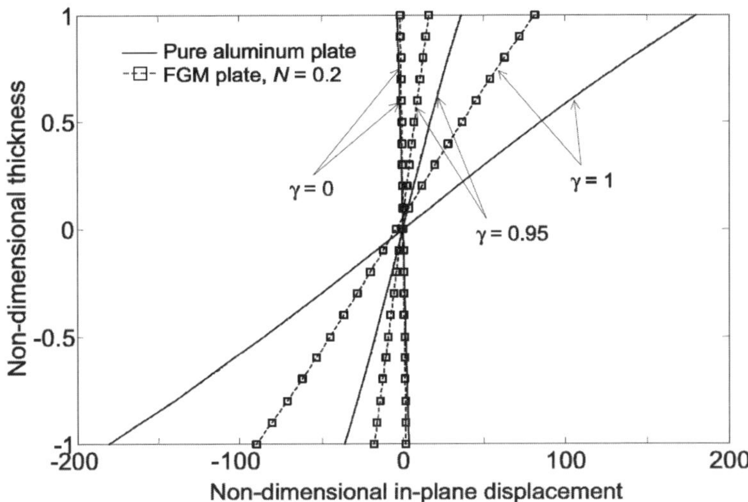

Figure 3.40. Through the thickness variation of the non-dimensional in-plane displacement for a pure aluminum plate and a FG plate with $N = 0.2$ for different values of γ

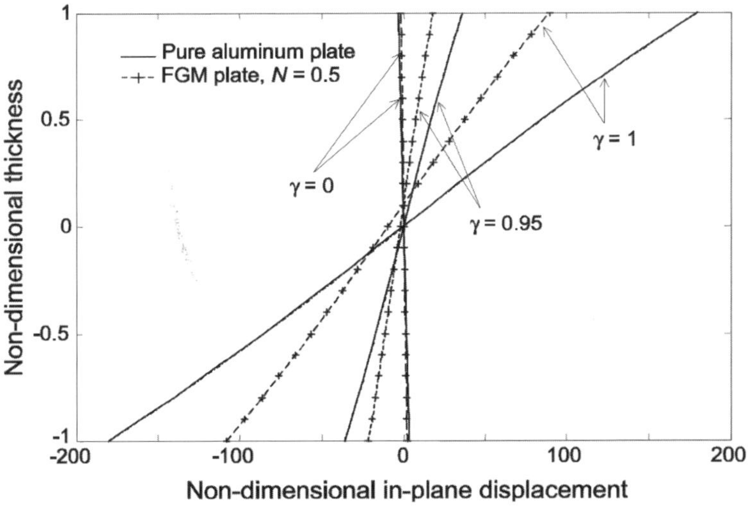

Figure 3.41. Through the thickness variation of the non-dimensional in-plane displacement for a pure aluminum plate and a FG plate with $N = 0.5$ for different values of γ

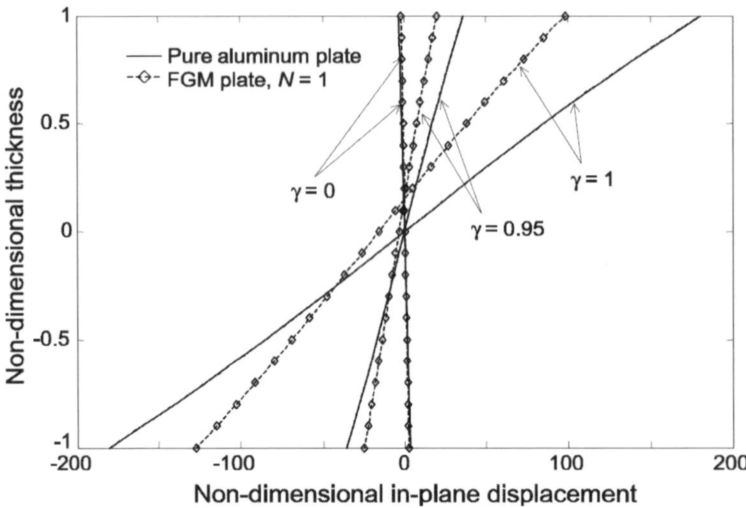

Figure 3.42. Through the thickness variation of the non-dimensional in-plane displacement for a pure aluminum plate and a FG plate with $N = 1$ for different values of γ

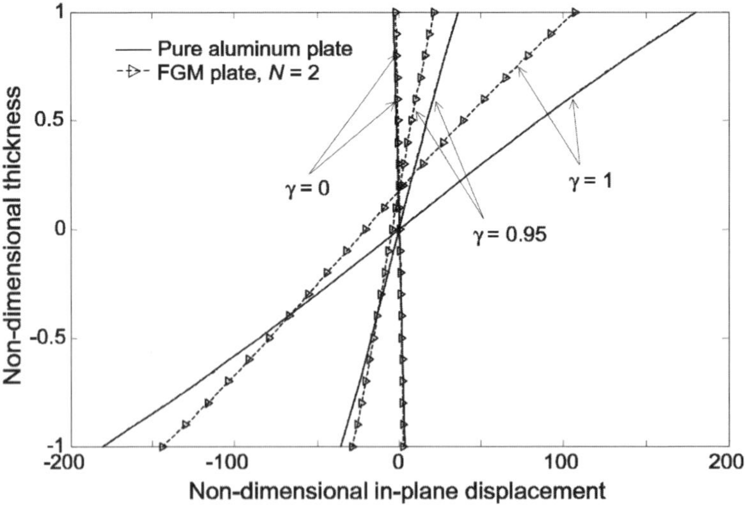

Figure 3.43. Through the thickness variation of the non-dimensional in-plane displacement for a pure aluminum plate and a FG plate with $N = 2$ for different values of γ

FG plate share the same in-plane displacement is moving towards the bottom surface of the plate.

This kind of result is expected because the more N becomes large the more the ceramic content is low and the FG plate approaches the case of the fully metallic plate.

3.6 Some remarks

A three-dimensional elasticity solution to the problem of forced, damped vibrations of simply supported and clamped rectangular plates made of FGMs has been presented. The material properties of the constituents, such as modulus of elasticity, mass density and damping coefficient are assumed to vary in the thickness direction in accordance with a power-law. The solution is obtained by the Ritz method that has been generalized in order to include material damping and the work done by the external force. Chebyshev polynomials multiplied by suitable boundary functions have been employed as admissible functions. From the present method, natural frequencies can be determined setting the load and dissipation terms equal zero in the expressions of the generalized energy functional.

The numerical results are supported by appropriate comparisons with results available in the literature for the case of simply supported and clamped homogeneous plates. No comparisons with analytical results for forced damped vibrations of FG plates could be made, apparently since they have not been available previously. Parametric studies have been performed for varying the power-law exponent and thickness side ratio. The convergence of vibration frequencies has been examined and frequency parameters have been presented for different thickness ratios (h/a) and power-law exponents (N).

Natural frequencies calculated for different values of power-law exponent are observed not to lie always in-between the natural frequencies of the plate made of pure ceramic and pure metal. Further, natural frequencies of FG plates decrease rapidly as h/a ratio increases. Since no simplifying assumptions have been made on the displacement fields, the method is capable of providing accurate frequency solutions for the vibration analysis of thick FG plates. Numerical results that are presented can serve as a reference for establishing the validity of approximate theories as well as checking numerical solutions. The formulation is readily extendible to other kinds of boundary conditions and laws of variation of the constituents. From the numerical results, it can be seen that when the excitation frequency is close or equal to the fundamental natural frequency, the gradients in material properties play an important role in determining the dynamic response of the FGM plates.

This means that the property gradients can be appropriately tailored in order to control the magnitude of the dynamic response at the resonance condition.

Chapter 4

Static Analysis of Plates Made of Functionally Graded Materials

In this chapter, a three-dimensional solution for the problem of transversely loaded, all round clamped rectangular plates of arbitrary thickness is presented within the linear, small deformation theory of elasticity. The energy functional defined in Chapter 3 is simplified in order to derive the governing equation of the functionally graded (FG) plate. Young's modulus is assumed to vary in the thickness direction.

Several investigators devoted themselves to the study of simply supported rectangular plates by means of different theories. Venini and Della Croce (2004), developed a hierarchic family of finite elements according to the Reissner–Mindlin theory, while Reddy (2000), Yang and Shen (2003) dealt with higher-order shear deformation plate theories. A finite element method for the static analysis and nonlinear vibrations of FG plates was developed by Praveen and Reddy (1998). Three-dimensional analysis were carried out by Cheng and Batra (2000) for FG elliptic plates, Reddy and Cheng (2001), Vel and Batra (2003) for transient thermal stresses of FG simply supported thick plates and Mian and Spencer (1998) gave a class of exact solutions of the three-dimensional elasticity equations in cylindrical polar coordinates.

As it was mentioned, previous studies are confined to simply supported plates. Herein, the other extreme case, namely, that of all-round clamped plate, is considered since realistic boundary conditions are in-between these two bounds.

The effects of the variation of the volume fractions of the constituent materials and thickness-to-side ratio on the through-the-thickness deflections, in-plane displacements and axial stress distributions are studied in detail. Preliminary results for static analysis of FG plates have been presented in Elishakoff and Gentilini (2005a), Gentilini (2005).

4.1 Statement of the problem and solution methodology

The geometry and dimensions of the rectangular plate made of functionally graded materials under consideration are represented in Fig. 4.1. The plate is subjected to a transversal loading $q(x, y)$ applied downward on the top surface.

In the non-dimensional coordinate system $(O; \xi, \eta, \zeta)$, the work done by the transverse loading $q(\xi, \eta)$ is given by:

$$W(\xi, \eta) = \frac{ab}{4} \iint_{-1}^{1} q(\xi, \eta) w(\xi, \eta, 1) d\xi d\eta \tag{4.1}$$

that is similar to Eq. (3.19), except that the dependence by time here is not considered. For convenience, the arbitrary load $q(\xi, \eta)$ is represented in the

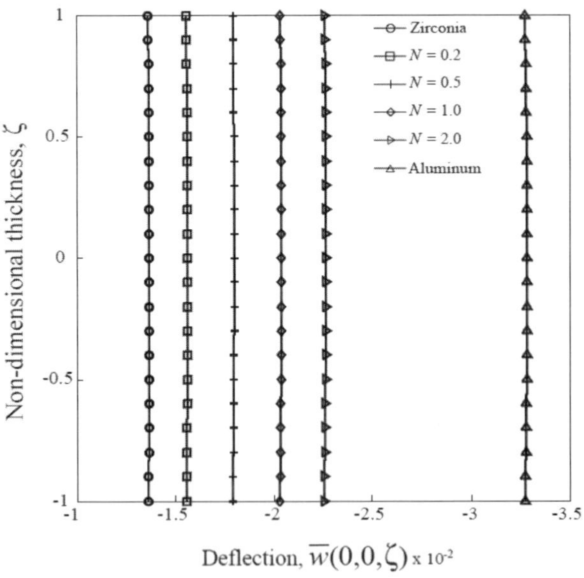

Figure 4.1. Case $h/a = 0.05$ — Non-dimensional deflection w in the thickness direction

general form of a double series:

$$q(\xi, \eta) = \sum_{\alpha} \sum_{\beta} Q_{\alpha\beta} P_\alpha(\xi) P_\beta(\eta) \tag{4.2}$$

where $P_\alpha(\xi)$ and $P_\beta(\eta)$ are Chebyshev polynomials.

Having studied dynamic response of FG plates in the previous chapter, static analysis is straightforward obtained. The energy functional Π_{max}, Eq. (3.30), is simplified to perform the static analysis:

$$\Pi_{max} = \Phi_{max} - W_{max} \tag{4.3}$$

where Φ_{max} and W_{max} have the expressions in Eqs. (3.33) and (3.34), respectively.

Thus, the following governing equation is obtained:

$$KC = \frac{a^2}{2h\sqrt{E_1 E_2}} Q$$

in which $\mathbf{K}$, $\mathbf{C}$ and $\mathbf{Q}$ are the stiffness matrix, the vector of the displacement coefficients and the load vector, respectively. The modulus of elasticity $E(\zeta)$ is considered varying through the thickness according to the power-law Equation (3.4). The above equation is the simplification of the governing equation (3.36). The explicit form of the elements in the stiffness sub-matrices $\mathbf{K}ij$ with $i,\ j = u,\ v,\ w$ in Eq. (4.3) are given in Eq. (3.41). The element $\mathbf{Q}^z$ of the load vector $\mathbf{Q}$ in Eq. (4.3) reads as in Eq. (3.55), reported here for sake of completeness:

$$\mathbf{Q}^z = \sum_{\alpha} \sum_{\beta} Q_{\alpha\beta} \int_{-1}^{1} P_\alpha(\xi) P_p(\xi) f_w^1(\xi) d\xi \int_{-1}^{1} P_\beta(\eta) P_q(\eta) f_w^2(\eta) d\eta\, P_r(1) \tag{4.4}$$

where $f_w^1(\xi)$ and $f_w^2(\eta)$ are boundary functions given in Table 4.1. From the governing equation (4.3), the coefficient vector $\mathbf{C}$ can be found. As stated before, the load is represented in the general form of a double series in terms of the Chebyshev polynomials, Eq. (4.2). To calculate any particular coefficient $Q_{\alpha\beta}$ of the series, both sides of Eq. (4.2) are multiplied by $P_{\alpha'}(\xi) P_{\beta'}(\eta) \varpi_\xi(\xi) \varpi_\eta(\eta)$ where $\varpi_\xi(\xi),\ \varpi_\eta(\eta)$ are weighting functions,

Table 4.1. Comparison of displacements and axial stresses for the homogeneous aluminum plate with thickness-to-side ratio equal to 0.2

Method of solution	$2\bar{w}\mu/q_0$ at $\xi = \eta = \zeta = 0$	$\sigma_{xx}a^2/h^2$ at $\xi = 1, \eta = 0, \zeta = 0$
Srinivas and Rao (1973)	−11.208	−3.9135
Liew *et al.* (2001)	−11.185	−3.9950
Present 3-D solution	−11.1604	−3.8134

Eq. (3.28), and integrating twice from -1 to 1:

$$\iint_{-1}^{1} q(\xi, \eta) P_{\alpha'}(\xi) P_{\beta'}(\eta)\varpi_\eta(\eta)d\xi d\eta$$

$$= \sum_\alpha \sum_\beta Q_{\alpha\beta} \int_{-1}^{1} P_\alpha(\xi) P_{\alpha'}(\xi)\varpi_\xi(\xi)d\xi \int_{-1}^{1} P_\beta(\eta) P_{\beta'}(\eta)\varpi_\eta(\eta)d\eta$$

$$(4.5)$$

In the case of a uniformly distributed load over the entire upper surface of the plate, we have

$$q(\xi, \eta) = q_0 \tag{4.6}$$

where q_0 is the intensity of the uniformly distributed load. Substituting Eq. (4.6) in Eq. (4.5), the following relation is obtained:

$$q_0 \int_{-1}^{1} P_{\alpha'}(\xi)\varpi_\xi(\xi)d\xi \int_{-1}^{1} P_{\beta'}(\eta)\varpi_\eta(\eta)d\eta$$

$$= \sum_\alpha \sum_\beta Q_{\alpha\beta} \int_{-1}^{1} P_\alpha(\xi) P_{\alpha'}(\xi)\varpi_\xi(\xi)d\xi \int_{-1}^{1} P_\beta(\eta) P_{\beta'}(\eta)\varpi_\eta(\eta)d\eta$$

$$(4.7)$$

From the orthogonality property, Eq. (3.29), the integrals in left hand side of Eq. (4.7) are different from zero, and in particular equal to π, only when $\alpha_0 = \beta_0 = 1$, that means $P_{\alpha 0}(\xi) = P_{\beta 0}(\eta) = 1$. As a consequence, also α and β have to be equal to 1, that means $P_\alpha(\xi) = P_\beta(\eta) = 1$. Based on these considerations, from Eq. (4.7) the following relation is obtained:

$$Q_{11} = q_0 \tag{4.8}$$

Thus, the component $\mathbf{Q}^z$ of the load vector $\mathbf{Q}$, Eq. (4.4), is given by:

$$\mathbf{Q}^z = q_0 \int_{-1}^{1} P_p(\xi) f_w^1(\xi)d\xi \int_{-1}^{1} P_q(\eta) f_w^2(\eta)d\eta\, P_r(1) \tag{4.9}$$

4.2 Numerical results

In this section, the proposed method is realized on FG all round clamped square ($a/b = 1$) plates and thickness-to-side ratios $h/a = 0.05, 0.2$, and 0.3. The analysis is performed for different values of the volume fraction exponent $N = 0.2, 5, 1$ and 2 Zirconia/aluminum FG plates are considered with material properties are listed in Table 3.2. Modulus of elasticity is assumed to vary through the thickness.

The numerical results are presented in terms of the following non-dimensional displacements and stresses:

$$\bar{u} = \frac{u}{a}, \quad \bar{w} = \frac{w}{h}, \quad \bar{\sigma}_{xx} = \frac{h^2 \sigma_{xx}}{a^2 |q_0|} \tag{4.10}$$

where h, a and q_0 are the plate thickness, the length and the applied load, respectively.

Results of comparison

In order to verify the proposed solution method, the displacement and axial stress of a square plate made of pure aluminum with thickness to width ratio equals to 0.2, is first solved. The results are listed in Table 4.1 and compared with the three-dimensional formulations provided by Srinivas and Rao (1973) and Liew *et al.* (2001). As is seen, a good agreement is established.

For the plate with thickness-to-side ratio equal to 0.5, Fig. 4.1 shows the non-dimensional central deflection w in the thickness direction.

The non-dimensional central deflection, w, of the purely metallic plate was found to be the largest magnitude and that of the purely ceramic plate, of the smallest magnitude. All the plates with intermediate properties undergo corresponding intermediate values. This is expected because the metallic plate is one with lower stiffness than the ceramic plate.

Figure 4.2 shows the non-dimensional in-plane displacement u, evaluated at $\xi = 1/2$ and $\eta = 1/2$, across the thickness. The homogeneous aluminum plate exhibits the largest in-plane displacements in absolute value at the top and bottom surfaces, while the plate made only by zirconia has the smallest ones. This can be expected because the Young's modulus for the ceramic is much greater than that for the metal.

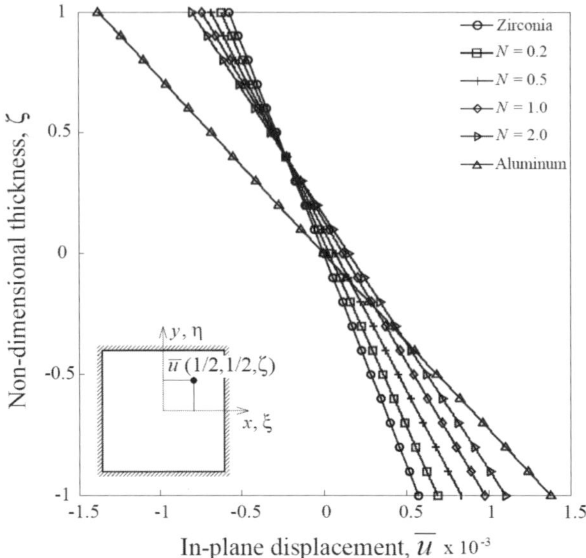

Figure 4.2. Case $h/a = 0.05$ — Non-dimensional in-plane displacement $\bar{u}$ in the thickness direction

As is seen from Fig. 4.2 at the bottom surface, $\zeta = -1$ and its vicinity, the displacements of the FG plate are in-between those of isotropic constituents. The same takes place at the upper surface $\zeta = 1$ and its vicinity.

Remarkably, across the thickness, there is a central region where the FG plates with different values of the exponent N exhibit a response that is not intermediate to the fully metallic and fully ceramic plates. Besides, there is a point at $\zeta \simeq 0.4$, where the plates for different values of N share the same displacement.

Figure 4.3 represents the plots of the non-dimensional axial stress, σ_{xx}, evaluated at $\xi = 1/2$, $\eta = 1/2$, through the thickness of the plate under the uniform load applied on the top surface. The axial stress due to the application of the pressure loading, is compressive at the top surface and tensile at the bottom surface. For the several volume fraction exponents chosen, the plate corresponding to $N = 2$ yielded the maximum compressive stress at the top surface, $\zeta = 1$, while the purely ceramic or purely metallic plates experience the maximum tensile stress at the bottom surface, $\zeta = -1$.

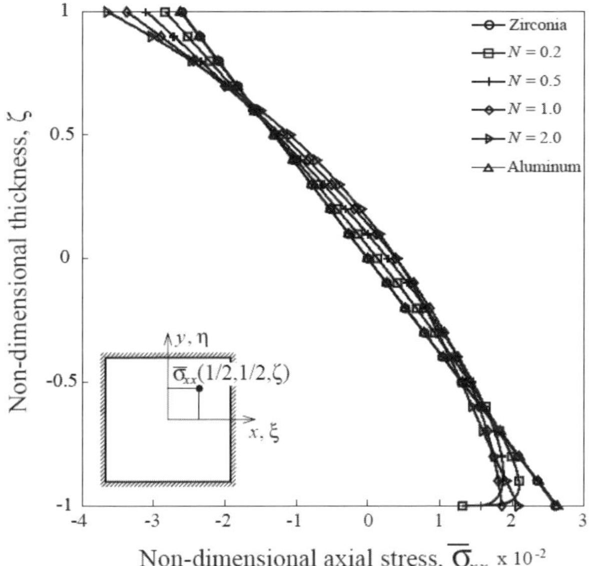

Figure 4.3. Case $h/a = 0.05$ — Non-dimensional axial stress $\bar{\sigma}_{xx}$ in the thickness direction

It is worth noting that the behavior of the axial stresses of the fully ceramic and fully metallic plates are coincident, because Poisson's ratio for aluminum and zirconia components is the same.

The stress profiles, σ_{xx}, in the clamped edge ($\xi = 1$, $\eta = 0$) across the thickness are presented in Fig. 4.4. In the clamped edge, there is tension at the top surface and compression at the bottom. Plates with different values of thickness-to-side ratio have also been investigated.

In Figs. 4.5 and 4.6, the behavior of the non-dimensional central deflection and in-plane displacement, w and u, for a plate with $h/a = 0.2$ are represented. Figures 4.7 and 4.8 illustrate the stress profiles at $\xi = 0.5$, $\eta = 0.5$ and in the clamped edge ($\xi = 1$, $\eta = 0$) across the thickness, respectively.

All the profiles are qualitatively similar to those of the $h/a = 0.05$ plate.

In Figs. 4.9 and 4.10, the central deflection and in-plane displacement of a moderately thick plate with thickness-to-side ratio equal to 0.3 are portrayed.

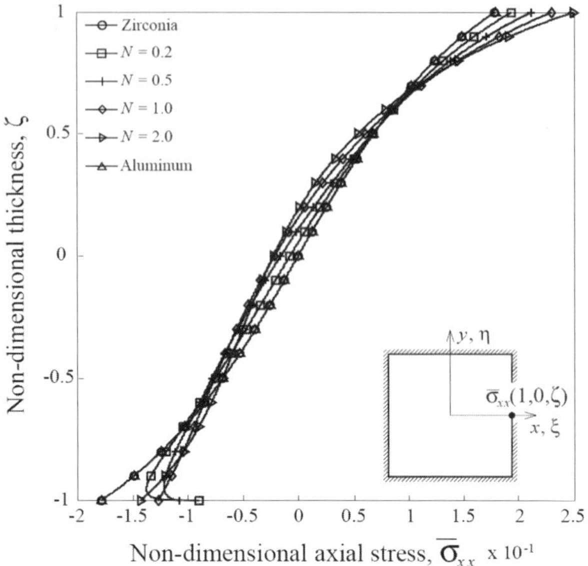

Figure 4.4. Case $h/a = 0.05$ — Non-dimensional axial stress $\bar{\sigma}_{xx}$ in the clamped edge in the thickness direction

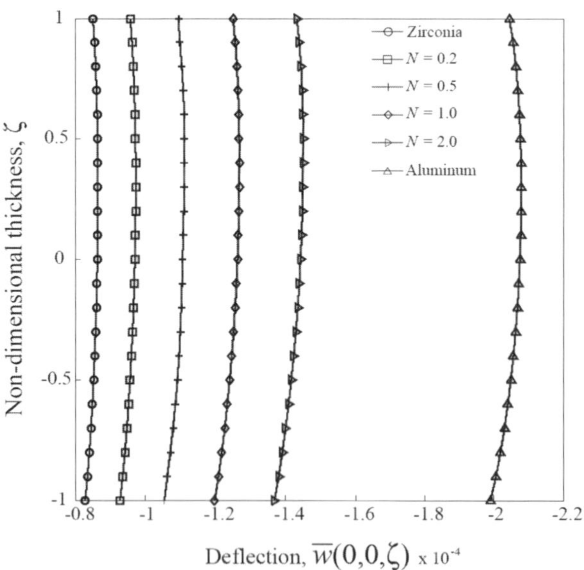

Figure 4.5. Case $h/a = 0.2$ — Non-dimensional deflection $\bar{w}$ in the thickness direction

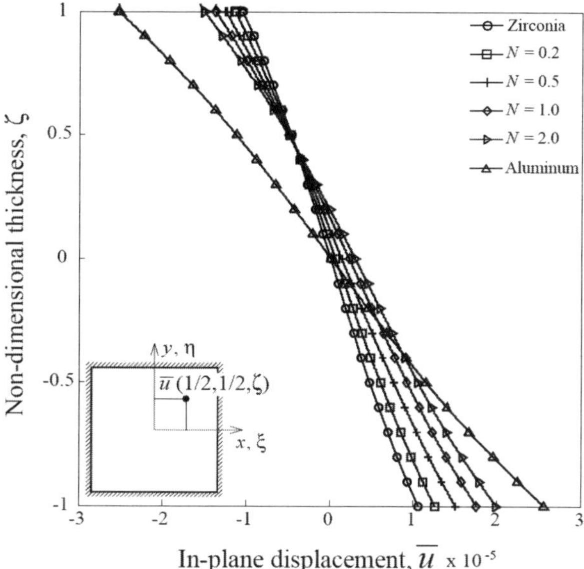

Figure 4.6. Case $h/a = 0.2$ — Non-dimensional in-plane displacement $\bar{u}$ in the thickness direction

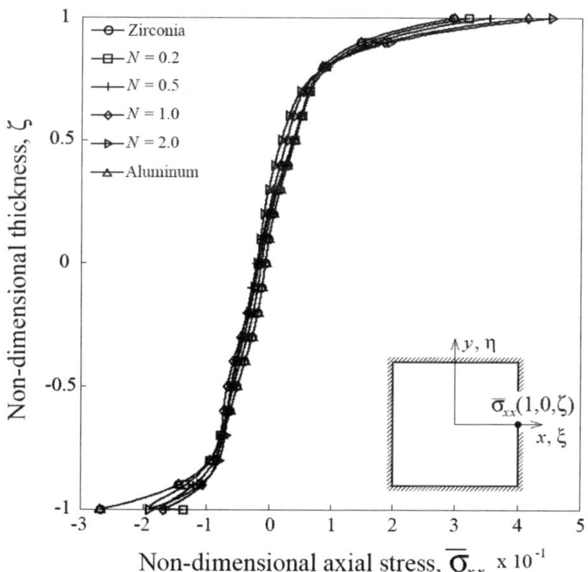

Figure 4.7. Case $h/a = 0.2$ — Non-dimensional axial stress $\bar{\sigma}_{xx}$ in the clamped edge in the thickness direction

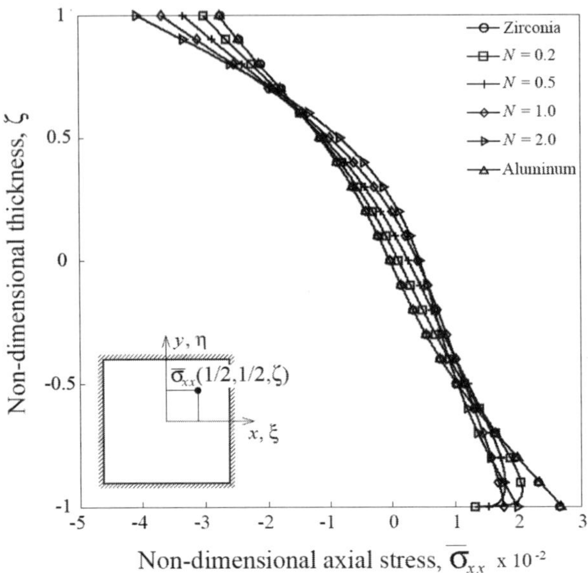

Figure 4.8. Case $h/a = 0.2$ — Non-dimensional axial stress $\bar{\sigma}_{xx}$ in the thickness direction

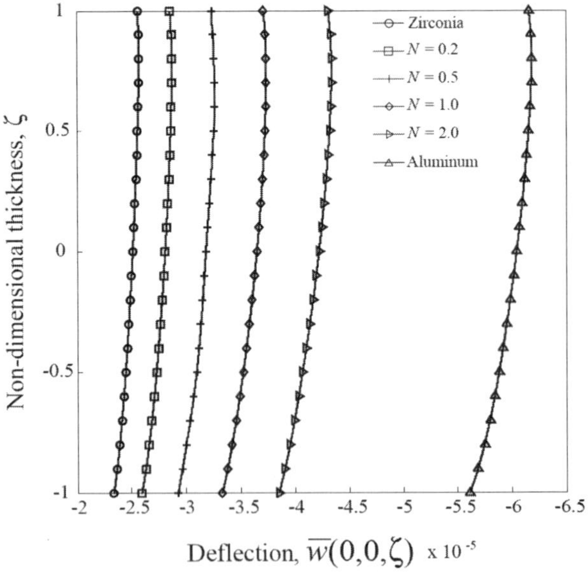

Figure 4.9. Case $h/a = 0.3$ — Non-dimensional deflection $\bar{w}$ in the thickness direction

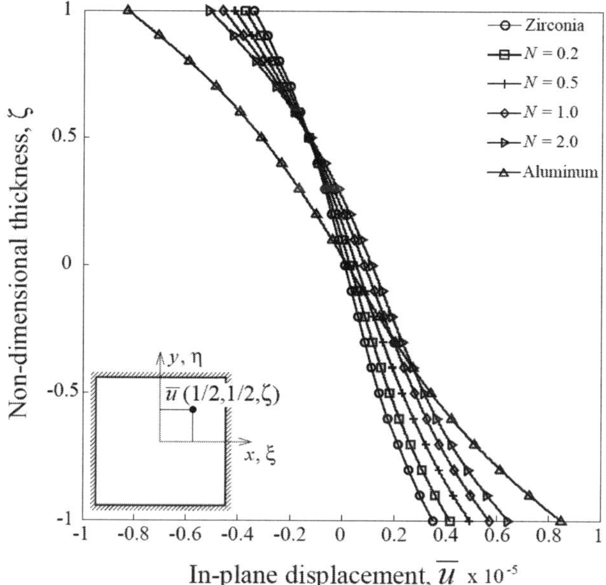

Figure 4.10. Case $h/a = 0.3$ — Non-dimensional in-plane displacement $\bar{u}$ in the thickness direction

For this plate, the non-symmetric behavior with respect to the midplane is more pronounced than in the case of smaller thickness side ratios.

Lack of symmetry in the displacement profiles is attributable to the fact that there is no symmetry in the load itself with respect to the midplane of the plate; in addition, the mechanical properties vary through the thickness of the plate in accordance with the non-symmetric power law, Eq. (3.4).

The stress profiles of the $h/a = 0.3$ plate are represented in Figs. 4.11 and 4.12. In particular, in Fig. 4.11, the through-the-thickness profiles of the stress calculated in the point $\xi = 0.5$, $\eta = 0.5$ are shown.

Figure 4.12 presents the distribution of the non-dimensional axial stress $\bar{\sigma}_{xx}$, in the clamped edge ($\xi = 1$, $\eta = 0$) versus the thickness coordinate. The stress profiles for the different values of the power-law exponent are much more clustered than inside domain of the plate.

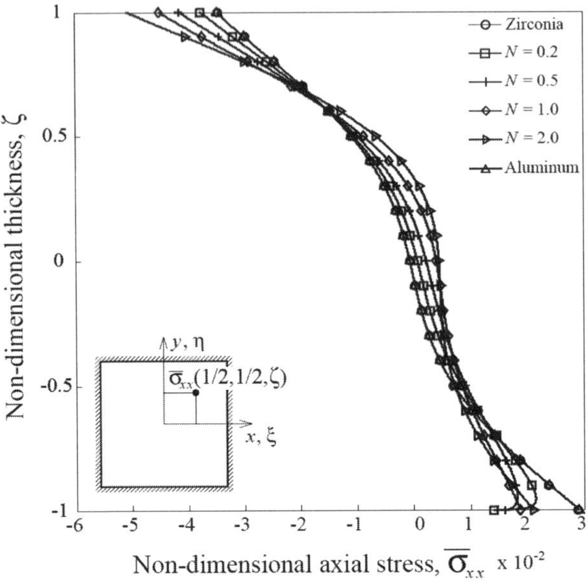

Figure 4.11.   Case $h/a = 0.3$ — Non-dimensional axial stress $\bar{\sigma}_{xx}$ in the thickness direction

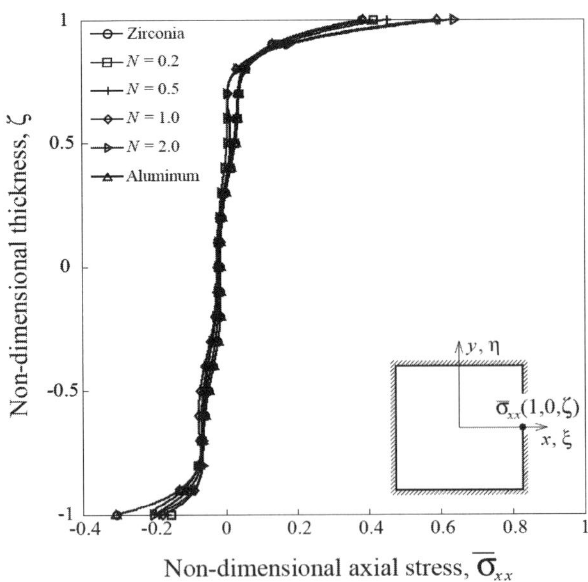

Figure 4.12.   Case $h/a = 0.3$ — Non-dimensional axial stress $\bar{\sigma}_{xx}$ in the clamped edge in the thickness direction

4.3 Some remarks

In this chapter, static analysis of all-round clamped FG plates is performed. The plate is made of a two-component (zirconia and aluminum) graded material with a power law variation of the volume fraction of the constituents through the thickness. The effects of varying the volume fraction and the side-thickness ratio on the non-dimensional displacements and stresses have been presented. From the reported results, it can be concluded that displacements and axial stresses in an arbitrary point of a FG plate do not necessarily lie between those associated with purely ceramic or metal plates.

Part II

Vibration Tailoring of Inhomogeneous Beams
and Circular Plates

In this part, we deal with the vibrational behavior of inhomogeneous beams and circular plates, utilizing the semi-inverse method developed by the first author and extensively discussed in his monograph (Elishakoff, 2005). The main thread of his methodology is that the knowledge of the mode shape encountered in vibration or buckling problem is postulated. The candidate mode shapes can be adopted from relevant static, dynamic or buckling problems. In this study, the exact mode shapes are sought as polynomial functions, in the context of vibration *tailoring*, i.e. designing the structure that possesses the pre-specified value. Apparently, for the first time in the literature, several closed-form solutions for vibration tailoring have been derived for vibrating inhomogeneous beams and circular plates. Twelve new closed-form solutions for vibration tailoring have been derived for an inhomogeneous polar orthotropic plate that is either clamped or simply supported around its circumference. Also, the vibration tailoring of a polar orthotropic circular plate with translational spring is analyzed. There is considerable potential of utilizing the developed method for the design of functionally graded materials.

Chapter 5

Beams Made of Functionally Graded Material

Analysis and design of beams engages a significant place in the field of Mechanical, Civil, Aerospace and Ocean Engineering. A beam is a structural element that carries load mainly in bending. The loads carried by a beam are transferred to supports, which in turn transfer the force to adjacent structural members.

Beams are characterized by the shape of their cross-section, the length, and the material. One of the most common types of beam is the I-beam which is widely used in steel-frame buildings and bridges. The I-beam is so common because it makes efficient use of material for carrying loads in bending. It maximizes material at the top and bottom of the beam where the bulk of the load is carried. The idea of I-beam was generalized later to the sandwich structure that has thin and strong facing and a thick soft core. Presently, this idea is further advanced in terms of functionally graded materials (FGMs) where the modulus of elasticity changes so as to meet some preselected requirements.

The primary tool for structural analysis of beams is the Euler–Bernoulli beam equation which allows quick calculation of the load-carrying capacity and deflection of structural elements.

In FGMs the microstructure varies continuously in thickness. Our motivation, in this book is to use materials that can be graded axially for the beams or radially for the circular plates. It is anticipated that the best functionally grading will combine that both in axial/radial and thickness directions.

5.1 Euler-Bernoulli beam equation

The Euler–Bernoulli governing differential equation for the inhomogeneous beam reads

$$\frac{d^2}{dx^2}\left[E(x)I(x)\frac{d^2w(x)}{dx^2}\right] = p(x) \tag{5.1}$$

where $w(x)$ is the transverse displacement, $p(x)$ is the distributed loading, $E(x)$ is the Young's modulus of elasticity, and $I(x)$ is the moment of inertia. For the homogeneous case, where E and I do not vary with x along the length of beam, then the Euler–Bernoulli equation simplifies to,

$$EI\frac{d^4w(x)}{dx^4} = p(x) \tag{5.2}$$

5.2 Method of solution of uniform beam vibrations

The governing differential equation of a uniform beam vibration was first written by Bernoulli and Euler in 18th century:

$$EI\frac{\partial^4w(x,t)}{\partial x^4} + \rho A\frac{\partial^2w(x,t)}{\partial t^2} = 0 \tag{5.3}$$

where $E =$ Young's modulus of elasticity, $I =$ moment of inertia, $\rho =$ material density, $A =$ cross-section area, $\omega(x,t) =$ transverse displacement, $x =$ spatial coordinate and $t =$ time. In order to determine the natural frequencies and vibration mode shape of a beam, Eq. (5.3) must be solved by using the method of separation of variables:

$$w(x,t) = W(x)T(t) \tag{5.4}$$

where $W(x)$ is a mode shape of vibration as a function of the axial coordinate, and $T(t)$ is a function of time. Substitution of Eq. (5.4) into (5.3) gives

$$T(t)EI\frac{d^4W(x)}{dx^4} + W(x)\rho A\frac{d^2T(t)}{dt^2} = 0 \tag{5.5}$$

or can be rewritten in the form

$$\frac{EI\frac{d^4W(x)}{dx^4}}{\rho A W(x)} = -\frac{\frac{d^2T(t)}{dt^2}}{T(t)} = \text{const} \tag{5.6}$$

In view of the fact that the left-hand side depends only on x, and the right-hand side only on t, these ratios are equal to be constant, symbolized by ω^2. From this consideration, the following two equations are obtained

$$\frac{d^2 T(t)}{dt^2} + \omega^2 T(t) = 0 \tag{5.7}$$

and

$$EI\frac{d^4 W(x)}{dx^4} - \rho A \omega^2 W(x) = 0 \tag{5.8}$$

Solution of Eq. (5.7) reads

$$T(t) = D_1 e^{i\omega t} + D_2 e^{-i\omega t} \tag{5.9}$$

The constants of integration D_1 and D_2 can be determined from the initial conditions set at $t = 0$. The constant ω is sought as the natural frequency of vibration. It can be obtained from Eq. (5.8) and by applying the boundary conditions at the two ends of the beam.

By introducing a frequency parameter

$$\alpha^4 = \frac{\rho A \omega^2}{EI} \tag{5.10}$$

then Eq. (5.8) is simplified to

$$\frac{d^4 W(x)}{dx^4} - \alpha^4 W(x) = 0 \tag{5.11}$$

Assume a solution of $W(x)$ in the form, for $\alpha \neq 0$,

$$W(x) = Ce^{rx} \tag{5.12}$$

where C is a constant, and r is the characteristic exponent to be determined. Substitution of Eq. (5.12) into Eq. (5.11) gives the following characteristic equation

$$r^4 - \alpha^4 = 0 \tag{5.13}$$

with roots

$$r_{1,2} = \pm\alpha, \quad r_{3,4} = \pm i\alpha \tag{5.14}$$

Then, solution of Eq. (5.11) can be written in the following form

$$W(x) = C_1 e^{\alpha x} + C_2 e^{-\alpha x} + C_3 \cos(\alpha x) + C_4 \sin(\alpha x) \tag{5.15}$$

Using the following hyperbolic functions

$$\cosh(\alpha x) = \frac{e^{\alpha x} + e^{-\alpha x}}{2} \tag{5.16}$$

$$\sinh(\alpha x) = \frac{e^{\alpha x} - e^{-\alpha x}}{2} \tag{5.17}$$

$$\cosh(\alpha x) + \sinh(\alpha x) = e^{\alpha x} \tag{5.18}$$

$$\cosh(\alpha x) - \sinh(\alpha x) = e^{-\alpha x} \tag{5.19}$$

we get

$$W(x) = C_1[\cosh(\alpha x) + \sinh(\alpha x)] + C_2[\cosh(\alpha x) - \sinh(\alpha x)]$$
$$+ C_3 \cos(\alpha x) + C_4 \sin(\alpha x) \tag{5.20}$$

or

$$W(x) = \sinh(\alpha x)(C_1 - C_2) + \cosh(\alpha x)(C_1 + C_2)$$
$$+ C_3 \cos(\alpha x) + C_4 \sin(\alpha x) \tag{5.21}$$

The final form of mode shape is

$$W(x) = B_1 \sinh(\alpha x) + B_2 \cosh(\alpha x) + B_3 \cos(\alpha x) + B_4 \sin(\alpha x) \tag{5.22}$$

where

$$B_1 = C_1 - C_2, \quad B_2 = C_1 + C_2, \quad B_3 = C_3, \quad B_4 = C_4 \tag{5.23}$$

5.2.1   Determination of natural frequencies and mode shapes

The above method is applied in this section to determine the natural frequencies and mode shapes of simply supported and clamped–clamped beam vibrations. First, lets take a simply supported beam of length L as shown in Fig. 5.1.

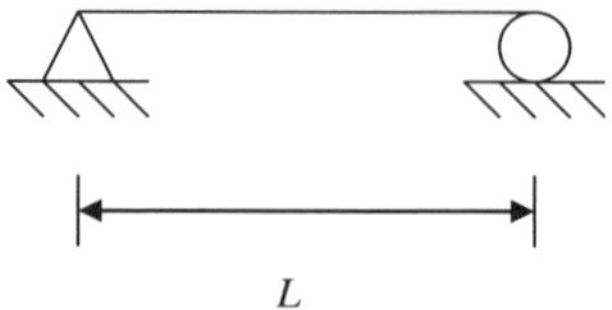

Figure 5.1.   Simply supported beam

By applying the boundary conditions, we get at $x = 0$:

$$W(0) = 0 \rightarrow B_2 + B_4 = 0 \tag{5.24}$$

$$W'(x) = B_1 \alpha \cosh(\alpha x) + B_2 \alpha \sinh(\alpha x) + B_3 \alpha \cos(\alpha x) - B_4 \alpha \cos(\alpha x)$$

$$W''(x) = B_1 \alpha^2 \sinh(\alpha x) + B_2 \alpha^2 \cosh(\alpha x) - B_3 \alpha^2 \sin(\alpha x) - B_4 \alpha^2 \cos(\alpha x)$$

$$W''(0) = 0 \rightarrow B_2 = B_4 \tag{5.25}$$

Solution of Eqs. (5.24) and (5.25) results

$$B_2 = 0 \tag{5.26}$$

$$B_4 = 0 \tag{5.27}$$

and at $x = L$

$$W(L) = 0 \rightarrow B_1 \sinh(\alpha L) + B_3 \sin(\alpha L) = 0 \tag{5.28}$$

$$W''(L) = 0 \rightarrow \alpha^2 B_1 \sinh(\alpha L) - \alpha^2 B_3 \sin(\alpha L) = 0 \tag{5.29}$$

Then, the determinant of coefficients in front of B_1 and B_3 must be equal to zero

$$\begin{vmatrix} \sinh(\alpha L) & \sin(\alpha L) \\ \alpha^2 \sinh(\alpha L) & -\alpha^2 \sin(\alpha L) \end{vmatrix} = 0 \tag{5.30}$$

or

$$\sinh(\alpha L) \sin(\alpha L) = 0 \tag{5.31}$$

In this case, we are looking for nontrivial solutions. For $\alpha \neq 0$ and $\sinh(\alpha L) \neq 0$, the nontrivial solutions of $\sin(\alpha L) = 0$ are

$$\alpha L \approx j\pi \tag{5.32}$$

where $j = 1, 2, 3, \ldots$.

The first four roots obtained from Fig. 5.2 are as follows:

$$(\alpha L)_j = 3.1416 \quad 6.2832 \quad 9.4248 \quad 12.5664 \tag{5.33}$$

The natural frequency is

$$\omega_j = (\alpha L)_j^2 \sqrt{\frac{EI}{\rho A L^4}} \tag{5.34}$$

The final form of mode shape is

$$W(x) = B_3 \left[\cos(\alpha x) - \frac{\sin(\alpha L)}{\sinh(\alpha L)} \right] \tag{5.35}$$

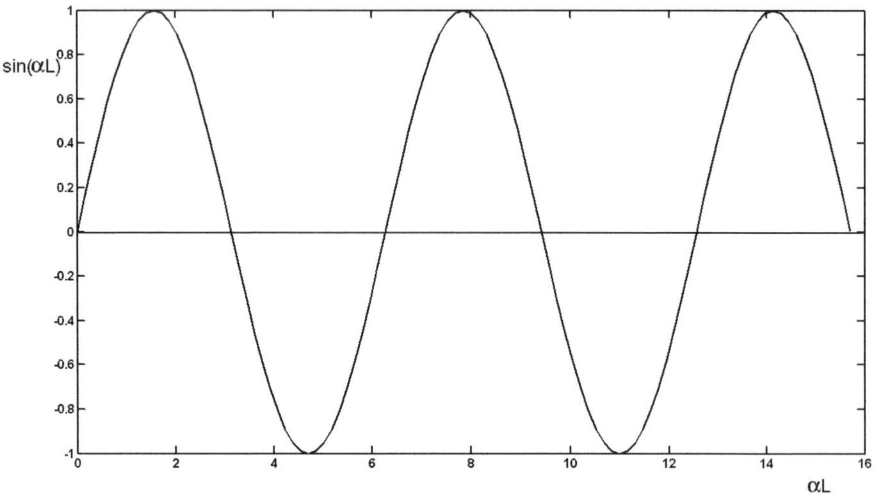

Figure 5.2. Variation of $\sin(\alpha L)$ versus αL

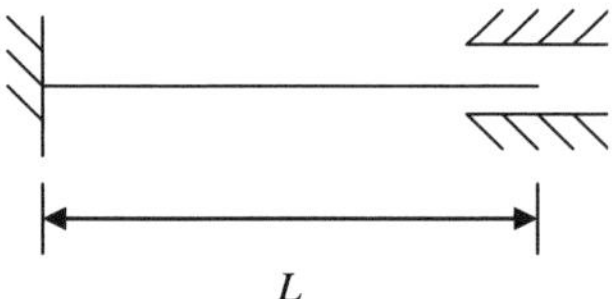

Figure 5.3. Clamped–clamped beam

where

$$\alpha = \frac{j\pi}{L} \tag{5.36}$$

The same method of solution is applied for the clamped–clamped beam, which is shown in Fig. 5.3.

By applying the boundary conditions, we get at $x = 0$:

$$W(0) = 0 \rightarrow B_4 = -B_2 \tag{5.37}$$

$$W'(x) = B_1\alpha \cosh(\alpha x) + B_2\alpha \sinh(\alpha x) + B_3\alpha \cos(\alpha x) - B_4\alpha \cos(\alpha x)$$

$$W'(0) = 0 \rightarrow B_3 = -B_1 \tag{5.38}$$

and at $x = L$

$$W(L) = 0 \rightarrow B_1[\sinh(\alpha L) - \sin(\alpha L)]$$
$$+ B_2[\cosh(\alpha L) - \cos(\alpha L)] = 0 \qquad (5.39)$$
$$W'(L) = 0 \rightarrow \alpha B_1[\cosh(\alpha L) - \cos(\alpha L)]$$
$$+ \alpha B_2[\sinh(\alpha L) + \sin(\alpha L)] = 0 \qquad (5.40)$$

Then, the determinant of coefficients in front of $B_1 \& B_2$ must be equal to zero

$$\begin{vmatrix} \sinh(\alpha L) - \sin(\alpha L) & \cosh(\alpha L) - \cos(\alpha L) \\ \alpha\cosh(\alpha L) - \alpha\cos(\alpha L) & \alpha\sinh(\alpha L) + \alpha\sin(\alpha L) \end{vmatrix} = 0 \qquad (5.41)$$

or

$$\cosh(\alpha L)\cos(\alpha L) - 1 = 0 \qquad (5.42)$$
$$\cosh(\alpha L)\cos(\alpha L) = 1 \qquad (5.43)$$

The nontrivial solutions of Eq. (5.43) are

$$\alpha L \approx \left(j + \frac{1}{2} \right) \pi \qquad (5.44)$$

where $j = 1, 2, 3, \ldots$.

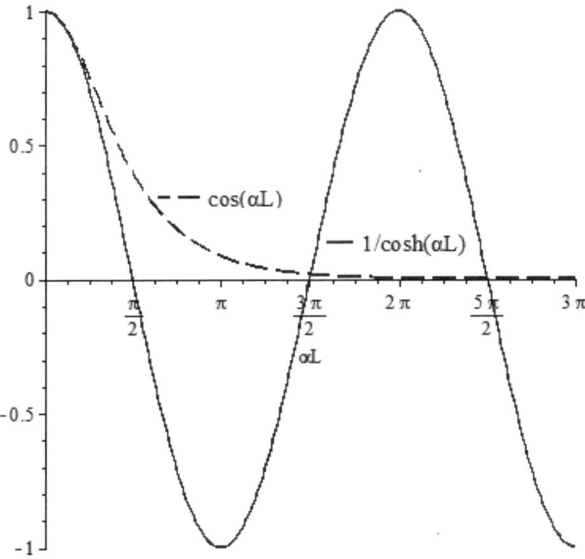

Figure 5.4. Variation of $y = \cos(\alpha L)$ versus $d = 1/\cosh(\alpha L$ versus αL

The first four roots obtained from Fig. 5.4 are as follows:

$$(\alpha L)_j = 4.7 \quad 7.85 \quad 10.99 \quad 14.137 \tag{5.45}$$

The natural frequency is

$$\omega_j = (\alpha L)_j^2 \sqrt{\frac{EI}{\rho A L^4}} \tag{5.46}$$

The mode shape reads

$$W(x) = B_2 \left[[\sinh(\alpha x) - \sin(\alpha x)] \left[-\frac{\cosh(\alpha L) - \cos(\alpha L)}{\sinh(\alpha L) - \sin(\alpha L)} \right] \right.$$

$$\left. + [\cosh(\alpha x) - \cos(\alpha x)] \right] \tag{5.47}$$

where

$$\alpha = \left(j + \frac{1}{2} \right) \frac{\pi}{L} \tag{5.48}$$

Chapter 6

Vibration Tailoring of Inhomogeneous Elastically Restrained Vibrating Beams

In this chapter, we consider free vibration of inhomogeneous Bernoulli–Euler beam which is elastically restrained at one end and clamped at the other, as shown in Fig. 6.1. The closed-form solution is obtained for the beam of constant material density and constant cross-section but of modulus of elasticity, which varies in a polynomial manner. The semi-inverse method is utilized; namely the fundamental mode of vibration is postulated as a polynomial. It turns out that such a formulation leads to infinite number of solutions, one can obtain a unique solution by introducing an additional requirement. It is shown that if in addition to the fundamental mode shape the natural frequency is specified also, the unique solution is derived.

6.1 Background

The free vibration of uniform and non-uniform beams attracted many investigators since Bernoulli and Euler derived the governing differential equation in the 18th century. Beams with end springs have been dealt with by many investigators. Handbook by Karnovsky and Lebel (2000) summarizes analytical and numerical results. Due to the numerous papers, it is virtually impossible to do justice to the accumulated literature. Therefore, only representative works will be cited. The studies by Liu and Chen (1989), Hibbeler (1975), Maurizi *et al.* (1991), Laura and Guttierez (1986), Lee and Kuo (1992) and Lizarev (1959) appear to be in the need of mentioning. The above papers dealt with direct problems, i.e., the ones in which the flexural

97

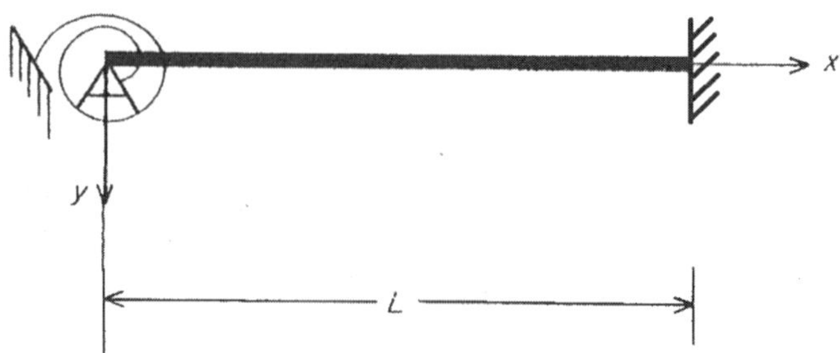

Figure 6.1. Beam elastically restrained at one end and clamped at the other

rigidity and the inertial coefficient are specified, and one needs to determine the natural frequencies and mode shapes. Inverse vibration problems were attacked by Barcilon (1976; 1979), Lowe (1993) and other investigators. In these problems, one deals with construction of Euler–Bernoulli beam from the spectral data. In these circumstances, the natural question arises on how can one get the spectra that serve as inputs for the construction problem.

Elishakoff and Candan (2001) dealt with a semi-inverse problem of more modest objective than that of Barcilon (1976; 1979), and Lowe (1993). Elishakoff and Candan (2001) dealt with a situation when only the fundamental mode shape is specified in the form of a simple polynomial; one assumes the inertial coefficient as given and seeks the polynomial flexural rigidity that is compatible with the mode shape and attendant postulated flexural rigidity. For the inhomogeneous beams with ideal boundary conditions, including pinned, clamped or free ends, Elishakoff and Candan (2001) constructed Bernoulli–Euler beams that correspond to the provided information. It turned out that there are infinite amount of beams that possess the specified natural mode. This finding opens a possibility to impose some additional requirement that allows then the unique solution.

In this case, the results of Elishakoff and Candan (2001) are generalized to include an inhomogeneous beam with an end spring. According to Einstein, the theories and methods ought to be as simple as possible, but not simpler; it appears that here simplest possible solution is presented for the titled problem. The results show the extreme usefulness of such a formulation: It turns out that by postulating the mode shape and setting a natural frequency at the preselected level, one obtains a unique solution.

6.2 Analysis

The governing differential equation for the inhomogeneous beam reads

$$\frac{\partial^2}{\partial x^2}\left[D(x)\frac{\partial^2 w}{\partial x^2}\right] + R(x)\frac{\partial^2 w}{\partial t^2} = 0 \tag{6.1}$$

where $w(x,t)$ is the transverse displacement, $D(x) = E(x)I(x)$ is the flexural rigidity, $E(x) =$ modulus of elasticity, $I(x) =$ moment of inertia, $R(x) = \rho(x)A(x)$ inertial coefficient, $\rho(x) =$ mass density, $A(x) =$ cross-sectional area, $x =$ axial coordinate, and $t =$ time. We set $R(x) =$ const, namely that $\rho(x) =$ const $= \rho_0$, $A(x) =$ const $= A_0$, $I(x) =$ const $= I_0$. The only function that varies along the beam's axis is the modulus of elasticity, as a result of which the flexural rigidity is a function of x.

We introduce a non-dimensional axial coordinate

$$\xi = x/L \tag{6.2}$$

where $L =$ length of the beam. The inertial coefficient $R(x)$ is considered to be a constant

$$R(x) = \rho A = \text{const} \tag{6.3}$$

so that Eq. (6.1) reduces to

$$\frac{d^2}{d\xi^2}\left[D(\xi)\frac{d^2 W}{d\xi^2}\right] - \rho A L^4 \omega^2 W = 0 \tag{6.4}$$

where $\omega =$ sought natural frequency and $W(\xi) =$ mode shape. Equation (6.4) is obtained from Eq. (6.1) by substituting

$$w(x,t) = W(x)\sin\omega t \tag{6.5}$$

We study a beam with a rotational spring at the left end of the beam and clamped at the right end that the boundary conditions are

$$W(\xi) = 0 \text{ at } \xi = 0 \tag{6.6}$$

$$kL\frac{dW(\xi)}{d\xi} = D(\xi)\frac{d^2 W(\xi)}{d\xi^2} \text{ at } \xi = 0 \tag{6.7}$$

$$W(\xi) = 0 \text{ at } \xi = 1 \tag{6.8}$$

$$\frac{dW(\xi)}{d\xi} = 0 \text{ at } \xi = 1 \tag{6.9}$$

where k is the spring stiffness.

Simplest function for the mode shape is a fourth-order polynomial

$$W(\xi) = a_0 + a_1\xi + a_2\xi^2 + a_3\xi^3 + a_4\xi^4 \tag{6.10}$$

The enforcement of the condition (6.6) yields

$$a_0 = 0 \tag{6.11}$$

Prior to satisfaction of the boundary condition in Eq. (6.7), we need a specific form of the flexural rigidity $D(\xi)$. We are solving a semi-inverse problem of determining the flexural rigidity in such a manner that the function in Eq. (6.10) represents an exact mode shape. The function for $D(\xi)$ that is compatible with Eq. (6.10) is a fourth-order polynomial

$$D(\xi) = b_0 + b_1\xi + b_2\xi^2 + b_3\xi^3 + b_4\xi^4 \tag{6.12}$$

Equations (6.10) and (6.12) are substituted into Eq. (6.7) to result in

$$kLa_1 = 2b_0a_2 \tag{6.13}$$

The boundary conditions in Eqs. (6.8) and (6.9) lead to

$$a_2 = -2a_3 - 3a_4 \tag{6.14}$$

$$a_1 = a_3 + 2a_4 \tag{6.15}$$

we get

$$a_2 = \frac{kL}{4b_0 + kL}a_4 \tag{6.16}$$

In view of Eq. (6.13), a_1 becomes

$$a_1 = \frac{2b_0}{4b_0 + kL}a_4 \tag{6.17}$$

Substituting Eq. (6.17) into (6.15) leads to

$$a_3 = -2\frac{(3b_0 + kL)}{4b_0 + kL}a_4 \tag{6.18}$$

Thus, the mode shape becomes

$$W(\xi) = a_4\left(\frac{2b_0}{4b_0 + kL}\xi + \frac{kL}{4b_0 + kL}\xi^2 - \frac{(6b_0 + 2kL)}{4b_0 + kL}\xi^3 + \xi^4\right) \tag{6.19}$$

where a_4 is an arbitrary constant. We fix it at unity. The expression of the mode shape reads

$$W(\xi) = \frac{2b_0}{4b_0 + kL}\xi + \frac{kL}{4b_0 + kL}\xi^2 - \frac{6b_0 + 2kL}{4b_0 + kL}\xi^3 + \xi^4 \tag{6.20}$$

The result of substitution of Eqs. (6.12) and (6.20) into Eq. (6.4) is

$$\sum_{j=0}^{4} C_j \xi^j = 0 \tag{6.21}$$

where

$$C_0 = 24b_0 kL - 24b_1 kL + 4b_2 kL + 96b_0^2 - 72b_1 b_0 \tag{6.22}$$

$$C_1 = -72b_2 kL + 72b_1 kL - 2\rho A\omega^2 L^4 b_0 - 216b_2 b_0$$
$$+ 288b_1 b_0 + 12b_3 kL \tag{6.23}$$

$$C_2 = 144b_2 kL - 144b_3 kL + 576b_0 b_2 - 432b_3 b_0$$
$$+ 24b_4 kL - \rho A L^5 \omega^2 k \tag{6.24}$$

$$C_3 = 960b_3 b_0 - 720b_4 b_0 + 240b_3 kL + 2\rho A L^5 \omega^2 k$$
$$+ 6\rho A\omega^2 L^4 b_0 - 240b_4 kL \tag{6.25}$$

$$C_4 = 360b_4 kL - 4\rho A L^4 \omega^2 b_0 + 1440b_4 b_0 - \rho A\omega^2 kL^5 \tag{6.26}$$

Equation (6.26) results in

$$\omega^2 = 360b_4/\rho A L^4 \tag{6.27}$$

It should be noted that although Eq. (6.26) depends on coefficient kL, the expression for ω^2 does not depend on it. Remarkably, formula (6.27) coincides with its counterpart valid for the beam with classical boundary conditions, as reported in Elishakoff and Candan (2001) (see also the monograph of Elishakoff (2005)). Whereas the relation between ω^2 and b_4 remains unchanged, the rest of the coefficients in $D(\xi)$ depend upon the stiffness of the end spring. Substituting Eq. (6.27) into (6.25) gives

$$b_3 = \frac{-2(3b_0 + kL)}{4b_0 + kL} b_4 \tag{6.28}$$

Eqs. (6.23) and (6.24) yield

$$b_2 = \frac{k^2 L^2 - 8b_0 kL - 54b_0^2}{3(4b_0 + kL)^2} b_4 \tag{6.29}$$

$$b_1 = \frac{2(k^3 L^3 - 9b_0^2 kL^5 + 87b_0^2 kL + 159b_0^3)}{3(4b_0 + kL)^3} b_4 \tag{6.30}$$

From Eq. (6.22), we get

$$b_4 = \frac{18b_0 \left(4b_0 + kL\right)^4}{5256kLb_0^3 + 1706k^2L^2b_0^2 + 232b_0k^3L^3 + 5724b_0^4 + 11k^4L^4}$$

(6.31)

The flexural rigidity $D(\xi)$ is obtained by substituting Eqs. (6.28)–(6.31) into Eq. (6.12). Note that when $k = 1$, we get $b_3 = -1.5b_4$ as found by Elishakoff (2005) for the pinned–clamped inhomogeneous beam. When k tends to infinity, b_3 approaches $-2b_4$ as for the clamped–clamped inhomogeneous beam as derived by Elishakoff (2005).

It is seen that when the end spring stiffness k is fixed, we get infinite amount of beams since the coefficient b_0 is arbitrary. For example, for $kL = 1$ and $b_0 = 1$, we have

$$D(\xi) = 1 + \frac{15780}{12929}\xi - \frac{9150}{12929}\xi^2 - \frac{18000}{12929}\xi^3 + \frac{11250}{12929}\xi^4$$

$$\approx 1 + 1.2205\xi - 0.70771\xi^2 - 1.3922\xi^3 + 0.87014\xi^4 \quad (6.32)$$

For $kL = 10$ and $b_0 = 2$, we get

$$D(\xi) = 2 + \frac{26856}{12281}\xi - \frac{7452}{12281}\xi^2 - \frac{46656}{12281}\xi^3 + \frac{26244}{12281}\xi^4$$

$$\approx 2 + 2.1868\xi - 0.60679\xi^2 - 3.799\xi^3 + 2.137\xi^4 \quad (6.33)$$

For $kL = 100$ and $b_0 = 3$

$$D(\xi) = 3 + \frac{7222900131543161}{2251799813685248}\xi - \frac{7365930465590695}{9007199254740992}\xi^2$$

$$- \frac{296257640492991}{35184372088832}\xi^3 + \frac{4870584328104953}{1125899906842624}\xi^4$$

$$\approx 3 + 3.2076\xi - 0.81778\xi^2 - 8.4201\xi^3 + 4.3259\xi^4 \quad (6.34)$$

These are shown in the Figs. 6.2–6.4. It may appear at the first glance that the Fig. 6.2 is symmetric with respect to the middle cross-section $\xi = 0.5$. However, it is not a symmetric function with respect to $\xi = 0.5$, because the presence of the end spring introduces a lack of symmetry. For $k = 0$, the figure is symmetric because of symmetric pinned–clamped boundary conditions (Elishakoff, 2005). At low values of kL, the mode shape may

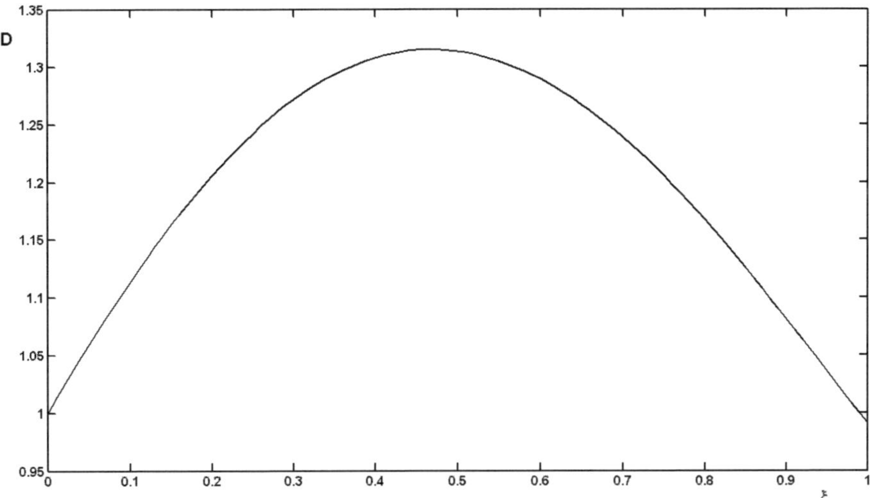

Figure 6.2.   Variation of $D(\xi)$ for $kL = 1$ and $b_0 = 1$, $\xi \in [0; 1]$

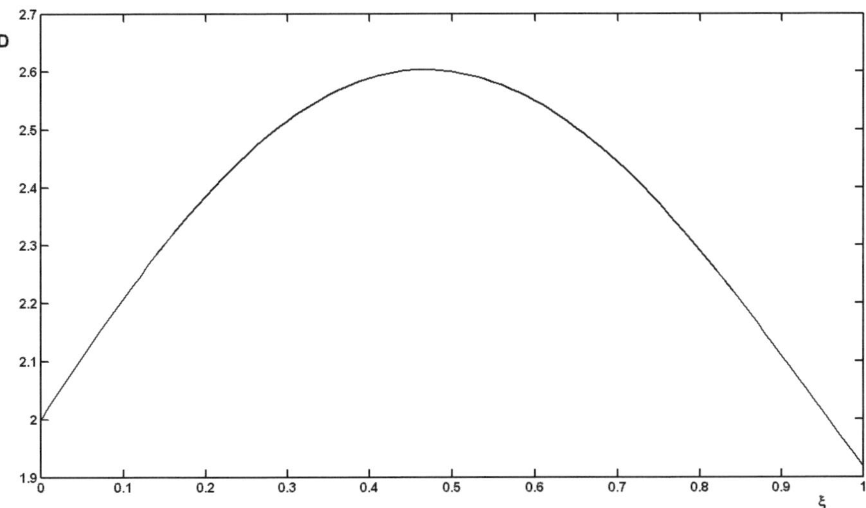

Figure 6.3.   Variation of $D(\xi)$ for $kL = 10$ and $b_0 = 2$, $\xi \in [0; 1]$

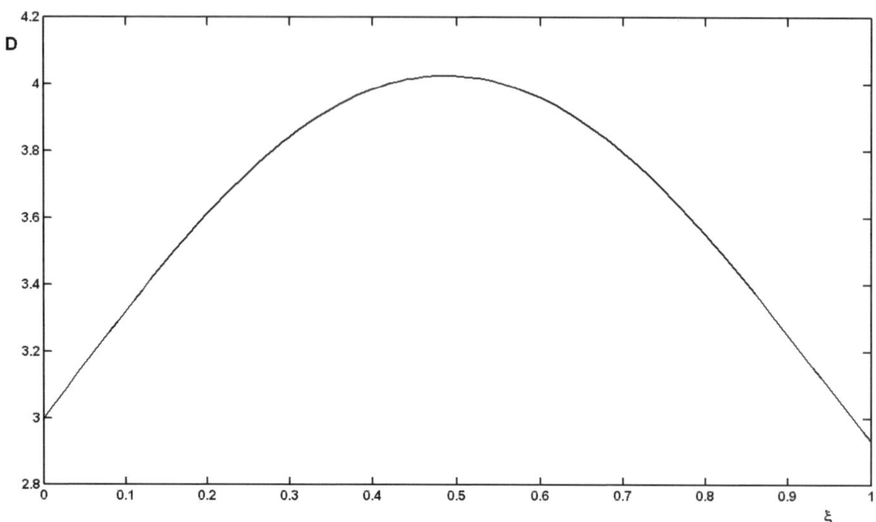

Figure 6.4. Variation of $D(\xi)$ for $kL = 100$ and $b_0 = 3$, $\xi \in [0; 1]$

appear symmetric. But this appearance is not present for a beam with end springs of larger stiffness. In Fig. 6.3, kL is set at 10, whereas in Fig. 6.4, $kL = 100$. In these figures, the lack of symmetry is apparent.

Presence of infinite amount of solutions is a favorable consequence of the present formulation of the semi-inverse problem. It allows the designer to introduce an additional requirement. For example, if the design requires the beam to possess a specified natural frequency Ω, then from Eq. (6.27) we get the expression for b_4 by setting $\omega = \Omega$:

$$b_4 = \rho A L^4 \Omega^2 / 360 \tag{6.35}$$

Then, by equating the left hand sides of Eqs. (6.31) and (6.35), we get the following equation for b_0:

$$\frac{18b_0(4b_0 + kL)^4}{5256kLb_0^3 + 1706k^2L^2b_0^2 + 232b_0k^3L^3 + 5724b_0^4 + 11k^4L^4} = \frac{\rho A L^4 \Omega^2}{360} \tag{6.36}$$

which is a cubic equation for b_0. Solution of Eq. (6.36) for specified kL and Ω and substitution of this b_0 into Eqs. (6.28)–(6.30) yields the coefficients b_1, b_2 and b_3, resulting in a unique beam.

6.3 Comparison between closed-form solutions of inhomogeneous vibrating beam and homogeneous beam

It is instructive to compare the above closed-form solution for a class of polynomially inhomogeneous beams to that for a uniform, homogeneous beam. In the latter case, the mode shape reads, Rao (2004),

$$W(x) = B_1 \sinh(\alpha x) + B_2 \cosh(\alpha x) + B_3 \sin(\alpha x) + B_4 \cos(\alpha x) \quad (6.37)$$

where

$$\alpha^4 = \rho A \omega^2 / EI \quad (6.38)$$

Satisfaction of the boundary conditions in Eqs. (6.6)–(6.9) leads to the following transcendental equation:

$$\frac{kL}{EI}\left(1 + \frac{1}{\cosh(\alpha L)\cos(\alpha L)}\right) = \tan(\alpha L) - \tanh(\alpha L) \quad (6.39)$$

Note that when k vanishes, the above equation reduces to

$$\tan(\alpha L) = \tanh(\alpha L) \quad (6.40)$$

which is frequency equation for the pinned–clamped beam. For kL approaching infinity

$$\cos(\alpha L)\cosh(\alpha L) + 1 = 0 \quad (6.41)$$

recovering the frequency equation of the clamped–clamped beam. For general kL, the fundamental natural frequencies are, for example, for $kL/EI = 0.01$, $\alpha L \approx 0.527$; for $kL/EI = 0.1$, $\alpha L \approx 0.6759$; for $kL/EI = 1$, $\alpha L \approx 1.305$; for $kL/EI = 10$, $\alpha L \approx 1.793$; for $kL/EI = 100$, $\alpha L \approx 1.877$. Finally, for $kL/EI \to \infty$, $\alpha L \approx 1.885$. The results are obtained by the numerical solution of the characteristic equation (6.39).

It is worth noting that while the solution for the uniform, homogeneous beam necessitates the *numerical* tackling of the *transcendental* equation, the semi-inverse method in the present setting furnishes the solution in a closed, polynomial form. There are also closed-form solutions available for non-polynomial variation of the mode shape and/or the flexural rigidity (Caliò and Elishakoff, 2005; Gilat, Caliò, Elishakoff, 2010).

6.4 Vibration tailoring: Numerical example

Let us imagine we need a beam with the following properties: $\omega = 100\,\text{Hz}$, $\rho = 7800\,\text{kg/m}^3$, $k = 1$, $L = 15\,\text{m}$, and $A = 0.18\,\text{m}^2$. Substitution of these given values into Eq. (6.27), yields

$$b_4 = 87.75 \times 10^5 \tag{6.42}$$

Substituting Eq. (6.42) into Eq. (6.31), we get a fifth-order equation for b_0:

$$-\frac{67869140625}{64} - \frac{381712449375}{256}b_0 - \frac{23390852625}{32}b_0^2$$

$$-\frac{60053875}{4}b_0^3 - \frac{174402885}{16}b_0^4 + b_0^5 = 0 \tag{6.43}$$

Solution of Eq. (6.43) yields five roots for b_0. The only positive root is $b_0 = 10900194.0861$. Therefore, the flexural rigidity, Eq. (6.12), turns out to be

$$D(\xi) = 1.2422 + 1.6562\xi - 1.125\xi^2 - 1.5\xi^3 + \xi^4 \tag{6.44}$$

Chapter 7

Some Intriguing Results Pertaining to Functionally Graded Columns

Some intriguing results are reported in conjunction with closed-form solutions obtained for a clamped-free vibrating inhomogeneous column under an axial concentrated load using the semi-inverse method. Fourth-order polynomial is postulated for both the vibration mode shape and buckling displacement. Solution is provided for the flexural rigidity and the natural frequency. It is shown that for each level of axial loading, there may exist up to five flexural rigidities satisfying the governing differential equation and boundary conditions.

7.1 Background

In their study, Elishakoff and Endres (2005) studied the semi-inverse method as applied to eigenvalue problems for axially graded functionally graded (FG) vibrating column subject to a concentrated load at its end. They showed that by postulating the polynomial mode shape of a FG structure, one can find the flexural rigidity function $D(x)$ as well as the associated eigenvalue. Specifically, Elishakoff and Endres (2005) investigated the case of a clamped-free column under an axial load in static setting. Unexpectedly, the same solution for the buckling load was derived, namely $P_{cr} = -12b_2/L^2$, as for other boundary conditions, L being the length, b_2 being an arbitrary negative coefficient. Remarkably, three different mode shapes were shown to be associated with the same buckling load in the

107

clamped-free column case whereas a single mode shape was presented with any other boundary condition. Ayadoğlu (2008), Ece *et al.* (2007), Caliò and Elishakoff (2005), Li (2009), Maróti (2011), Caliò *et al.* (2011) obtained other novel results inspired by the closed-form solution by Elishakoff and Endres (2005).

In this chapter, which follows closely paper by Elishakoff and Miglis (2012), we extend Elishakoff and Endres' (2005) work to vibrating inhomogeneous clamped-free beam under an axial load. Interestingly, we find up to five-mode shapes when the beam is subjected to both vibration and buckling.

We also demonstrate that there is a linear relationship between buckling load and natural frequency. Finally, we are discussing the limiting case $P = 0$ and associated implications. It appears that the remarkable findings that are reported in this study will inspire more research in functionally graded structures.

7.2 Formulation of the problem

We consider a FG beam in both vibration and buckling settings. The Young's modulus depends on x and is denoted $E(x)$. The cross-sectional area A is constant. We denote the flexural rigidity of the beam $E(x)I = D(x)$. In this case, the governing differential equation reads

$$\frac{d^2}{dx^2}\left(D(x)\frac{d^2}{dx^2}[w(x)]\right) + P\frac{d^2}{dx^2}[w(x)] - \rho A\omega^2 w(x) = 0 \qquad (7.1)$$

ρ is the density of the material, A is the cross-sectional area, ω is the natural frequency of the beam, $w(x)$ is the mode shape and P is the axial load. Equation (7.1) can be written in a more convenient manner introducing the non-dimensional coordinate $\xi = x/L$

$$\frac{d^2}{d\xi^2}\left(D(\xi)\frac{d^2}{d\xi^2}[w(\xi)]\right) + PL^2\frac{d^2}{d\xi^2}[w(\xi)] - \Omega^2 w(\xi) = 0, \quad \Omega^2 = \rho A\omega^2$$

$$(7.2)$$

We solve hereinafter a semi-inverse problem postulating the form of $w(\xi)$ and finding the grading function $E(\xi)$ that is compatible with the specified function $w(\xi)$. We take $w(\xi)$ as a fourth-order polynomial.

For a clamped-free beam, the boundary conditions read

$$w(\xi) = 0, \quad \frac{dw}{d\xi} = 0 \quad \text{at } \xi = 0 \tag{7.3}$$

$$D(\xi)\frac{d^2 w}{d\xi^2} = 0 \quad \text{at } \xi = 1 \tag{7.4}$$

$$\frac{d}{dx}\left[D(\xi)\frac{d^2 w(\xi)}{d\xi^2}\right] + PL^2\frac{dw(\xi)}{d\xi} = 0 \quad \text{at } \xi = 1 \tag{7.5}$$

At this juncture, we see that the satisfying boundary conditions necessitates the knowledge of flexural rigidity $D(\xi)$. We look for the mode shape in the polynomial form. In order function $w(\xi)$ to serve as a mode shape it has to satisfy boundary conditions.

$$w(\xi) = a_0 + a_1\xi + a_2\xi^2 + a_3\xi^3 + \xi^4 \tag{7.6}$$

In these circumstances, $D(x)$ has to be also a fourth-order polynomial in the order of the first and the third term to match in Eq. (7.2). For the rigidity of the beam, we write

$$D(\xi) = b_0 + b_1\xi + b_2\xi^2 + b_3\xi^3 + b_4\xi^4 \tag{7.7}$$

In Eq. (7.6) we have to determine 4 unknowns a_j. Introducing Eq. (7.6) and Eq. (7.7) into Eq. (7.2) yields

$$(360b_4 - \Omega^2)\xi^4 + (240b_3 - \Omega^2 a_3 + 120b_4 a_3)\xi^3 + (12PL^2$$

$$+ 144b_2 - \Omega^2 a_2 + 72b_3 a_3 + 24b_4 a_2)\xi^2 + (6PL^2 a_3 + 36b_2 a_3$$

$$+ 12b_3 a_2 + 72b_1 - \Omega^2 a_1)\xi + 4b_2 a_2 + 12b_1 a_3 + 2PL^2 a_2$$

$$+ 24b_0 - \Omega^2 a_0 = 0 \tag{7.8}$$

Introducing Eq. (7.6) and Eq. (7.7) into Eqs. (7.3)–(7.5) gives the following values for $a_i, i = (0, 1, 2, 3)$

$$a_0 = 0 \tag{7.9}$$

$$a_1 = 0 \tag{7.10}$$

$$a_2 = \frac{2(PL^2 - 6b_2 - 6b_1 - 6b_3 - 6b_4 - 6b_0)}{PL^2 - 2b_2 - 2b_1 - 2b_3 - 2b_4 - 2b_0} \tag{7.11}$$

$$a_3 = -\frac{8}{3}\frac{PL^2 - 3b_2 - 3b_1 - 3b_3 - 3b_4 - 3b_0}{PL^2 - 2b_2 - 2b_1 - 2b_3 - 2b_4 - 2b_0} \tag{7.12}$$

We eliminate the terms a_0 and a_1 in Eq. (7.8) to get

$$(360b_4 - \Omega^2)\xi^4 + (240b_3 - \Omega^2 a_3 + 120b_4 a_3)\xi^3$$
$$+ (12PL^2 + 144b_2 - \Omega^2 a_2 + 72b_3 a_3 + 24b_4 a_2)\xi^2$$
$$+ (6PL^2 a_3 + 36b_2 a_3 + 12b_3 a_2 + 72b_1)\xi + 4b_2 a_2$$
$$+ 12b_1 a_3 + 2PL^2 a_2 + 24b_0 = 0 \tag{7.13}$$

Equation (7.13) has to be satisfied for any ξ. Therefore, the coefficients in front of each power of ξ have to vanish. The term in front of ξ^4 has to be equal to 0. Thus,

$$b_4 = \frac{1}{360}\Omega^2 \tag{7.14}$$

The term in front of ξ^3, must also be equal to zero. We get

$$240b_3 - \Omega^2 a_3 + 120b_4 a_3 = 0 \tag{7.15}$$

Introducing Eqs. (7.11), (7.12) and (7.14) into Eq. (7.15) yields

$$240b_3 - \Omega^2 a_3 + 120b_4 a_3$$
$$= \frac{8}{3(-360b_1 - \Omega^2 + 180PL^2 - 360_{b_2} - 360_{b_3} - 360_{b_0})}$$
$$\times [-360\Omega^2 b_2 - 450\Omega^2 b_3 - \Omega^4 + 120\Omega^2 PL^2 - 360\Omega^2 b_0$$
$$- 360\Omega^2 b_1 - 32400 b_3 b_1 + 16200 b_3 PL^2 - 32400 b_3 b_2$$
$$- 32400 b_3^2 - 32400 b_3 b_0] = 0 \tag{7.16}$$

Therefore, we have to solve the following equation

$$-360\Omega^2 b_2 - 450\Omega^2 b_3 - \Omega^4 + 120\Omega^2 PL^2 - 360\Omega^2 b_0 - 360\Omega^2 b_1$$
$$- 32400 b_3 b_1 + 16200 b_3 PL^2 - 32400 b_3 b_2 - 32400 b_3^2 - 32400 b_3 b_0$$
$$= 0$$
$$\tag{7.17}$$

Expressing b_0 from Eq. (7.17) yields

$$b_0 = \frac{1}{360} \frac{\begin{array}{c} -360\Omega^2 b_1 - 360\Omega^2 b_2 - 450\Omega^2 b_3 - \Omega^4 + 120\Omega^2 PL^2 \\ +16200b_3 PL^2 - 32400b_3 b_1 - 32400b_3 b_2 - 32400b_3^2 \end{array}}{\Omega^2 + 90b_3}$$

$$(7.18)$$

Now, equating the term in front of ξ^2 in Eq. (7.13) to zero results in

$$12PL^2 + 144b_2 - \Omega^2 a_2 + 72b_3 a_3 + 24b_4 a_2 = 0 \qquad (7.19)$$

Introducing Eqs. (7.11), (7.12), (7.14), (7.18) into Eq. (7.19) one obtains

$$\frac{4}{5} \frac{15\Omega^2 PL^2 + 7\Omega^4 + 1260\Omega^2 b_3 + 32400b_3^2 + 180\Omega^2 b_2}{\Omega^2} = 0 \qquad (7.20)$$

Solving Eq. (7.20) for b_2 yields

$$b_2 = -\frac{1}{180} \frac{15\Omega^2 PL^2 + 7\Omega^4 + 1260\Omega^2 b_3 + 32400b_3^2}{\Omega^2} \qquad (7.21)$$

The coefficient in front of ξ in Eq. (7.13) must also be equal to zero. Thus,

$$6PL^2 a_3 + 36b_2 a_3 + 12b_3 a_2 + 72b_1 = 0 \qquad (7.22)$$

Replacing b_2, Ω^2, a_2, a_3, b_0, by their values given in Eqs. (7.11), (7.12), (7.18) and (7.21) and introducing them into Eq. (7.22) yields

$$\frac{72\left(15b_3 PL^2 \Omega^2 + \Omega^4 b1 - 8\Omega^4 b_3 - 1440b_3^2 \Omega^2 - 32400b_3^3\right)}{\Omega^4} = 0 \qquad (7.23)$$

From Eq. (7.23), we obtain the value of b_1 as a function of b_3 and Ω^2

$$b_1 = -\frac{b_3\left(15\Omega^2 PL^2 - 8\Omega^2 - 1440\Omega^2 b_3 - 32400b_3^2\right)}{\Omega^4} \qquad (7.24)$$

In these circumstances, we express the factor in front of ξ^0 in Eq. (7.13)

$$4b_2 a_2 + 12b_1 a_3 + 2PL^2 a_2 + 24b_0 = 0 \qquad (7.25)$$

as a function of b_3 and Ω^2:

$$4b_2 a_2 + 12b_1 a_3 + 2PL^2 a_2 + 24b_0$$

$$= -\frac{9}{5\Omega^6(\Omega^2 + 90b_3)}\left[600b_3 PL^2 \Omega^6 + 108000b_3^2 \Omega^4 PL^2\right.$$

$$+ 3240000b_3^3 PL^2\Omega^2 - 250\Omega^8 b_3 - 36000b_3^2\Omega^6 - 5400000\Omega^4 b_3^3$$

$$- 388800000\Omega^2 b_3^4 - \Omega^{10} - 69984000000b_3^5]$$

$$= 0 \tag{7.26}$$

Expressing P from Eq. (7.26) yields

$$P = \frac{1}{600} \frac{(77760000b_3^4 + 3456000\Omega^2 b_3^3 + 21600b_3^2\Omega^4 + 160\Omega^6 b_3 + \Omega^8)(\Omega^2 + 90b_3)}{\Omega^2 b_3 L^2(\Omega^4 + 180\Omega b_3 + 5400b_3^2)} \tag{7.27}$$

In the numerator of P, we have the factor $\Omega^2 + 90b_3$. Therefore, b_3 and Ω^2 ought to have the same dimension. As a consequence, we propose a possible linear relationship between b_3 and Ω^2:

$$b_3 = k\Omega^2 \quad k \in \mathrm{R} \tag{7.28}$$

with k being a proportionality parameter. Later, we will show that this representation is useful. In these circumstances, we can rewrite b_0, b_1, b_2 and P as follows:

$$b_0 = \frac{1}{360(1 + 90k)}[150PL^2 + 13\Omega^2 + 450k\Omega^2 - 518400k^2\Omega^2$$

$$+ 24300kPL^2 - 52488000k^3\Omega^2 + 486000k^2 PL^2$$

$$- 1049760000k^4\Omega^2] \tag{7.29}$$

$$b_1 = -k(15PL^2 - 8\Omega^2 - 1440k\Omega^2 - 32400k^2\Omega^2) \tag{7.30}$$

$$b_2 = -\frac{1}{12}PL^2 + \Omega^2\left(-\frac{7}{180} - 7k - 180k^2\right) \tag{7.31}$$

$$P = \frac{1}{600} \frac{(36000k^2 + 5400000k^3 + 388800000k^4 + 69984000000k^5 + 1 + 250k)\Omega^2}{k(5400k^2 + 1 + 180k)L^2} \tag{7.32}$$

From Eq. (7.32), we also get the expression of Ω^2 as a function of P

$$\Omega^2 = \frac{600k(5400k^2 + 1 + 180k)PL^2}{6998400000k^5 + 250k + 36000k^2 + 5400000k^3 + 388800000k^4 + 1} \tag{7.33}$$

For a beam, if one specifies the applied load, the beam's length and the parameter k, one can find the rigidity $D(\xi)$ and the displacement $w(\xi)$. We first write the expression of the rigidity:

$$
\begin{aligned}
D(\xi) &= \frac{5}{3} \frac{k(1 + 180k + 5400k^2)PL^2}{(1 + 90k)(77760000k^4 + 3456000k^3 + 21600k^2 + 160k + 1)}\xi^4 \\
&\quad + \frac{600k^2(1 + 180k + 5400k^2)PL^2}{(1 + 90k)(77760000k^4 + 3456000k^3 + 21600k^2 + 160k + 1)}\xi^3 \\
&\quad - \frac{1}{12}\frac{(530k + 136800k^2 + 17280000k^3 + 894240000k^4 + 1 + 13996800000k^5)PL^2}{(1 + 90k)(77760000k^4 + 3456000k^3 + 21600k^2 + 160k + 1)}\xi^2 \\
&\quad + \frac{15k(70k - 1 + 7992000k^3 + 155520000k^4 + 79200k^2)PL^2}{(77760000k^4 + 3456000k^3 + 21600k^2 + 160k + 1)(1 + 90k)}\xi \\
&\quad + \frac{5}{12}\frac{(5598720000k^5 + 318816000k^4 + 5421600k^3 + 57240k^2 + 374k + 1)PL^2}{(1 + 90k)(77760000k^4 + 3456000k^3 + 21600k^2 + 160k + 1)}
\end{aligned} \tag{7.34}
$$

The attendant mode shape reads

$$w(\xi) = (-6 - 1080k + 360k\xi + \xi^2)\xi^2 \tag{7.35}$$

Following question arises: How to determine the value of the parameter k? One observes that in the expression of $D(\xi)$ in Eq. (7.4) and Eq. (7.34), k is the only unknown parameter since the rest of the coefficients b_0, b_1 and b_2 are expressed via k in Eqs. (7.29)–(7.31), respectively. Likewise, since P is a pre-specified parameter, Eq. (7.33) provides us with values of squared natural frequency once k is known.

We demonstrate hereinafter that if the value of the flexural rigidity is known in any specified cross-section, say at the origin, namely if $D(0)$ at $\xi = 0$ is specified, we can find the value of the parameter k. Indeed, from Eq. (7.34), one arrives at the following equation to be solved for k

$$\frac{5}{12} \frac{(5598720000k^5 + 318816000k^4 + 5421600k^3 + 57240k^2 + 374k + 1)\,PL^2}{(1 + 90k)(77760000k^4 + 3456000k^3 + 21600k^2 + 160k + 1)}$$

$$= D(0) = b_0 \tag{7.36}$$

Equation (7.36) is obtained by substituting $\xi = 0$ in Eq. (7.34) and equating the result with $D(0)$. Thus, we have to solve the following fifth-order polynomial with respect to k

$$[2332800000\,PL^2 - 6998400000\,D(0)]k^5 + [132840000\,PL^2$$

$$- 388800000\,D(0)]k^4 + [2259000\,PL^2 - 5400000\,D(0)]k^3$$

$$+ [23850\,PL^2 - 36000\,D(0)]k^2$$

$$+ \left[\frac{935}{6}PL^2 - 250D(0)\right]k + \left[\frac{5}{12}PL^2 - D(0)\right] = 0 \tag{7.37}$$

It is convenient to non-dimensionalize Eq. (7.37). Dividing each side by PL^2 and denoting b_0/PL^2 as Θ yields

$$f(k) = (2332800000 - 6998400000\Theta)k^5 + (132840000$$

$$- 388800000\Theta)k^4 + (132840000 - 388800000\Theta)k^3$$

$$+ (23850 - 36000\Theta)k^2 + \left(\frac{935}{6} - 250\Theta\right)k + \left(\frac{5}{12} - \Theta\right) = 0$$

$$\tag{7.38}$$

We are interested in determining the number of real roots of Eq. (7.38). These roots can be either positive or negative. First, we investigate the number of positive roots.

We resort to the Descartes' Rule of signs in order to determine the number of positive roots of Eq. (7.38) knowing that the value of $D(x)$, and hence $D(0)$ must be positive. According to this rule, one has to count the number of changes in sign in front of coefficients of the polynomial. If there are no changes in sign, the polynomial has no positive roots. If there are n changes

Table 7.1. Values of Θ where the sign of the coefficient vanishes

	k^5	k^4	k^3	k^2	k^1	k^0
Θ	$\dfrac{1}{3}$	$\dfrac{41}{120}$	$\dfrac{251}{600}$	$\dfrac{53}{80}$	$\dfrac{187}{300}$	$\dfrac{5}{12}$

in sign, n being even, there are n, or $n-2$, or $n-4$, …, or 0 positive roots. If there are $n+1$ changes in sign, n being an even number, there are then $n+1$, or $n-1$, or $n-3$, …, or 1 positive root. In our case, it is instructive to observe where the coefficients in front of k^j change their signs. Finally, we can determine whether in some regions we have real roots for our equation. Table 7.1 denotes values of Θ at which the sign of the coefficients in front of k^j change from positive to negative.

Two specific cases must be considered first. If $\Theta = 1/3$, the term in front of k^5 vanishes and we get a fourth-order polynomial

$$3240000k^4 + 459000k^3 + 11850k^2 + \frac{145}{2}k + \frac{1}{12} = 0 \qquad (7.39)$$

One can observe that there are no changes in sign and according to Descartes' Rule, Eq. (7.39) does not have any positive root.

When $\Theta = 5/12$, the constant term vanishes, and Eq. (7.38) reduces to the following fourth-order polynomial since k must be non zero:

$$-583200000k^4 - 29160000k^3 + 9000k^2 + 8850k + \frac{155}{3} = 0 \quad (7.40)$$

In this case, we have one change in sign corresponding to one positive root.

Consider now the case $\Theta < 1/3$. The sign sequence is $-, -, -, -, -$. Therefore, Eq. (7.40) does not have a positive root. If $1/3 < \Theta < 5/12$, there is one change in signs in the coefficient sequence, and hence Eq. (7.40) has one positive root. On the other hand, if $5/12 < \Theta < 53/80$, the sign sequence has two changes in sign, resulting in two positive roots. We conclude that in this range Descartes rule of signs is inconclusive, and one has to resort to other means (for example, numeric ones). Finally, we observe that if $\Theta > 53/80$, we do not have any change in sign. Thus, in this region, there are no positive roots.

Since k in Eq. (7.28) must be real, we also have to investigate the number of negative roots. Concerning the negative roots, we first note that the number of negative roots of $f(k)$ equals to the number of positive roots equals the number of positive roots of the polynomial equation $f(-v)$. We get the polynomial

$$f(-v) = (-2332800000 + 69984000000\Theta)v^5 + (132840000$$

$$- 388800000\Theta)v^4 + (-132840000 + 388800000\Theta)v^3$$

$$+ (23850 - 360000\Theta)v^2 + \left(-\frac{935}{6} + 2500\Theta\right)v$$

$$+ \left(\frac{5}{12} - \Theta\right) = 0 \tag{7.41}$$

When $\Theta < 1/3$, the sign sequence is $-, +, -, +, -, +$. Therefore, the polynomial in Eq. (7.41) has 5, 3 or 1 positive roots. Thus, Eq. (7.41) surely has one positive root and, hence Eq. (7.38) has at least one negative root. It can be shown that in the range $5/12 < \Theta < 53/80$ one cannot determine the number of negative roots of Eq. (7.38) using Descartes' rule of signs. For $\Theta > 53/80$, Descartes rule is again inconclusive.

Results of such an analysis are summarized in Table 7.2.

Table 7.2. Results of application of Descartes' rule of signs for analytical determination of number or real roots

Regions	Value of Θ	Number of roots
I	$\Theta < \dfrac{1}{3} \approx 0.3333$	No positive root, at least one negative root
II	$\dfrac{1}{3} \approx 0.3333 < \Theta < \dfrac{41}{120} \approx 0.34166$	At least one positive root
III	$\dfrac{41}{120} \approx 0.34166 < \Theta < \dfrac{5}{12} \approx 0.4166$	
IV	$\dfrac{5}{12} \approx 0.41666 < \Theta < \dfrac{251}{600} \approx 0.41833$	At least one negative root
V	$\dfrac{251}{600} \approx 0.41833 < \Theta < \dfrac{187}{300} \approx 0.62333$	
VI	$\dfrac{187}{300} \approx 0.62333 < \Theta < \dfrac{53}{80} = 0.6625$	
VII	$\Theta > \dfrac{53}{80} = 0.6625$	At least one negative root

We perform *numerical* analysis for $5/12 < \Theta < 53/80$ (Regions IV–VI in Table 7.2). Numerical solution of Eq. (7.38) yields 5 real roots, 2 positive and 3 negative ones if $5/12 < \Theta < 0.5302083333$ (the latter value being in Region V). For example, if $\Theta = 0.475$ (Region V), the five roots are -0.04141101588, -0.01573492764, -0.006766827430, 0.001291278483, and 0.01033391077. The mode shape and the rigidity associated with these solutions are depicted in Fig. 7.1. Moreover, if Θ is greater than 0.5302083333 (but less than 0.6625) there are no positive roots; one has three negative roots. For example, if $\Theta = 13/24 = 0.541666$ (Region V), we obtain three negative roots as follows -0.04046658146, -0.10470211335, and -0.006766059379.

In conclusion, one can state that if the value of Θ is between $1/3$ and 0.5302083333, one has 5 roots, for Eq. (7.39), and thus the needed parameter k can be determined if $D(0)$ is specified. Outside this region, the equation has 3 negative roots. These results are summarized in Table 7.3.

Therefore, we want to follow the change of each root with variation of the axial load P. The variation of the roots k_i of Eq. (7.38) as a function of $PL^2/D(0) = \Theta^{-1}$ is depicted in Fig. 7.1.

It is seen from Fig. 7.1 that for $PL^2/D(0) \leq 1.89$, Eq. (7.38) has 3 roots, two of which are depicted in Fig. 7.1(a), whereas the variation of the root k_2

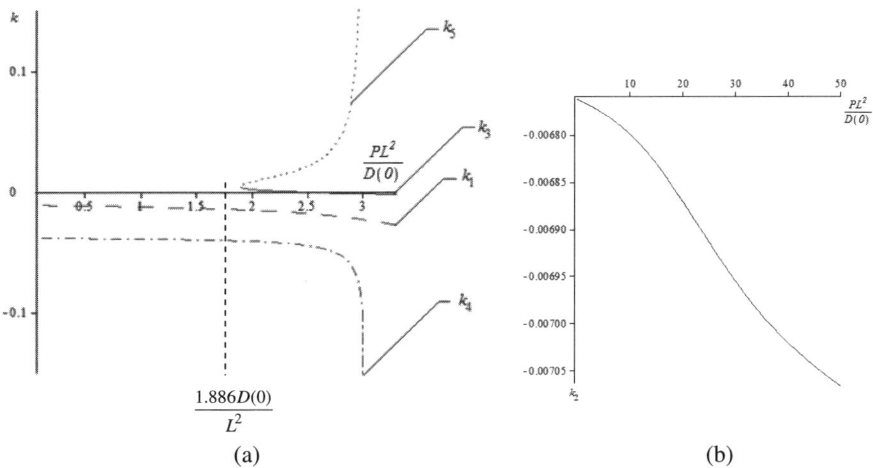

Figure 7.1. Variation of the roots of Eq. (7.38) with axial load P; (a) roots k_1, k_3, k_4, k_5; (b) root k_2

Table 7.3. Numerically determined number of roots of Eq. (7.38)

Value of Θ	Number of roots
$0 < \Theta < \dfrac{1}{3} \approx 0.3333$	Three negative roots
$\dfrac{1}{3} \approx 0.3333 \leq \Theta \leq 0.53$	Five roots real roots, at least three negatives and one positive
$0.53 < \Theta < \infty$	Three negative roots

is portrayed in Fig. 7.1(b). In the region $1.89 \leq PL^2/D(0) \leq 3$, Eq. (7.38) has 5 roots, four of which are depicted in Fig. 7.1(a), with the remaining 1 is presented in Fig. 7.1(b). Note that value $PL^2/D(0) = 1.89$ corresponds to $\Theta = 0.530208333$ in Table 7.3. For $PL^2/D(0) \geq 3$ (or $\Theta < 1/3$), Eq. (7.38) has 3 roots, 2 roots out of 5 in the previous region, namely k_4 and k_5, not appearing once the value $PL^2/D(0)$ is equal to or greater than 3 (or $\Theta \leq 1/3$).

Figure 7.2 depicts the variation of the five possible flexural rigidities and the associated mode shapes for $\Theta = 0.475$ or $PL^2/D(0) = 2.105263$.

7.3 Dependence between Ω^2 and P

In Eq. (7.33), we observe that there is a postulated dependence between the natural frequency Ω^2 and the load P. It is interesting to investigate the nature of this dependence for each of the 5 fundamental mode shapes. In order to plot the dependence $\Omega^2(P)$, we first determine the values of roots k of Eq. (7.38) for a sufficiently large number of values of Θ. For each specified value of Θ^j, we obtain 3 or 5 solutions for k, in accordance with Table 7.3. The roots are denoted by k_i^j, where i varies between 1 and 3 (for $\Theta < 1/3$ or $\Theta > 0.5302083333$) or between 1 and 5 for ($1/3 < \Theta < 0.5302083333$). The second step consists in collecting all k_i^j values together, along with the associated values of Θ^j. In other words, for an assigned value of $\Theta = \Theta^a$, we get the roots $k_1^a, k_2^a \ldots k_m^a$, where $m = 3$ or 5. For another value $\Theta = \Theta^b$, we get another set of roots $k_1^b, k_2^b \ldots k_m^b$. However, we know from Fig. 7.1 and Table 7.3 that some roots k_i are not existent for all Θ values (or all $PL^2/D(0)$). Numerically, as shown in Fig. 7.1 and Table 7.4, the roots k_1

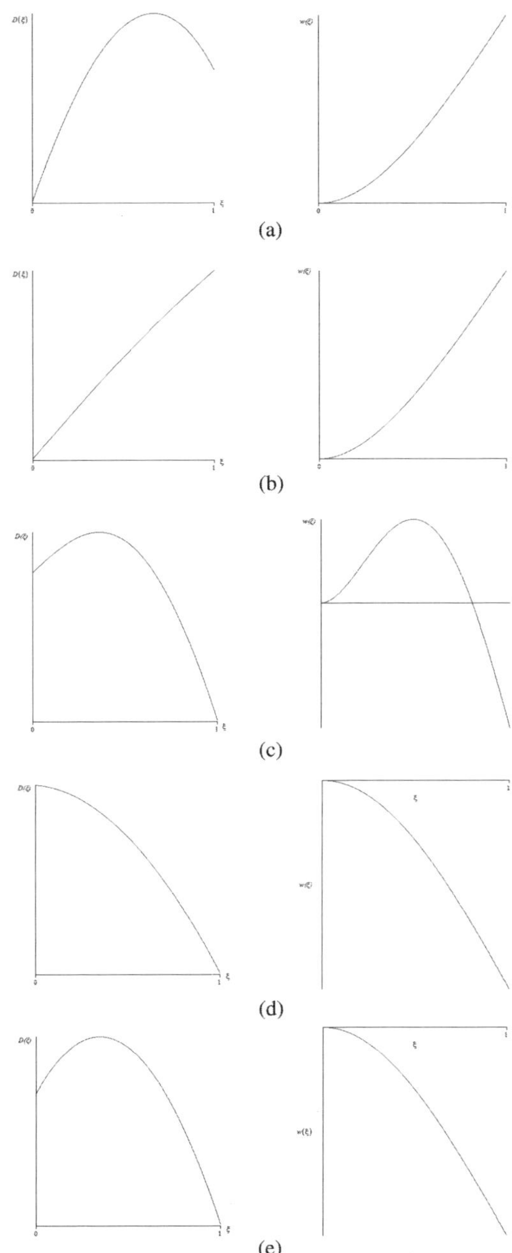

Figure 7.2. Flexural rigidities and associated mode shapes corresponding to the roots (a) $k_4 = -0.041411$, (b) $k_1 = -0.01573492764$, (c) $k_2 = -0.006766827430$, (d) $k_3 = 0.001291278483$, (e) $k_5 = 0.01033391077$

Table 7.4. Region of existence of the roots

Regions	k_1	k_2	k_3	k_4	k_5
$0 < \Theta < 1/3$	X	X	X	Not existent	Not existent
$1/3 < \Theta < 0.5302$	X	X	X	X	X
$0.5302 < \Theta < \infty$	X	X	Not existent	X	Not existent

and k_2 exist on the entire interval of variation of $\Theta \in [0, \infty[$. We compose the following two sets that are valid for any Θ

$$S_1 = [\{\Theta^a, k_1^a\}, \{\Theta^b, k_1^b\}, \ldots, \{\Theta^j, k_1^j\}, \ldots]$$
$$S_2 = [\{\Theta^a, k_2^a\}, \{\Theta^b, k_2^b\}, \ldots, \{\Theta^j, k_2^j\}, \ldots] \tag{7.42}$$

Concerning the root k_3, it exists when Θ varies from zero to $\Theta \approx 0.53$. We construct the set S_3 with Θ^j having value in this interval

$$S_3 = [\{\Theta^a, k_3^a\}, \{\Theta^b, k_3^b\}, \ldots, \{\Theta^j, k_3^j\}, \ldots] \tag{7.43}$$

The root k_4 exists when preselected values belong to interval $\Theta^j \in [1/3, \infty[$

$$S_4 = [\{\Theta^a, k_4^a\}, \{\Theta^b, k_4^b\}, \ldots, \{\Theta^j, k_4^j\}, \ldots] \tag{7.44}$$

Finally, the root k_5 exists for $\Theta^i \in [1/3, 0.53[$. The set S_5 reads

$$S_5 = [\{\Theta^a, k_5^a\}, \{\Theta^b, k_5^b\}, \ldots, \{\Theta^j, k_5^j\}, \ldots] \tag{7.45}$$

The results in Table 7.4 summarize Eqs. (7.42)–(7.45), where the sign X indicating the presence of the corresponding root in the corresponding region of variation of Θ.

Concerning the order of the roots, for $0 < \Theta < 1/3$ we have $k_1 < k_2 < k_3$. Then, in the region $1/3 \leq \Theta \leq 0.53$, $k_4 < k_1 < k_2 < k_3 < k_5$. In the region $\Theta > 0.53$, we have $k_4 < k_1 < k_2$.

Recalling that $\Theta = D(0)/PL^2 = b_0/PL^2$, where b_0 is fixed and the only varying parameter is the load P, we deduce that

$$P^a = \frac{b_0}{\Theta^a L^2} \tag{7.46}$$

In each set S_j, through the values Θ^a we determine the corresponding values of P^a via Eq. (7.46); likewise values of $k^a, k^b \ldots$ allow us to determine the associated squared frequencies $(\Omega^a)^2$, $(\Omega^2)^b$ and so on. The results of above substitution yields sets $\mathfrak{R}_j$ as follows

$$\mathfrak{R}_1 = \left[\left\{ \frac{P^a L^2}{D(0)}, (\Omega_1^a)^2 \right\}, \left\{ \frac{P^b L^2}{D(0)}, (\Omega_1^b)^2 \right\}, \ldots, \left\{ \frac{P^j L^2}{D(0)}, \left(\Omega_1^j\right)^2 \right\}, \ldots \right]$$

$$\mathfrak{R}_2 = \left[\left\{ \frac{P^a L^2}{D(0)}, (\Omega_2^a)^2 \right\}, \left\{ \frac{P^b L^2}{D(0)}, (\Omega_2^b)^2 \right\}, \ldots, \left\{ \frac{P^j L^2}{D(0)}, \left(\Omega_2^j\right)^2 \right\}, \ldots \right]$$

$$\mathfrak{R}_3 = \left[\left\{ \frac{P^a L^2}{D(0)}, (\Omega_3^a)^2 \right\}, \left\{ \frac{P^b L^2}{D(0)}, (\Omega_3^b)^2 \right\}, \ldots, \left\{ \frac{P^j L^2}{D(0)}, \left(\Omega_3^j\right)^2 \right\}, \ldots \right]$$

$$\mathfrak{R}_4 = \left[\left\{ \frac{P^a L^2}{D(0)}, (\Omega_4^a)^2 \right\}, \left\{ \frac{P^b L^2}{D(0)}, (\Omega_4^b)^2 \right\}, \ldots, \left\{ \frac{P^j L^2}{D(0)}, \left(\Omega_4^j\right)^2 \right\}, \ldots \right]$$

$$\mathfrak{R}_5 = \left[\left\{ \frac{P^a L^2}{D(0)}, (\Omega_5^a)^2 \right\}, \left\{ \frac{P^b L^2}{D(0)}, (\Omega_5^b)^2 \right\}, \ldots, \left\{ \frac{P^j L^2}{D(0)}, \left(\Omega_5^j\right)^2 \right\}, \ldots \right]$$

$$(7.47)$$

The sets in Eq. (7.47) allow one to plot the dependence of Ω^2 as a function of the load P for the five possible fundamental mode shapes. Figure 7.3 depicts the five possible dependencies of the squared natural frequency Ω^2 versus the load P.

As Eq. (7.47) implies, one gets five different relationships, associated with the five different mode shapes found previously. The relationship between the squared natural frequency Ω^2 and the load P depicted in Figs. 7.3(a)–7.3(d) is nearly linear. However, for other cases the relationship is not linear.

It should be noted that the curves in Figs. 7.2(a)–7.2(c) intersect with the P-axis. These points are associated with zero natural frequency, and thus the points of intersection with P-axis constitute buckling loads.

Remarkably, the curves in Figs. 7.2(d) and 7.2(e) do not intersect with the P-axis. Intriguingly, this means that the axially graded column may possess three buckling loads but 5 vibration modes. In Fig. 7.2(d), the maximum value P can take is 3, associated with $\Theta = 1/3$ between Regions I and II. At $P = 3$, the squared natural frequency does not vanish, it equals $\Omega^2 =$

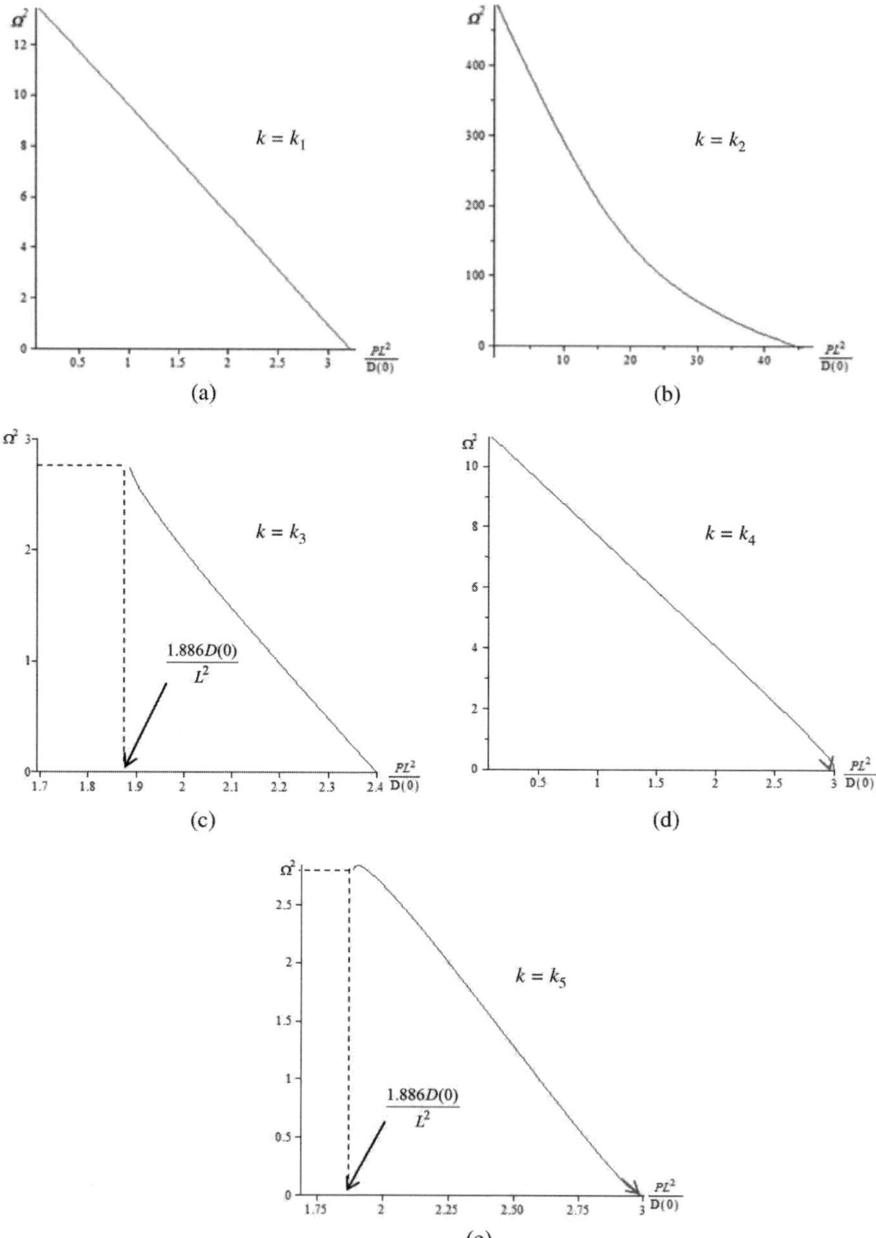

Figure 7.3. Dependence of the squared natural frequency versus load (a) $k = k_4$, (b) $k = k_1$, (c) $k = k_2$, (d) $k = k_3$, and (e) $k = k_5$

$8.436 \cdot 10^{-4}$. In Fig. 7.3(e), maximum possible value for P is 3 at which $\Omega^2 = 2.27 \cdot 10^3$.

It is interesting to observe that for $k = k_3$, for $P \leq 1.886D(0)/L^2$, the natural frequency is not determined. Analogous conclusion takes place in case $k = k_5$; specifically for $P \leq 1.886D(0)/L^2$, natural frequency is not determined.

To summarize, an unexpected phenomenon takes place: There are five vibration modes associated with an axial load; only three of these are associated with a buckling load. Thus, only three of five vibration modes coalesce with the buckling modes. This is in accordance with the study of Elishakoff and Endres (2005) who found three distinct buckling modes in their static counterpart in this chapter.

Let us now deal with the exact values of buckling load. The value of the buckling load P_{cr} is obtained from the condition $\Omega^2 = 0$. Equation (7.33) shows that Ω^2 vanishes if $k(5400k^2 + 1 + 180k) = 0$. The roots of the latter equation are

$$k_{cr} = \left\{ k_{cr_1} = 0, \; k_{cr_2} = -\frac{1}{60} + \frac{1}{180}\sqrt{3}, \; k_{cr_3} = -\frac{1}{60} - \frac{1}{180}\sqrt{3} \right\}$$

$$(7.48)$$

Substituting values of k into Eq. (7.36), we find P as a function of $D(0) = b_0$ and L^2. We get three different values of P_{cr}, associated with the 3 mode shapes

$$P_{cr_1} = \frac{12b_0}{5L^2} = \frac{2.4b_0}{L^2},$$

$$P_{cr_2} = -\frac{12(\sqrt{3} - 1)(-7 + 4\sqrt{3})b_0}{(-71 + 41\sqrt{3})L^2} = \frac{44.78454607b_0}{L^2}, \qquad (7.49)$$

$$P_{cr_3} = \frac{12(1 + \sqrt{3})(7 + 4\sqrt{3})}{(71 + 41\sqrt{3})L^2}b_0 = \frac{3.215390310b_0}{L^2}$$

The above buckling loads are in accordance with the values observed on the figures. Moreover, the two other relationships given by k_4 and k_5 do not intersect with. Moreover, Elishakoff and Endres (2005) found that for a clamped free beam, the buckling load corresponds to $P_{cr} = -12b_2/L^2$. Introducing the values of k_{cr_i}, in Eq. (7.45), into the expression of the coefficient b_2 in front of ξ^2 in Eq. (7.34) yields, for all the different values of k_{cr}, a unique value

$$P_{cr} = -\frac{12b_2}{L^2} \tag{7.50}$$

in accordance with Elishakoff and Endres (2005). Furthermore, this result is in accordance with the expression of b_2 in Eq. (7.31) when $\Omega^2 = 0$.

The last interesting case is when the load P is set at zero value. The numerator in Eq. (7.32) is equal to 0. It yields

$$(77760000k^4 + 3456000k^3 + 21600k^2 + 160k + 1)(1 + 90k)\Omega^2 = 0 \tag{7.51}$$

We know that Ω^2 and P cannot be 0 at the same time, therefore $\Omega^2 \neq 0$ and

$$(77760000k^4 + 3456000k^3 + 21600k^2 + 160k + 1) = 0$$

or

$$(1 + 90k) = 0 \tag{7.52}$$

The roots of the first equation are

$$k_1 = -0.03837569383, \quad k_2 = -0.006760973389 \tag{7.53}$$

and the root of the second equation is

$$k_3 = -\frac{1}{90} \tag{7.54}$$

It would mean that one could get 3 mode shapes when $P = 0$. However, the third derivative of Eq. (7.35) has to be 0 when $\xi = 1$, because $P = 0$. The third derivative at $\xi = 1$ reads

$$w'''(\xi) = 2160k + 24 = 0 \tag{7.55}$$

and the only acceptable value of k is given in Eq. (7.54). Therefore, the values reported in Eq. (7.53) are not acceptable.

We have 3 values for k when $0 < P < 1.89$. For vanishingly small values of P, one has three roots. However, when P coincides with zero one gets only one root.

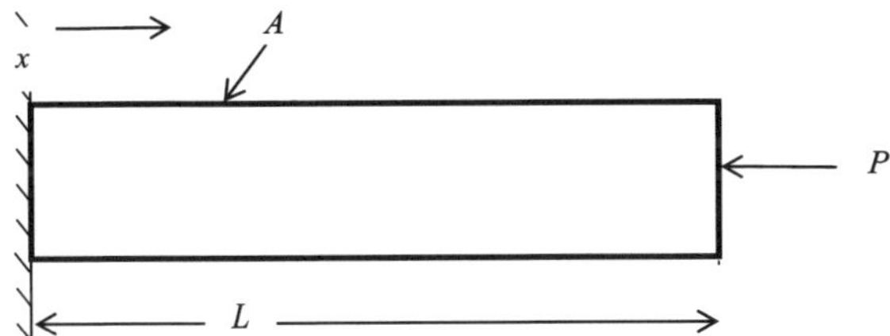

Figure 7.4. Vibrating clamped-free beam loaded at its tip

7.4 Numerical example

For illustrative purposes, let us consider a clamped-free beam, with a length $L = 2$ m, and a concentrated load at the top $P = 20$ kN

The density of the beam is set at $\rho = 2400$ kg/m^3, the cross-sectional area A is a square with dimension 0.2 m $\times$ 0.2 m $= 0.04$ m^2 and the moment of inertia is $I = (0.2\,\text{m})^4/12 = 1^{1/3}10^{-4}$ m^4. The Young's modulus E_0 at $x = 0$ is $11.25 \cdot 10^9$ Pa. Finally, the rigidity $D(0)$ at $x = 0$ is $E_0 I = 11.25 \cdot 10^9 \cdot 1/\frac{1}{3}10^{-4} = 1.5 \cdot 10^6$ N $\cdot$ m^2.

Given these numerical values, the value of Θ becomes

$$D(0)/PL^2 = 1.5 \cdot 10^6/(20000 \cdot 2^2) = 25 \tag{7.56}$$

This value of Θ indicates that we are in the Region 3 (Table 7.3), with three negative roots. The first step consists in finding the different values of k that satisfies Eq. (7.38). Introducing the value of Θ in Eq. (7.51) into Eq. (7.38) gives the following values for k and considering that $\Theta > 0.53$ we get 3 different values of k

$$k_1 = -0.3839987231 \cdot 10^{-1}, \quad k_2 = -0.1116894106 \cdot 10^{-1},$$

$$k_3 = -0.6761111363 \cdot 10^{-2} \tag{7.57}$$

With the above values of k_i, we find the associated values of Ω^2 by plugging k_i into Eq. (7.33). Thus, the following values for natural frequency parameter are obtained

$$\Omega_1^2 = 1.668364561 \cdot 10^7, \quad \Omega_2^2 = 2.043467897 \cdot 10^7,$$

$$\Omega_3^2 = 7.297832976 \cdot 10^8 \tag{7.58}$$

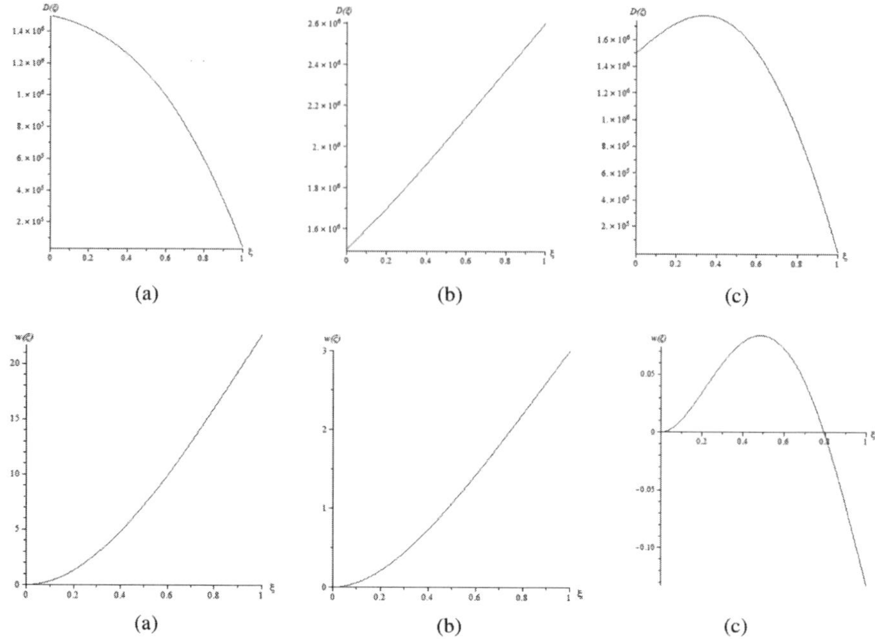

Figure 7.5. Five possible flexural rigidities and displacements for (a) k_1, (b) k_2, (c) k_3, (d) k_4, and (e) k_5

From Eq. (7.2), we determine the natural frequencies

$$\omega_1 = 416.8788494 \, \text{rad/s}, \quad \omega_2 = 533.8324848 \, \text{rad/s},$$

$$\omega_3 = 2757.156026 \, \text{rad/s}, \tag{7.59}$$

Figure 7.5 portrays the curve of five possible flexural rigidity variation with ξ and displacement $w(\xi)$

It appears instructive to compare the natural frequencies of an axially graded beam and the associated beam with constant flexural rigidity where the value of the flexural rigidity of the associated beam is the spatial average of the rigidity of the FG beam. For each of five columns associated with various k_j, the associated beam is defined as

$$D_{ave} = \frac{1}{L} \int_0^L D(x)\,dx \tag{7.60}$$

It gives, for each column

$$D_{ave,1} = 1.018821504 \cdot 10^6$$

$$D_{ave,2} = 2.034683653 \cdot 10^6 \tag{7.61}$$

$$D_{ave,3} = 1.353846115 \cdot 10^6$$

From these values, and using the result given by Bokaian (1998), we can compute the natural frequencies of the beam under a load 90 kN and a flexural rigidity given in Eq. (7.62). We get

$$\omega_{1,ave} = 363.58 \text{ rad/s}, \quad \omega_{2,ave} = 513.81 \text{ rad/s}, \quad \omega_{3,ave} = 605.37 \text{ rad/s} \tag{7.62}$$

The percentage-wise error between the values found by the two methods are

$$e_1 = 11.53\%, \quad e_2 = 10.13\%, \quad e_3 = 355\% \tag{7.63}$$

Moreover, FEM software RDM6© allows one to check that for each beam the applied load P is smaller than the critical buckling load. Performing a buckling analysis on a 10 finite elements' beam yields the following critical loads, the subscript i being associated with the subscript of the root k

$$P_{cr_1} = 53,494 \text{ kN}, \quad P_{cr_2} = 93810 \text{ kN}, \quad P_{cr_3} = 77,196 \text{ kN} \tag{7.64}$$

Since the percentage-wise error in Eq. (7.63) between average and actual quantities varies between 10.13% and 355%, one concludes that FGM might modify either slightly or significantly the natural frequency of the beam.

7.5 Conclusion

This chapter deals with free vibrations of an inhomogeneous clamped-free beam subjected to a compressive force at the end. We generalize previous study by Elishakoff and Endres, devoted to static buckling load of such a column. The extension reveals some extremely interesting and unexpected phenomena: Whereas in the buckling case there could be, due to inhomogeneity, three different buckling modes, in the vibratory setting the column may possess three or five fundamental mode shapes.

Here the following remark appears to be relevant: One can notice that the equations in this study have large numbers as coefficients (see, for example, Eq. (7.34)). These large numbers result from expanding characteristic equations. One may have an intuitive concern on whether the unexpected phenomena that this study claims to have discovered are the result of a combination of numerical instabilities in association with the idealized structural model. Naturally, the large coefficients occur due to the evaluation of the quantities by the symbolic algebra that performs analytical calculations. Can these phenomena occur in reality?

It appears that the behavior of inhomogeneous, structures made of FGMs is much richer than the associated behavior of homogeneous structures. Therefore, the behavior of such structures may show some unexpected results.

It is hoped that the above intriguing results will trigger much interest in the applied mechanics and mathematics community, and hopefully many more remarkable facts — still uncovered — will be derived for FGM structures.

Chapter 8

Design of Heterogeneous Polar–Orthotropic Clamped Circular Plates with Specified Fundamental Natural Frequency

The titled problem is solved by the aid of the semi-inverse formulation. At the first glance, a surprising demand is posed: Design of a radially heterogeneous polar orthotropic circular plate whose fundamental mode shape coalesces with the static deflection of a homogeneous circular plate of the same radius and thickness. Compatible polynomial variations of the radial and circumferential flexural rigidities are introduced. It turns out that there is an infinite number of polar–orthotropic circular plates that solve the posed semi-inverse problem, depending on the free parameter represented by one of the coefficients of flexural rigidities. This allows us to provide a unique solution to the titled problem: There is a single plate that possesses the specified fundamental natural frequency.

8.1 Background

Several papers are devoted to the free vibration of homogeneous orthotropic circular plates. Method of Frobenius was utilized by Akasaka and Takagishi (1958), Borsuk (1960) and Minkarah and Hoppman (1964) formulating the solution in terms of the hypergeometric functions. Power series expansion methodology was utilized by Pandalai and Patel (1965). The Garlerkin method was applied by Prathap and Varadan (1976), while the Rayleigh–Ritz method was employed by Elishakoff (1987). Undetermined non-integer power method was used by Grossi *et al.* (1986). In the above papers, both the thickness and modulus of elasticity in radial and circumferential direction were fixed at constant values.

Circular polar–orthotropic plates with variable properties were addressed in several investigations. Gupta *et al.* (1990) studied polar orthotropic radially tapered circular plates.

In this study, we consider heterogeneous circular polar-orthotropic plates. As Leissa (1981) mentions "a plate is heterogeneous if its material properties vary from point to point within it." The candidates for the variation are the material density and/or the elastic moduli. The case of the varying material density, albeit for the isotropic case, was studied by Ran (1976). Here, we deal with the variation of the elastic moduli. Semi-inverse vibration problem is formulated in which the mode shape is specified while the natural frequency is sought. The postulated mode shape is taken in the form of the static deflection of the isotropic circular plate published by Timoshenko and Woinowsky-Krieger (1959).

Immediately one can pose a provocative question: "Why should polar orthotropic plate possess the mode shape that coincides with the static deflection? Secondly, why should the polar orthotropic plate share the mode shape of an isotropic plate?" The answer, obviously, is yet another question "Why not?" We show that heterogeneity of the plate allows for this phenomenon to take place.

8.2   Derivation of governing differential equation

The relationship between the radial and circumferential bending moments with the transverse displacement w reads, in the polar coordinate system, in axisymmetric setting (Lekhnitskii, 1968)

$$M_r(r) = -D_r \left(w'' + \frac{\nu_\theta}{r} w' \right) \tag{8.1}$$

$$M_\theta(r) = -D_\theta \left(\nu_r w'' + \frac{1}{r} w' \right) \tag{8.2}$$

where M_r = radial bending moment, M_θ = circumferential bending moment, D_r and D_θ are, respectively, radial and circumferential flexural rigidities, ν_r = Poisson's ratio, $'$ denotes differentiation with respect to radial coordinate r. Equations (8.1) and (8.2) were derived by Lekhnitskii (1968).

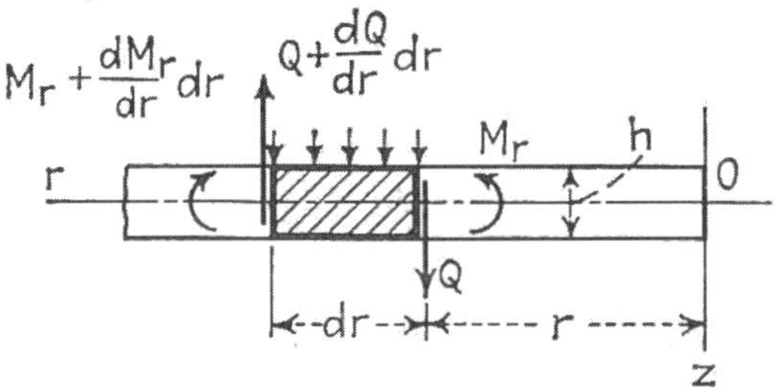

Figure 8.1. Element of the plate

The equilibrium equation of an element of the plate, as shown in Fig. 8.1, is

$$M_r + \frac{dM_r}{dr} r - M_\theta + Q_r r = 0 \tag{8.3}$$

Following Timoshenko and Woinowsky-Krieger (1959), the shearing force Q_r is expressed as:

$$2\pi r Q_r = \int_0^r \int_0^{2\pi} q(\rho, \theta) \rho \, d\rho \, d\theta \tag{8.4}$$

where $q(r, \theta)$ is the distributed transverse load. We are considering the problem in axisymmetric setting, hence $q(r, \theta) = q(r)$. From Eq. (8.4), we get

$$Q_r = \int_0^r q(\rho) \rho \, d\rho \tag{8.5}$$

After substitution of Eqs. (8.1), (8.2) and (8.5) into (8.3) we arrive at

$$rD_r w''' - \left(-D_r - r\frac{dD_r}{dr} - v_\theta D_r + v_r D_\theta \right) w''$$

$$- \left(-v_\theta \frac{dD_r}{dr} + \frac{1}{r} D_\theta \right) w' = \int_0^r q(r) r \, dr \tag{8.6}$$

Differentiating the result with respect to r, we obtain

$$rD_r w^{IV} + \left(2D_r + 2r\frac{dD_r}{dr} + v_\theta D_r - v_r D_\theta \right) w'''$$

$$- \left[\frac{d}{dr} \left(-D_r - r\frac{dD_r}{dr} - v_\theta D_r + v_r D_\theta \right) - v_\theta \frac{dD_r}{dr} + \frac{1}{r} D_\theta \right] w''$$

$$- \left[\frac{d}{dr} \left(-v_\theta \frac{dD_r}{dr} + \frac{1}{r} D_\theta \right) \right] w' = rq(r) \tag{8.7}$$

To derive the equation of motion, one applies D'Alembert's principle and substitutes

$$q(r) = -\rho h \partial^2 \omega / \partial t^2 \qquad (8.8)$$

into Eq. (8.7). In the resulting equation, we let $w(r, t) = W(r) \sin \omega t$, where $W(r)$ is the mode shape, $t =$ time, $\rho =$ material density, $h =$ thickness, and $\omega =$ sought natural frequency. The final governing equation reads

$$r D_r W^{IV} + \left(2D_r + 2r \frac{dD_r}{dr} + v_\theta D_r - v_r D_\theta \right) W'''$$

$$- \left[\frac{d}{dr} \left(-D_r - r \frac{dD_r}{dr} - v_\theta D_r + v_r D_\theta \right) - v_\theta \frac{dD_r}{dr} + \frac{1}{r} D_\theta \right] W''$$

$$- \left[\frac{d}{dr} \left(-v_\theta \frac{dD_r}{dr} + \frac{1}{r} D_\theta \right) \right] W' = r \rho h \omega^2 W(r) \qquad (8.9)$$

For an isotropic plate $D_r = D_\theta = D$, $v_r = v_\theta = v$, and Eq. (8.9) reduces to Eq. (11.1) in Elishakoff's (2005) monograph.

8.3   Semi-inverse method of solution

We postulate the following mode shape

$$W(r) = (R^2 - r^2)^2 \qquad (8.10)$$

that is proportional to the static displacement of the isotropic homogeneous circular plate of constant thickness, as obtained by Timoshenko and Woinowsky-Krieger (1959).

In this section, we consider the case of the constant material density $\rho =$ const. The flexural rigidities are expressed as polynomials of fourth-order

$$D_r(r) = b_0 + b_1 r + b_2 r^2 + b_3 r^3 + b_4 r^4 \qquad (8.11)$$

$$D_\theta(r) = k^2 D_r(r) \qquad (8.12)$$

where k is taken as constant. Equation (8.12) signifies that the circumferential flexural rigidity is proportional to the radial flexural rigidity. In subsequent analysis, the Poisson's ratios v_r and v_θ are taken as constants.

The substitution of Eqs. (8.10)–(8.12) into Eq. (8.9) leads to the following polynomial equation

$$A_0 + A_1 r + A_2 r^2 + A_3 r^3 + A_4 r^4 + A_5 r^5 = 0 \qquad (8.13)$$

where A_j are following coefficients

$$A_0 = -8R^2 b_1 - 8v_\theta R^2 b_1 + 4R^2 v_r k^2 b_1 + 4R^2 k^2 b_1 \qquad (8.14)$$

$$A_1 = 72b_0 + 24v_\theta b_0 - 24v_r k^2 b_0 - 8k^2 b_0 - 24R^2 b_2 - 24R^2 v_\theta b_2$$
$$+ 8v_r k^2 R^2 b_2 + 8k^2 R^2 b_2 - \rho h \omega^2 R^4 \qquad (8.15)$$

$$A_2 = 144b_1 + 48v_\theta b_0 - 36v_r k^2 b_1 - 12k^2 b_1 - 48R^2 b_3 - 48R^2 v_\theta b_3$$
$$+ 12v_r k^2 R^2 b_3 + 12k^2 R^2 b_3 \qquad (8.16)$$

$$A_3 = 240b_2 + 80v_\theta b_2 - 48v_r k^2 b_2 - 16k^2 b_2 - 80R^2 b_4 - 80R^2 b_4$$
$$- 80v_\theta R^2 b_4 + 6v_r k^2 R^2 b_4 + 16k^2 R^2 b_4 + 2\rho h \omega^2 R^2 \qquad (8.17)$$

$$A_4 = 360b_3 + 120v_\theta b_3 - 60v_r k^2 b_3 - 20k^2 b_3 \qquad (8.18)$$

$$A_5 = 504b_4 + 168v_\theta b_4 - 72v_r k^2 b_4 - 24k^2 b_4 - \rho h \omega^2 \qquad (8.19)$$

From Eq. (8.19), we get the relationship between the natural frequency squared ω^2 and the coefficient b_4:

$$\omega^2 = (504 + 168v_\theta - 72v_r k^2 - 24k^2) b_4 / \rho h \qquad (8.20)$$

Equation (8.18) shows that b_3 vanishes identically. The equation resulting from substitution of Eq. (8.20) into Eq. (8.17), results in the formula for b_2, as related to b_4

$$b_2 = -2R^2 \frac{(k^2 - 8v_\theta - 29 + 4k^2 v_r)}{3k^2 v_r + k^2 - 5v_\theta - 15} b_4 \qquad (8.21)$$

Equation (8.16) leads to the conclusion that $b_1 = 0$. The equation resulting for substitution of Eq. (8.20) and (8.21) into Eq. (8.15) yields to the expression for b_0:

$$b_0 = \frac{Q_1 R^4}{(k^2 + 3k^2 v_r - 5v_\theta - 15)(k^2 - 9 - 3v_\theta + 3k^2 v_r)} b_4 \qquad (8.22)$$

where

$$Q_1 = -242k^2v_r - 68k^2v_rv_\theta + 408v_\theta - 44k^2b + 8k^4v_r$$
$$- 14k^2v_\theta + 57v_\theta^2 + 19k^4v_r^2 + k^4 + 771 \tag{8.23}$$

Equtaion (8.14) is automatically satisfied since $b_1 = 0$. Thus, we arrive at the following expression for the flexural rigidity:

$$D_r(r) = \left[\frac{Q_1 R^4}{(k^2 + 3k^2v_r - 5v_\theta - 15)(k^2 - 9 - 3v_\theta + 3k^2v_r)} \right.$$
$$\left. - \frac{2R^2(k^2 - 8v_\theta - 29 + 24k^2v_r)}{3k^2v_r + k^2 - 5v_\theta - 15}r^2 + r^4 \right] b_4 \tag{8.24}$$

or, in the view of the relationship

$$v_\theta = k^2v_r \tag{8.25}$$

we get

$$D_r(r) = \frac{1}{(2k^2v_r - k^2 + 15)(k^2 - 9)}[-8R^4k^4v_r^2 - 166R^4k^4v_r + 44R^4k^2$$

$$+ 6R^4k^4v_r - 771R^4 - R^4k^4 + (-76k^2R^2 + 2R^2k^4 + 522R^2$$

$$+ 72R^2k^2v_r - 8R^2k^4v_r)r^2 + (-18v_r + 2k^4v_r$$

$$+ 24k^2 - k^4 - 135)r^4]b_4 \tag{8.26}$$

Figure 8.2 shows the variation of the radial flexural rigidity coefficient

$$d = D_r(r)/b_4R^4 \tag{8.27}$$

with the non-dimensional radial coordinate for different values of k. Larger values of $d(r)$ are associated with greater anisotropy measure, namely k.

8.3.1 Parabolic mode shape

In this case, the mode shape represents a second-order polynomial

$$W(r) = (R - r)^2 \tag{8.28}$$

The result of substitution of Eq. (8.28) in conjunction of Eqs. (8.11) and (8.12) in Eq. (8.9) yields the equation

$$\sum_{j=0}^{5} B_j r^j = 0 \tag{8.29}$$

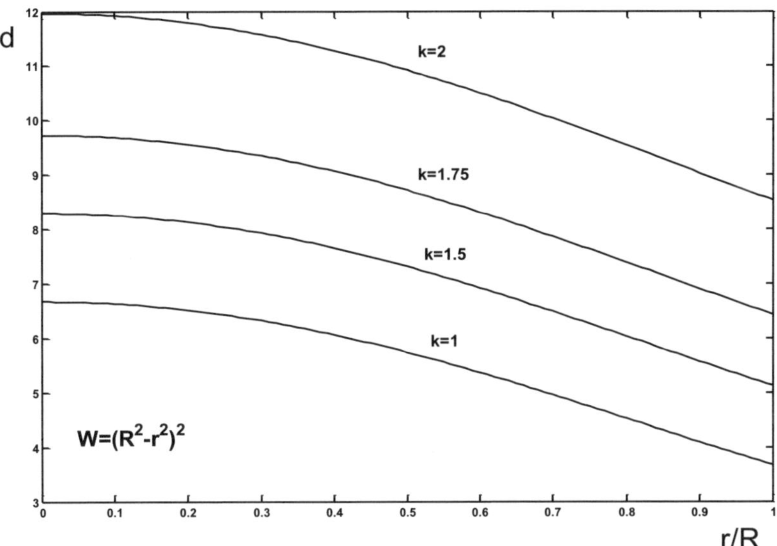

Figure 8.2. Variation of d versus non-dimensional radial coordinate r/R for various values of k, $v_r = 0.35$

where

$$B_0 = -2Rk^2 b_0 \tag{8.30}$$

$$B_1 = 0 \tag{8.31}$$

$$B_2 = 2k^2 b_2 R + 4v_\theta b_1 - 4Rv_\theta b_2 - 2k^2 b_1 + 4b_1 - 2v_r k^2 b_1 \tag{8.32}$$

$$B_3 = -4k^2 b_2 - 12v_\theta b_3 R + 4k^2 b_3 R + 12b_2 + 12v_\theta b_2$$
$$- 4v_r k^2 Rb_2 - \rho h \omega^2 R^2 \tag{8.33}$$

$$B_4 = -24b_4 v_\theta R - 6k^2 b_3 + 24b_3 - 6k^2 b_3 v_r + 24b_3 v_\theta$$
$$+ 6k^2 Rb_4 + 2\rho h \omega^2 R \tag{8.34}$$

$$B_5 = 40b_4 v_\theta - 8k^2 b_4 - 40b_4 - 8k^2 b_4 v_r - \rho h \omega^2 \tag{8.35}$$

From Eq. (8.35), we get the expression for the natural frequency squared:

$$\omega^2 = -8(k^2 v_r - 5v_\theta + k^2 - 5)b_4/\rho h \tag{8.36}$$

Substitution of Eq. (8.36) into Eq. (8.34) leads to

$$b_3 = -\frac{1}{3}\frac{(5k^2 + 8k^2 v_r - 28v_\theta - 40)}{(-4v_\theta + k^2 + k^2 v_r - 4)}Rb_4 \tag{8.37}$$

Substituting Eq. (8.36) and (8.37) into Eq. (8.33) we get

$$b_2 = \frac{1}{3(k^2 - 4v_\theta + k^2 v_r - 4)(k^2 - 3 - 3v_\theta + k^2 v_r)}$$
$$\times (-30k^2 v_r v_\theta + 4k^4 v_r + 6k^4 v_r^2 - 54k^2 v_r + 36v_\theta^2 - 11v_\theta k^2$$
$$+ 120v_\theta + k^4 - 14k^2 + 120)R^2 b_4 \tag{8.38}$$

Equations (8.32) and (8.38) yield

$$b_1 = \frac{Q_2}{3(k^2 - 4v_\theta + k^2 v_r - 4)(k^2 - 3 - 3v_\theta + k^2 v_r)(k^2 + k^2 v_r - 2v_\theta - 2)}$$
$$\times R^3 b_4 \tag{8.39}$$

where

$$Q_2 = -30k^2 v_r v_\theta + 4k^4 v_r + 6k^4 v_r^2 - 54k^2 v_r + 36v_\theta^2 - 11v_\theta k^2$$
$$+ 120v_\theta + k^4 - 14k^2 + 120 \tag{8.40}$$

From Eq. (8.30), we derive

$$b_0 = 0 \tag{8.41}$$

The flexural rigidity $D_r(r)$ is obtained by substituting Eqs. (8.37)–(8.41) into Eq. (8.11)

$$D_r(r) = \left[\frac{1}{\substack{3(k^2 - 4v_\theta + k^2 v_r - 4)(k^2 - 3v_\theta + k^2 v_r - 3) \\ (k^2 - 2v_\theta + k^2 v_r - 2)}}\right.$$
$$\times (-2v_\theta + k^2)(36v_\theta^2 - 11v_\theta k^2 - 30v_\theta v_r k^2 + 120v_\theta + 4k^4 v_r^2$$
$$+ 6k^4 v_r - 54k^2 v_r + k^4 - 14k^2 + 120)R^3 r$$

$$+\frac{\begin{array}{c}(36v_\theta^2 - 11v_\theta k^2 - 30v_\theta v_r k^2 + 120\theta_\theta 4k^4 v_r^2 + 6k^4 v_r \\ - 54k^2 v_r + k^4 - 14k^2 + 120)\end{array}}{3(k^2 - 4v_\theta + k^2 v_r - 4)(k^2 - 3v_\theta + k^2 v_r - 3)}R^2 r^2$$

$$\left. -\frac{(5k^2 - 28v_\theta + 8k^2 v_r - 40)}{3(k^2 - 4v_\theta + k^2 v_r - 4)}Rr^3 + r^4\right]b_4 \tag{8.42}$$

Due to the fact that $b_0 = 0$, $D_r(0)$ and $D_\theta(0)$ vanish. For $v_r = 0.35$, the minimum value of k resulting in non-negative D_r over $r/R \in [0; 1]$ equals 1.75412, as shown in Fig. 8.3. Figure 8.4 depicts the variation of the radial flexural rigidity coefficient $d_r = D_r(r)/b_4 R^4$ with the non-dimensional radial coordinate r/R for $k = 2, 2.5, 2.75, 3$ and $v_r = 0.35$. Note that when $k = 1$, for a Poisson's ratio that differs from $1/2$, the plate does not acquire the parabolic shape in Eq. (8.28), as demonstrated by Elishakoff (2005).

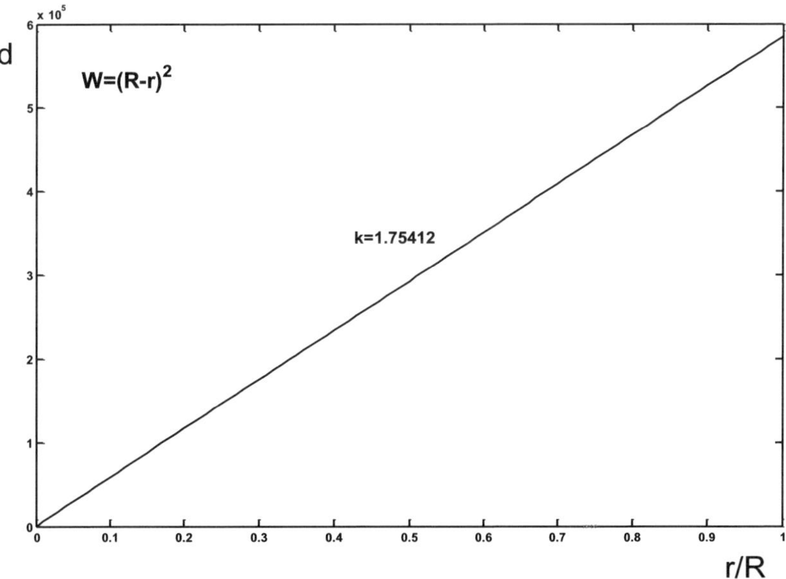

Figure 8.3. Variation of d versus non-dimensional radial coordinate r/R for $k = 1.75412$, $v_r = 0.35$

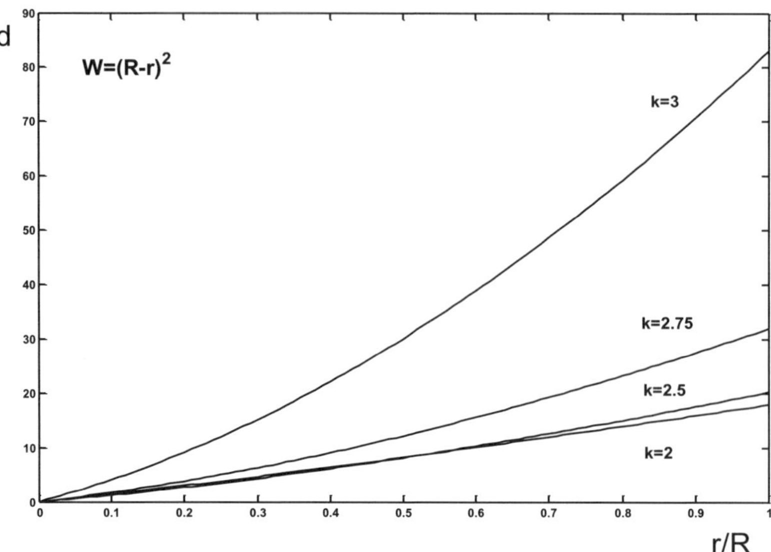

Figure 8.4. Variation of d versus non-dimensional radial coordinate r/R for various values of k, $v_r = 0.35$

8.3.2 Two cubic mode shapes

Now we check if the following expression of cubic mode shape

$$W(r) = (R - r)^3 \tag{8.43}$$

can be possessed by the polar orthotropic plate. We are following the same procedure as in the previous cases. The substitution of Eqs. (8.43), (8.11), and (8.12) into Eq. (8.9) leads to the equation

$$\sum_{j=0}^{6} C_j r^j = 0 \tag{8.44}$$

where

$$C_0 = -3R^2 k^2 b_0 \tag{8.45}$$

$$C_1 = 0 \tag{8.46}$$

$$C_2 = -6k^2 b_1 R + 6k^2 v_r b_0 + 3k^2 R^2 b_2 + 3k^2 b_0 - 6k^2 b_1 R v_r + 12 R b_1$$
$$\quad - 6b_2 v_\theta R^2 + 12 b_1 v_\theta R - 6 v_\theta b_0 - 12 b_0 \tag{8.47}$$

$$C_3 = -12k^2 R b_2 + 36 b_2 R + 36 v_\theta b_2 R - 18 b_3 v_\theta R^2 + 6k^2 b_3 R^2 - 36 b_1$$

$$- 18 b_1 v_\theta + 12 k^2 v_r b_1 - 12 k^2 b_2 R v_r + 6 k^2 b_1 - \rho h \omega^2 R^3 \qquad (8.48)$$

$$C_4 = 72 b_3 v_\theta R - 36 v_\theta R^2 b_4 + 72 b_3 R - 36 b_2 v_\theta - 72 b_2 + 9 k^2 b_2$$

$$- 18 k^2 b_3 R - 18 k^2 b_3 R v_r + 18 k^2 b_2 v_r + 9 k^2 R^2 b_4 + 3 \rho h \omega^2 R^2$$

$$\qquad (8.49)$$

$$C_5 = 120 b_4 R + 24 k^2 b_3 v_r - 24 k^2 b_4 R v_r - 60 b_3 v_\theta + 12 k^2 b_3$$

$$+ 120 b_4 v_\theta R - 120 b_3 - 24 k^2 b_4 R - 3 \rho h \omega^2 R \qquad (8.50)$$

$$C_6 = 30 k^2 b_4 v_r - 180 b_4 - 90 b_4 v_\theta + 15 k^2 b_4 + \rho h \omega^2 \qquad (8.51)$$

From Eq. (8.51) we get the expression of the natural frequency squared ω^2 with respect of the coefficient b_4:

$$\omega^2 = -15(2k^2 v_r - 6v_\theta + k^2 - 12) b_4 / \rho h \qquad (8.52)$$

The coefficients of flexural rigidity are as follows:

$$b_0 = -\frac{Q_3}{(k^2 - 5v_\theta + 2k^2 v_r - 10)(k^2 - 8 - 4v_\theta + 2k^2 v_r)(k^2 + 2k^2 v_r - 3v_\theta - 6)(k^2 - 2v_\theta - 4 + 2k^2 v_r)} R^4 b_4 \qquad (8.53)$$

$$b_1 = -\frac{Q_4}{4(k^2 - 5v_\theta + 2k^2 v_r - 10)(k^2 - 8 - 4v_\theta + 2k^2 v_r)(k^2 + 2k^2 v_r - 3v_\theta - 6)} R^3 b_4 \qquad (8.54)$$

$$b_2 = \frac{1}{2(k^2 - 5v_\theta + 2k^2 v_r - 10)(k^2 - 8 - 4v_\theta + 2k^2 v_r)}$$

$$\times (-66 k^2 v_r v_\theta + 7 k^4 v_r + 18 k^4 v_r^2 - 212 k^2 v_r + 60 v_\theta^2$$

$$- 14 v_\theta k^2 + 360 v_\theta + k^4 - 32 k^2 + 640) R^2 b_4 \qquad (8.55)$$

$$b_3 = -\frac{1}{4} \frac{(7k^2 + 22 k^2 v_r - 50 v_\theta - 140)}{(k^2 - 5v_\theta + 2k^2 v_r - 20)} R b_4 \qquad (8.56)$$

where

$$Q_3 = -2664k^2 v_r v_\theta + 816k^4 v_r + 544k^4 v_r^2 - 1776k^2 v_r + 2400v_\theta^2$$
$$- 708v_\theta k^2 + 4800v_\theta - 172k^4 - 1776k^2 + 1920 + 600v_\theta^3 - k^6$$
$$+ 25k^6 v_r - 152k^6 v_r^2 - 76k^6 v_r^3 - 35v_\theta k^4 - 218v_\theta^2 k^2 + 544k^4 v_r^2 v_\theta$$
$$+ 376k^4 v_r v_\theta - 1052k^2 v_r v_\theta^2 + 116k^4 v_r^2 v_\theta^2 + 65k^4 v_r v_\theta^2 - 142k^2 v_r v_\theta^3$$
$$- 45v_\theta k^6 v_r^2 - 38v_\theta k^6 v_r^3 - 30v_\theta^3 k^2 + 60v_\theta^4 + 10k^8 v_r^3 + 4k^8 v_r^4$$

(8.57)

$$Q_4 = -224k^2 v_r v_\theta + 316k^4 v_r - 136k^4 v_r^2 - 54k^2 v_r + 720v_\theta^2 - 580v_\theta k^2$$
$$+ 960v_\theta + 36k^4 - 1104k^2 - 1920 + 120v_\theta^3 - k^6 - 8k^6 v_r - 24k^6 v_r^2$$
$$+ 8k^6 v_r^3 + 17v_\theta k^4 - 102v_\theta^2 k^2 + 12k^4 v_r^2 v_\theta + 100k^4 v_r v_\theta - 116k^2 v_r v_\theta^2$$

(8.58)

with $v_\theta = k^2 v_r$ the final expression of the flexural rigidity becomes

$$D_r(r) = \left[-\frac{\begin{array}{c}(280k^4 v_r^2 + 3024k^2 v_r - 1776k^2 - 172k^4 + 108k^4 v_r \\ + 16k^6 v_r^3 - 10k^6 v_r + 6k^6 v_r^2)\end{array}}{\begin{array}{c}(k^2 - 10 - 3k^2 v_r)(k^2 - 2k^2 v_r - 8) \\ (k^2 - 6 - k^2 v_r)(k^2 - 4)\end{array}} R^4 \right.$$

$$- \frac{(12k^4 v_r^2 - 7k^4 v_r + 66k^2 v_r + k^4 - 14k^2 + 120)(-2k^2 v_r - 6)}{4(k^2 - 3k^2 v_r - 10)(k^2 - 8 - 2k^2 v_r)(k^2 - k^2 v_r - 6)} R^3 r$$

$$+ \frac{(12k^4 v_r^2 - 7k^4 v_r + 148k^2 v_r + k^4 - 32k^2 + 640)}{2(k^2 - 3k^2 v_r - 10)(k^2 - 8 - 2k^2 v_r)} R^2 r^2$$

$$\left. - \frac{(7k^2 - 28k^2 v_r - 140)}{4(k^2 - 3k^2 v_r - 10)} R r^3 + r^4 \right] b_4$$

(8.59)

It should be noted that not for all values of k the physically realizable solutions exist; namely, for $v_r = 0.35$ the minimum value of k resulting in positive D_r over $r/R \in [0; 1]$ equals 3.039, as shown in Fig. 8.5. In Fig. 8.6, the flexural rigidity coefficients are depicted for values $k = 3.5$, 4.2, 4.4 and 4.6. It is remarkable to note that at $r/R = 0.27$ the values of $d(r)$

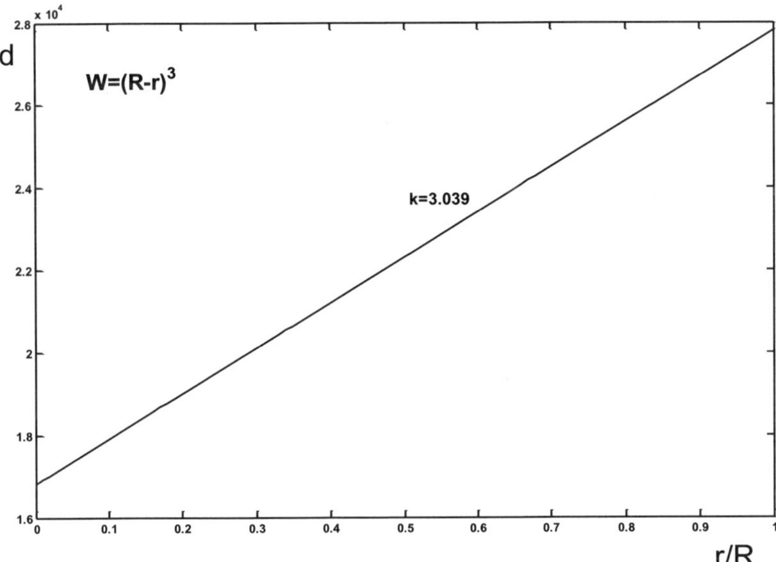

Figure 8.5. Variation of d versus non-dimensional radial coordinate r/R for $k = 3.039$, $v_r = 0.35$

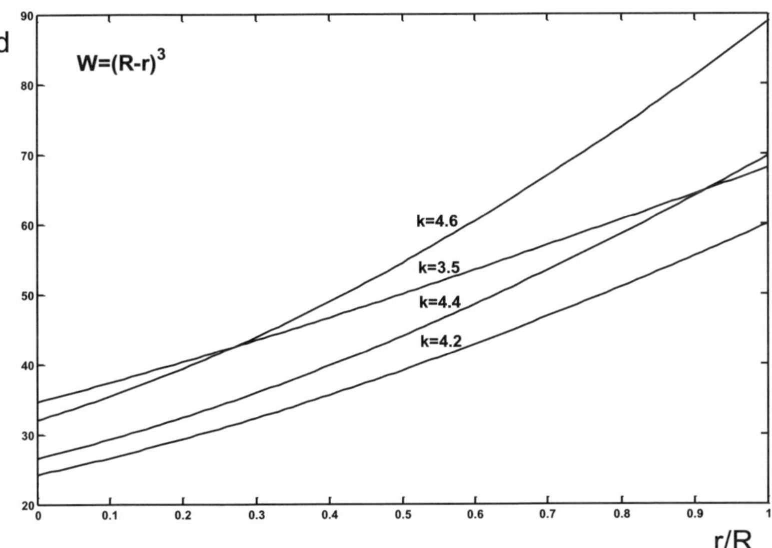

Figure 8.6. Variation of d versus non-dimensional radial coordinate r/R for various values of k, $v_r = 0.35$

corresponding to $k = 3.5$ and 4.6 intersect. Likewise, at $r/R = 0.92$ the $d(r)$ values associated with $k = 3.5$ and $k = 4.4$ intersect. This shows, that there is no monotonic dependence of the coefficient $d(r)$ with the rigidity ratio k.

Consider now an alternative cubic mode shape

$$W(r) = R^3 - 3r^2 R + 2r^3 \tag{8.60}$$

The substitution of Eqs. (8.60), (8.11), and (8.12) into Eq. (8.9) leads to the equation

$$\sum_{j=0}^{4} D_j r^j = 0 \tag{8.61}$$

where

$$\begin{aligned} D_0 = {} &6k^2 b_1 R - 12k^2 b_0 v_r + 24b_0 + 12b_0 v_\theta - 6k^2 b_0 \\ &+ 6k^2 b_1 v_r R - 12b_1 v_\theta R - 12b_1 R \end{aligned} \tag{8.62}$$

$$\begin{aligned} D_1 = {} &12k^2 b_2 v_r R - 36b_2 R + 36b_1 v_\theta - 12k^2 b_1 + 12k^2 b_2 R \\ &+ 72b_1 - 36b_2 v_\theta R - 24k^2 b_1 v_r - \rho h \omega^2 R^3 \end{aligned} \tag{8.63}$$

$$\begin{aligned} D_2 = {} &18k^2 b_3 v_r R - 18k^2 b_2 + 72v_\theta b_2 - 72Rb_3 + 18k^2 Rb_3 \\ &- 36k^2 b_2 v_r - 72b_3 v_\theta R + 144b_2 \end{aligned} \tag{8.64}$$

$$\begin{aligned} D_3 = {} &24k^2 v_r R + 240b_3 - 24k^2 b_3 - 48k^2 b_3 v_r - 120b_4 R \\ &+ 120b_3 v_\theta - 120b_4 R v_\theta + 24k^2 b_4 R + \rho h \omega^2 R \end{aligned} \tag{8.65}$$

$$D_4 = -30k^2 b_4 - 60k^2 b_4 v_r + 180b_4 v_\theta + 360b_4 - 2\rho h \omega^2 \tag{8.66}$$

From Eq. (8.66), we get the expression of the natural frequency squared ω^2:

$$\omega^2 = -15(k^2 + 2k^2 v_r - 6v_\theta - 12)b_4/\rho h \tag{8.67}$$

The coefficients of flexural rigidity are as follows

$$b_0 = \frac{Q_5}{\begin{array}{c} 8(k^2 - 5v_\theta + 2k^2 v_r - 10)(k^2 - 8 - 4v_\theta + 2k^2 v_r) \\ (k^2 + 2k^2 v_r - 3v_\theta - 6)(k^2 - 2v_\theta - 4 + 2k^2 v_r) \end{array}} R^4 b_4 \tag{8.68}$$

$$b_1 = \frac{Q_6}{8(k^2 - 5v_\theta + 2k^2 v_r - 10)(k^2 - 8 - 4v_\theta + 2k^2 v_r)} R^3 b_4$$
$$(k^2 + 2k^2 v_r - 3v_\theta - 6)$$

$$(8.69)$$

$$b_2 = -\frac{(7k^2 - 50v_\theta + 22k^2 v_r - 140)(k^2 - 4 + k^2 v_r - 4v_\theta)}{8(k^2 - 5v_\theta + 2k^2 v_r - 10)(k^2 - 8 - 4v_\theta + 2k^2 v_r)} R^2 b_4 \quad (8.70)$$

$$b_3 = -\frac{1}{8} \frac{(7k^2 + 22k^2 v_r - 50v_\theta - 140)}{(k^2 - 5v_\theta + 2k^2 v_r - 10)} R b_4 \quad (8.71)$$

where

$$Q_5 = (4676k^2 v_r + 1896k^2 - 10440v_\theta + 1896k^2 - 297k^4 v_\theta v_r$$
$$- 396v_r^2 v_\theta k^4 + 1462v_\theta k^2 - 4320v_\theta^2 - 717k^4 v_r - 906k^4 v_r^2$$
$$- 51v_\theta k^4 + 306k^2 v_\theta^2 + 24k^6 v_r + 69k^6 v_r^2 - 111k^4 + 3k^6$$
$$- 600v_\theta^3 + 866k^2 v_r v_\theta^2 + 58k^6 v_r^3 - 7920)(-2v_\theta - 2 + k^2 v_r + k^2)$$

$$(8.72)$$

$$Q_6 = 4062k^2 v_r v_\theta - 717k^4 v_r - 906k^4 v_r^2 + 4676k^2 v_r - 4320v_\theta^2$$
$$+ 1462v_\theta k^2 - 10440v_\theta - 111k^4 + 1896k^2 - 7920 - 600v_\theta^3$$
$$+ 3k^6 + 24k^6 v_r + 69k^6 v_r^2 + 58k^6 v_r^3 - 51v_\theta k^4 + 306v_\theta^2 k^2$$
$$- 396k^4 v_r^2 v_\theta - 297k^4 v_r v_\theta + 866k^2 v_r v_\theta^2 \quad (8.73)$$

with $v_\theta = k^2 v_r$ the final expression of the flexural rigidity is

$$D_r(r) = \left[\frac{\begin{array}{c} (-1164k^4 v_r^2 - 5764k^2 v_r + 1896k^2 - 111k^4 + 745k^4 v_r \\ -72k^6 v_r^3 - 27k^6 v_r + 78k^6 v_r^2 + 3k^6)(k^2 - k^2 v_r - 2) \end{array}}{8(k^2 - 10 - 3k^2 v_r)(k^2 - 2k^2 v_r - 8)(k^2 - 6 - k^2 v_r)(k^2 - 4)} R^4 \right.$$

$$\begin{aligned}
&+\frac{\begin{array}{c}(-1164k^4v_r^2 + 745k^4v_r - 5764k^2v_r - 111k^4 + 1896k^2 \\ -27k^6v_r + 78k^6v_r^2 - 72k^6v_r^3 - 7920)\end{array}}{8(k^2 - 3k^2v_r - 10)(k^2 - 8 - 2k^2v_r)(k^2 - k^2v_r - 6)}R^3r \\[2ex]
&-\frac{(7k^2 - 28k^2v_r - 140)(k^2 - 3k^2v - 4_r)}{8(k^2 - 3k^2v_r - 10)(k^2 - 8 - 2k^2v_r)}R^2r^2 \\[2ex]
&\left.-\frac{(7k^2 - 28k^2v_r - 140)}{8(k^2 - 3k^2v_r - 10)}Rr^3 + r^4\right] b_4
\end{aligned}$$
(8.74)

Figure 8.7 depicts the flexural rigidity coefficient d for three different values of k. Note that for k taking values 2, $\sqrt{6/(1 - v_r)}$, $\sqrt{8/(1 - 2v_r)}$ or $\sqrt{10/(1 - 3v_r)}$ the denominator in Eq. (8.68) becomes zero, and the solution must be modified for these particular cases. The case $k = 3$, $v_r = 0.35$ is shown in Fig. 8.8.

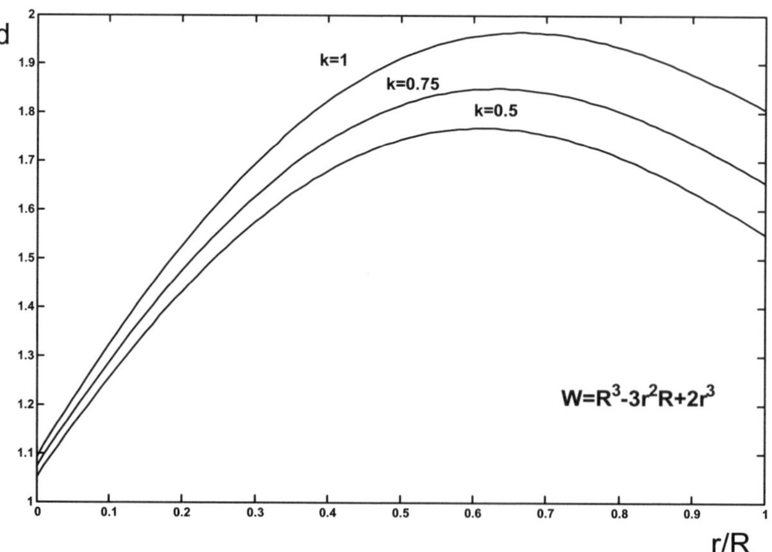

Figure 8.7. Variation of d versus non-dimensional radial coordinate r/R for various values of k, $v_r = 0.35$

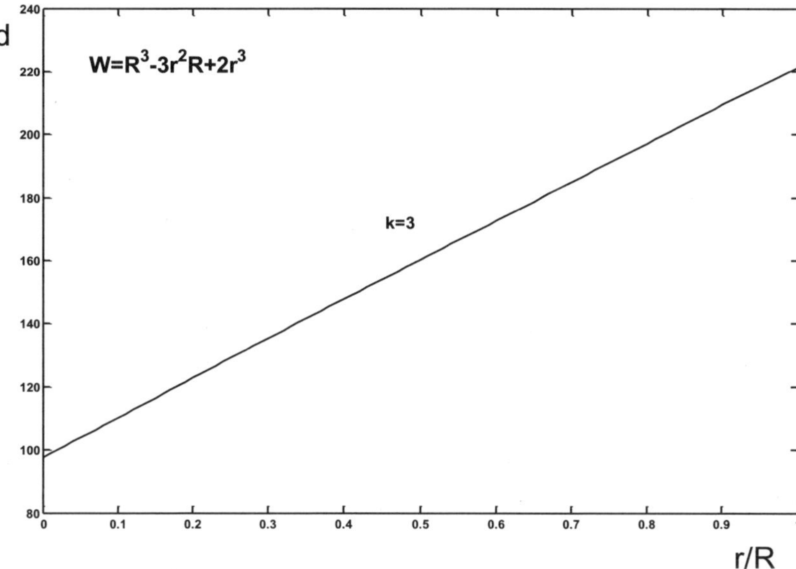

Figure 8.8. Variation of d versus non-dimensional radial coordinate r/R for $k = 3$, and $v_r = 0.35$

8.3.3 Two alternative quartic mode shapes

Note that the expression for the static displacement of the uniform homogeneous circular plate contains in it a quartic polynomial, given in Eq. (8.10). Natural question arises if other quartics may serve as exact mode shape. We investigate the following expression

$$W(r) = (R - r)^4 \tag{8.75}$$

as a candidate mode shape.

The substitution of Eqs. (8.75), (8.11), and (8.12) into Eq. (8.9) leads to the equation

$$\sum_{j=0}^{7} E_j r^j = 0 \tag{8.76}$$

where

$$E_0 = -4R^3 k^2 b_0 \tag{8.77}$$

$$E_1 = 0 \tag{8.78}$$

$$
\begin{aligned}
E_2 = {}& 4k^2 b_2 R^3 + 12k^2 b_0 R + 24k^2 v_r R b_0 - 48 R b_0 - 8R^3 b_2 v_\theta + 24 b_1 R^2 v_\theta \\
& - 24 v_\theta R b_0 + 24 k^2 b_0 R v_r - 48 b_0 R - 8 b_2 R^2 v_\theta + 24 b_1 R^2 v_\theta \\
& - 24 b_0 v_\theta R - 12 k^2 b_1 R^2 v_r - 12 k^2 b_1 R^2 + 24 b_1 R^2 \tag{8.79}
\end{aligned}
$$

$$
\begin{aligned}
E_3 = {}& 24 v_\theta b_0 + 48 v_r k^2 R b_1 - 24 k^2 b_2 R^2 v_r + 8 k^2 b_3 R^3 - 24 k^2 b_2 R^2 \\
& + 72 b_2 R^2 - 72 b_1 v_\theta R - 144 b_1 R - + 72 b_0 + 24 k^2 b_1 R - 8 k^2 b_0 \\
& + 72 b_2 R^2 v_\theta - 24 k^2 b_0 v_r - 24 b_3 R^3 v_\theta - \rho h \omega^2 R^4 \tag{8.80}
\end{aligned}
$$

$$
\begin{aligned}
E_4 = {}& 48 k^2 b_3 R + 80 b_2 v_\theta - 16 k^2 b_2 + 240 b_2 + 96 k^2 b_3 R v_r + 240 b_4 R^2 \\
& - 48 k^2 b_4 R^2 - 240 R v_\theta b_3 - 48 k^2 b_2 v_r - 48 k^2 b_4 R^2 v_r \\
& + 240 b_4 R^2 v_\theta - 480 b_3 R - 6 \rho h \omega^2 R^2 \tag{8.81}
\end{aligned}
$$

$$
\begin{aligned}
E_5 = {}& 360 b_3 + 120 b_3 v_\theta - 60 k^2 b_3 v_r - 360 k^2 R v_r b_4 + 120 k^2 R b_4 v_r \\
& - 710 b_4 R - 20 k^2 b_3 + 60 k^2 b_4 R + 4 \rho h R \omega^2 \tag{8.82}
\end{aligned}
$$

$$E_6 = 168 b_4 v_\theta - 72 k^2 v_r b_4 + 504 b_4 - 24 k^2 b_4 - \rho h \omega^2 \tag{8.83}$$

From Eq. (8.83), we get the expression of the natural frequency squared ω^2:

$$\omega^2 = -24(k^2 + 3k^2 v_r - 7 v_\theta - 21) b_4 / \rho h \tag{8.84}$$

The coefficients of flexural rigidity are as follows

$$
b_0 = -\frac{9 Q_7}{5(k^2 - 6 v_\theta + 3 k^2 v_r - 18)(k^2 - 15 - 5 v_\theta + 3 k^2 v_r)} R^4 b_4 \\
(k^2 + 3 k^2 v_r - 4 v_\theta - 12)(k^2 - 3 v_\theta - 9 + 3 k^2 v_r)
$$

$$\tag{8.85}$$

$$b_0 = -\frac{Q_8}{5(k^2 - 6v_\theta + 3k^2 v_r - 18)(k^2 - 15 - 5v_\theta + 3k^2 v_r)} R^4 b_4 \quad (8.86)$$
$$(k^2 + 3k^2 v_r - 4v_\theta - 12)$$

$$b_2 = \frac{3}{5(-6v_\theta + k^2 + 3k^2 v_r - 18)(-15 - 5v_\theta + 3k^2 v_r + k^2)}$$
$$\times (-144k^2 v_r v_\theta + 10k^4 v_r + 36k^4 v_r^2 - 522k^2 v_r + 90v_\theta^2 - 17v_\theta k^2$$
$$+ 780v_\theta + k^4 - 56k^2 + 1980) R^2 b_4 \quad (8.87)$$

$$b_3 = -\frac{3}{5}\frac{(3k^2 + 14k^2 v_r - 26v_\theta - 108)}{(k^2 - 6v_\theta + 3k^2 v_r - 18)} R b_4 \quad (8.88)$$

where

$$Q_7 = -8586k^2 v_r v_\theta + 2652k^4 v_r + 1989k^4 v_r^2 - 8586k^2 v_r + 7200v_\theta^2$$
$$- 2448v_\theta k^2 + 21600v_\theta - 399k^4 - 8586k^2 + 9720 + 1200v_\theta^3 - k^6$$
$$+ 42k^6 v_r - 360k^6 v_r^2 - 216k^6 v_r^3 - 54v_\theta k^4 - 472v_\theta^2 k^2 + 1326k^4 v_r^2 v_\theta$$
$$+ 838k^4 v_r v_\theta - 2292k^2 v_r v_\theta^2 + 191k^4 v_r^2 v_\theta^2 + 94k^4 v_r v_\theta^2 - 208k^2 v_r v_\theta^3$$
$$- 72v_\theta k^6 v_r^2 - 72v_\theta k^6 v_r^3 - 40v_\theta^3 k^2 + 80v_\theta^4 + 18k^8 v_r^3 + 9k^8 v_r^4$$
$$(8.89)$$

$$Q_8 = 1098k^2 v_r v_\theta + 972k^4 v_r - 1134k^4 v_r^2 + 8532k^2 v_r + 720v_\theta^2$$
$$- 1494v_\theta k^2 - 2160v_\theta + 66k^4 - 4770k^2 - 25920 + 120v_\theta^3 - k^6$$
$$- 12k^6 v_r - 54k^6 v_r^2 + 54k^6 v_r^3 + 21v_\theta k^4 - 158v_\theta^2 k^2 - 108k^4 v_r^2 v_\theta$$
$$+ 186k^4 v_r v_\theta - 42k^2 v_r v_\theta^2 \quad (8.90)$$

with $v_\theta = k^2 v_r$, the final expression of the flexural rigidity is

$$D_r(r) = \left[-\frac{9(-399k^4 - 8586k^2 + 204k^4 v_r + 603k^4 v_r^2 + 13014k^2 v_r + 9720 + 6k^6 v_r^2 - 12k^6 v_r + 18k^6 v_r^3)}{5(k^2 - 3k^2 v_r - 18)(k^2 - 15 - 2k^2 v_r)(k^2 - 12 - k^2 v_r)(k^2 - 9)} R^4 \right.$$

$$-\frac{\begin{array}{c}(66k^4 - 4770k^2 - 522k^4 v_r + 684k^4 v_r^2 + 6372k^2 v_r \\ -26k^6 v_r^2 + 9k^6 v_r + 24k^6 v_r^3 - k^6 - 25920)\end{array}}{5(k^2 - 3k^2 v_r - 18)(k^2 - 15 - 2k^2 v_r)(k^2 - 12 - k^2 v_r)}R^3 r$$

$$+\frac{3}{5}\frac{(-56k^2 + 12k^4 v_r^2 - 7k^4 v_r + k^4 + 258k^2 v_r + 1980)}{(k^2 - 3k^2 v_r - 18)(k^2 - 15 - 2k^2 v_r)}R^2 r^2$$

$$\left. -\frac{3}{5}\frac{(3k^2 - 12k^2 v_r - 108)}{(k^2 - 3k^2 v_r - 18)}Rr^3 + r^4 \right] b_4 \tag{8.91}$$

Figure 8.9 depicts the variation of d versus r/R. Note that for the isotropic case, $k = 1$ and $v_r = 0.35$ the function $d(r)$ takes negative values; this implies that for the isotropic case the Eq. (8.75) cannot serve the mode shape of the plate. For the polar orthotropic plate, at $v_r = 0.35$, it turns out that the minimum value of k for which the function D_r is non-negative is 1.459. Figure 8.9 shows $d(r)$ for $k = 1$; 1.459; 1.75; 2.25, 2; 2.5. Note that for $k = 3$, the solution in Eq. (8.75) is not valid because the denominator contains the factor $k^2 - 9$. Likewise, for any combination of k and v_r for

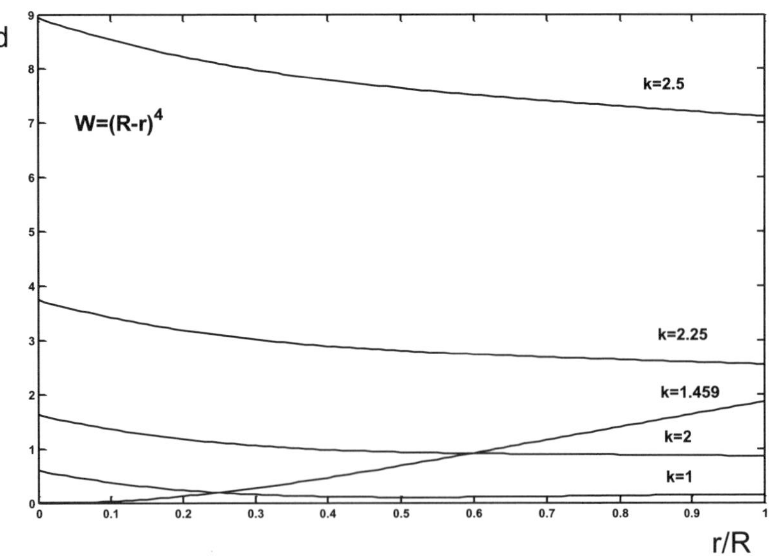

Figure 8.9. Variation of d versus non-dimensional radial coordinate r/R for various values of k, $v_r = 0.35$

which either of the following equalities

$$k^2 - 3k^2 v_r - 18 = 0;\ k^2 - 2k^2 v_r - 15 = 0;\ k^2 - k^2 v_r - 12 = 0 \tag{8.92}$$

holds, there is no solution in the form discussed.

Consider now the alternative quartic mode shape:

$$W(r) = R^4 - 4r^3 R + 3r^4 \tag{8.93}$$

Note that this function was utilized in Ref. 6 in the context of the Rayleigh–Ritz method. The substitution of Eqs. (8.93), (8.11), and (8.12) into Eq. (8.9) leads to the equation

$$\sum_{j=0}^{5} F_j r^j = 0 \tag{8.94}$$

where

$$F_0 = 24k^2 b_0 v_r R - 48 b_0 R - 24 b_0 v_\theta R + 12 k^2 b_0 R \tag{8.95}$$

$$F_1 = -72k^2 b_0 v_r - 144 b_1 R + 72 b_0 v_\theta - 24 k^2 b_0 + 25 k^2 b_1 R$$
$$\quad + 216 b_0 - 72 b_1 v_\theta R + 48 k^2 b_1 v_r R - \rho h \omega^2 R^4 \tag{8.96}$$

$$F_2 = 144 b_1 v_\theta - 144 b_2 v_\theta R + 432 b_1 + 36 k^2 b_2 R - 108 k^2 b_1 v_r$$
$$\quad + 72 k^2 R b_2 v_r - 36 b_1 k^2 - 288 b_2 R \tag{8.97}$$

$$F_3 = -240 v_\theta b_3 R - 144 k^2 b_2 v_r + 720 b_2 - 48 k^2 b_2 - 480 b_3 R$$
$$\quad + 240 b_2 v_\theta + 48 k^2 R b_3 + 96 k^2 b_3 v_r R \tag{8.98}$$

$$F_4 = -180 k^2 b_3 v_r - 720 b_4 R - 360 b_4 v_\theta R + 1080 b_3 + 360 b_3 v_\theta$$
$$\quad + 60 b_4 k^2 R - 60 k^2 b_3 + 120 k^2 b_4 v_r R + 4 \rho h \omega^2 R \tag{8.99}$$

$$F_5 = -216 k^2 b_4 v_r - 72 k^2 b_4 + 504 b_4 v_\theta + 1512 b_4 - 3 \rho h \omega^2 \tag{8.100}$$

From Eq. (8.100), we get the expression of the natural frequency squared ω^2 with respect of the coefficient b_4:

$$\omega^2 = -24(k^2 + 3k^2 v_r - 7 v_\theta - 21) b_4 / \rho h \tag{8.101}$$

The coefficients of flexural rigidity are as follows:

$$b_0 = \frac{Q_9}{5(k^2 - 6\nu_\theta + 3k^2\nu_r - 18)(k^2 - 15 - 5\nu_\theta + 3k^2\nu_r)} R^4 b_4 \quad (8.102)$$
$$(k^2 + 4k^2\nu_r - 4\nu_\theta - 12)(k^2 - 3\nu_\theta - 9 + 3k^2\nu_r)$$

$$b_1 = -\frac{(3k^2 - 26\nu_\theta + 14k^2\nu_r - 08)(k^2 - 10 + 2k^2\nu_r - 5\nu_\theta)}{5(k^2 - 6\nu_\theta + 3k^2\nu_r - 18)(k^2 - 15 - 5\nu_\theta + 3k^2\nu_r)} R^3 b_4$$
$$\frac{(k^2 + 2k^2\nu_r - 4\nu_\theta - 8)}{(k^2 + 4k^2\nu_r - 4\nu_\theta - 12)}$$

$$(8.103)$$

$$b_2 = -\frac{(3k^2 - 26\nu_\theta + 14k^2\nu_r - 08)(k^2 - 10 + 2k^2\nu_r - 5\nu_\theta)}{5(k^2 - 6\nu_\theta + 3k^2\nu_r - 18)(k^2 - 15 - 5\nu_\theta + 3k^2\nu_r)} R^2 b_4$$

$$(8.104)$$

$$b_3 = -\frac{1}{5}\frac{(3k^2 + 14k^2\nu_r - 26\nu_\theta - 108)}{(k^2 - 6\nu_\theta + 3k^2\nu_r - 18)} R b_4 \quad (8.105)$$

where

$$\begin{aligned}
Q_9 = (&-2090k^6\nu_r^3\nu_\theta + 5491k^4\nu_r^2\nu_\theta^2 + 3182k^4\nu_r\nu_\theta^2 - 522k^6\nu_\theta\nu_r \\
&+ 20780k^4\nu_\theta\nu_r - 64k^2 - 211062k^2\nu_r + 34201k^4\nu_r - 58778k^2\nu_\theta \\
&+ 4899k^4 + 169200\nu_\theta^2 + 56863k^4\nu_r^2 + 35386k^4\nu_r^2\nu_\theta \\
&- 197926k^2\nu_r\nu_\theta - 1698k^6\nu_r + 363360\nu_\theta + 34560\nu_\theta^3 \\
&+ 6702k^6\nu_r^3 - 5982k^6\nu_r^2 - 61162k^2\nu_r\nu_\theta^2 - 1766k^2\nu_\theta^2 288360)
\end{aligned}$$

$$(8.106)$$

with $\nu_\theta = k^2\nu_r$, the final expression of the flexural rigidity is

$$D_r(r) = \left[\frac{1}{5(k^2 - 3k^2\nu_r - 18)(k^2 - 15 - 2k^2\nu_r)(k^2 - 12 - k^2\nu_r)(k^2 - 9)}\right.$$
$$\times (152298k^2\nu_r + 28137k^4\nu_r^2 - 64386k^4\nu_r + 4899k^4$$

$$-2868k^6 v_r^2 + 2082k^6 v_r^3 + 1188k^6 v_r - 20k^8 v_r + 70k^8 v_r^2$$

$$-100k^8 v_r^3 - 24576k^4 v_r + 288360 - 150k^6 + 2k^8 + 48k^8 v_r^4)R^4$$

$$-\frac{(3k^2 - 12k^2 v_r - 108)(k^2 - 3k^2 v_r - 10)(k^2 - 2k^2 v_r - 8)}{5(k^2 - 3k^2 v_r - 18)(k^2 - 15 - 2k^2 v_r)(k^2 - 12 - k^2 v_r)}R^3 r$$

$$-\frac{1}{5}\frac{(3k^2 - 12k^2 v_r - 108)(k^2 - 3k^2 v_r - 10)}{(k^2 - 3k^2 v_r - 18)(k^2 - 15 - 2k^2 v_r)}R^2 r^2$$

$$\left. -\frac{1}{5}\frac{(3k^2 - 12k^2 v_r - 108)}{(k^2 - 3k^2 v_r - 18)}Rr^3 + r^4 \right] b_4 \tag{8.107}$$

Figure 8.10 presents the flexural rigidity for various values of k. It must be noted that the mode shape in Eq. (8.93) is proportional to the static displacement of the polar orthotropic circular plate, when it is specified at $k = 2$. Indeed, in his classic book Lekhnitskii (1968) derives the formula for the displacement of the polar orthotropic circular plate under uniform

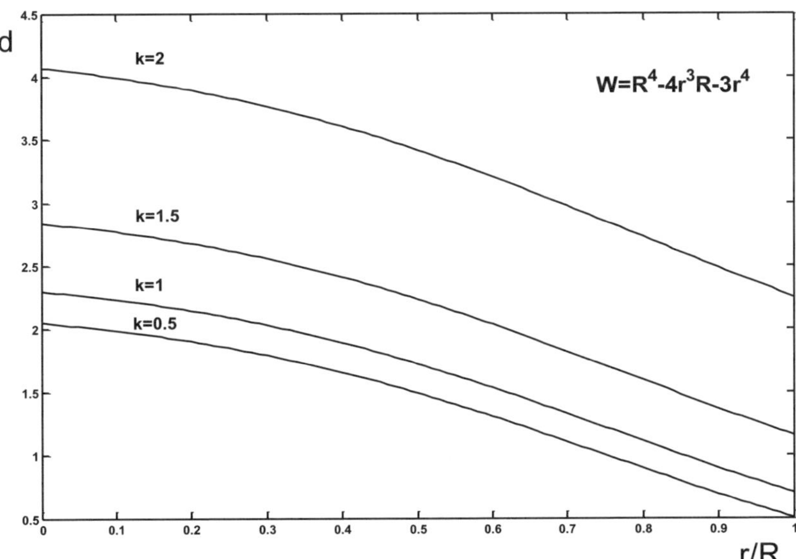

Figure 8.10. Variation of d versus non-dimensional radial coordinate r/R for various values of k, $v_r = 0.35$

loading q_0

$$w = \frac{q_0 R^4}{8(9 - k^2)(1 + k)D_r}\left[3 - k - 4\left(\frac{r}{R}\right)^{1+k} + (1 + k)\left(\frac{r}{R}\right)^4\right]$$

$$(8.108)$$

For $k = 2$, the expression in brackets reduces to Eq. (8.93). The remarkable fact is that the Lekhnitskii's static displacement function, when it is specified at $k = 2$, can serve as an exact mode shape of vibrating polar orthotropic plate both for $k = 2$ and $k \neq 2$.

8.4 Discussion

Apparently, for the first time, six closed-form solutions have been derived for the polar orthotropic plate that is clamped at its circumference. The appropriate mode shapes and the natural frequency expressions are summarized in Table 8.1. It must be stressed, that the expressions for the fundamental natural frequency corresponding to two cubic mode shapes, coincide with each other. However, the associated circular plates possess differing flexural rigidities, given in Eqs. (8.59) and (8.74), respectively. Likewise, the natural frequencies associated with two quartic mode shapes $(R - r)^4$ and $R^4 - 4r^3 R + 3r^4$ coincide. The corresponding flexural rigidities are given,

Table 8.1. Mode shapes, associated natural frequencies and flexural rigidities of polar orthotropic clamped circular plates

Cases	Mode shapes	Fundamental natural frequencies	Equation no. where flexural rigidity is given
1	$W(r) = (R^2 - r^2)^2$	$\omega^2 = (504 + 168v_\theta - 72v_r k^2 - 24k^2)$ $b_4/\rho h$	8.26
2	$W(r) = (R - r)^2$	$\omega^2 = -8(k^2 v_r - 5v_\theta + k^2 - 5)b_4/\rho h$	8.42
3	$W(r) = (R - r)^3$	$\omega^2 = -15(k^2 + 2k^2 v_r - 6v_\theta - 12)b_4/\rho h$	8.59
4	$W(r) = R^3 - 3r^2$ $R + 2r^3$	$\omega^2 = -15(k^2 + 2k^2 v_r - 6v_\theta - 12)b_4/\rho h$	8.74
5	$W(r) = (R - r)^4$	$\omega^2 = -24(k^2 + 3k^2 v_r - 7v_\theta - 21)b_4/\rho h$	8.91
6	$W(r) = R^4 - 4r^3$ $R + 3r^4$	$\omega^2 = -24(k^2 + 3k^2 v_r - 7v_\theta - 21)b_4/\rho h$	8.107

respectively, in Eqs. (8.91) and (8.107), and differ from each other, as they should be. Note that another quartic mode shape $(R^2 - r^2)^2$ does not lead to the same expression of the natural frequency.

The information contained in Table 8.1 allows one to tailor the plate so as to have a predetermined natural frequency Ω. Indeed, Eqs. (8.20), (8.36), (8.52), (8.67), (8.84), and (8.101) can be put in the form

$$\omega^2 = Ab_4 \tag{8.109}$$

where A depends on the problem parameters. For example, for mode shape in Eq. (8.10), the value A equals

$$A = (504 + 168v_\theta - 72v_r k^2 - 24k^2)/\rho h \tag{8.110}$$

as obtained from Eq. (8.20). From our demand $\omega = \Omega$, we get

$$b_4 = \Omega^2/A \tag{8.111}$$

Once this value is substituted into the flexural rigidity expressions, namely, in Eqs. (8.26), (8.42), (8.59), (8.74), (8.91) or (8.107), the unique plate is obtained, that possesses the predetermined natural frequency. Such a tailoring became extremely important in various fields of engineering. It appears remarkable that the vibrational tailoring problem is resolved here in a closed-form, analytical manner. Thus, the problem that is posed and solved herein, can serve as a benchmark case for tailoring problems.

8.5 Vibration tailoring: Numerical example

Consider a case, where the clamped–clamped inhomogeneous circular plate has a cubic mode shape, given by Eq. (8.43). The fundamental natural frequency is obtained, given by Eq. (8.52). We need to examine the behavior of this circular plate by specifying the following properties

$$\omega = 150\,\text{Hz}$$

$$\rho = 5000\,\text{kg m}^3$$

$$h = 0.01\,\text{m}$$

$$k = 3.5$$

$$v_r = 0.35$$

Substitution of the above values into Eq. (8.52), yields

$$b_4 = 4437.87 \tag{8.112}$$

Then, by substituting the known values into Eqs. (8.53)–(8.56), we get the values for the coefficients b_0, b_1, b_2, and b_3 respectively:

$$b_0 = 153377.98 \, R^4 \tag{8.113}$$

$$b_1 = 119303.85 \, R^3 \tag{8.114}$$

$$b_2 = 42810.91 \, R^2 \tag{8.115}$$

$$b_3 = -18221.19 \, R \tag{8.116}$$

Hence, the flexural rigidity becomes

$$\frac{D(r)}{R^4} = 153377.98 + 119303.85 \left(\frac{r}{R}\right) + 42810.91 \left(\frac{r}{R}\right)^2$$
$$- 18221.19 \left(\frac{r}{R}\right)^3 + 4437.87 \left(\frac{r}{R}\right)^4$$

$$\tag{8.117}$$

Figure 8.11 shows the variation of the radial flexural rigidity coefficient Eq. (8.117) with the non-dimensional radial coordinate.

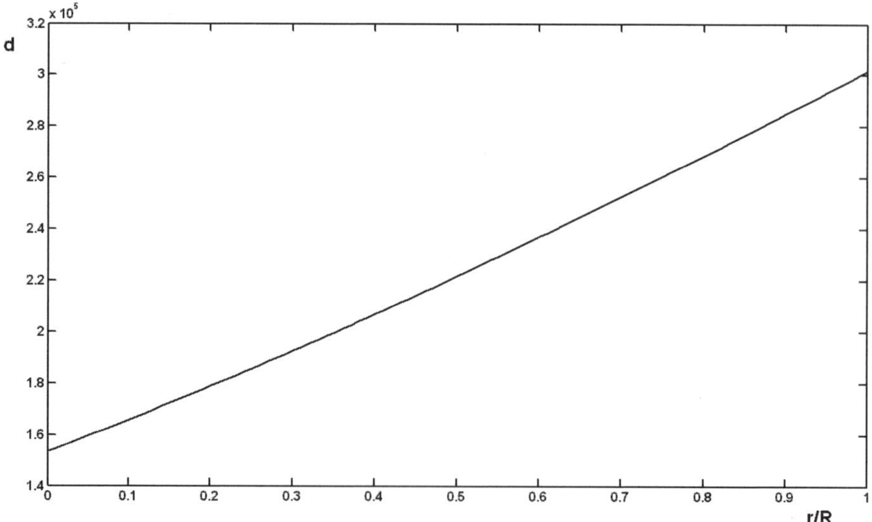

Figure 8.11.   Variation of d versus non-dimensional radial coordinate r/R

Chapter 9

Vibration Tailoring of Simply-Supported Polar Orthotropic Inhomogeneous Circular Plates

Semi-inverse problem of free vibration of simply-supported inhomogeneous polar orthotropic plates is studied. The mode shape is postulated in the form of the classic formula by Lekhnitskii, namely, the static deflection of the associated uniform polar-orthotropic circular plate under uniform loading. The ratios of the circumferential rigidity to the radial rigidity are identified as integer numbers, so that the candidate mode shapes constitute the polynomial functions. The flexural rigidities themselves are also represented by polynomials. The semi-inverse problem of identifying the coefficients of the flexural rigidities is solved analytically. It appears remarkable, that for numerous cases, the simple method developed in this study provides novel closed-form solutions for the design of the polar-orthotropic circular plates with pre-specified mode shapes.

9.1 Introduction

Since the study on vibrations of circular plates by Chladni (1787) and Poisson (1829), numerous studies have been conducted on this subject. The definitive reference in plate vibrations is the monograph by Leissa (1969). Closed-form solution for simply-supported isotropic circular plates was reported, apparently for the first time, by Elishakoff and Storch (2005).

Prior to Elishakoff and Storch (2005), various approximate methods have been utilized. Pertinent references for our purpose are those of Jones (1975) and Johns (1975). As Leissa (1977) mentions, "Jones (1975) devised a simple approximate formula for calculating fundamental (i.e. lowest)

155

frequencies if the static deflected shape of a uniformly loaded plate is known. Johns (1975) used a two-term solution for static deflection shapes to evaluate frequencies of clamped- and simply-supported circular plates and compares them with those of Jones (1975)". Another study that relates the vibration frequency of plates and their deflection was presented by Mazumdar (1971).

An unconventional application of static deflections of circular plates to the vibration problems was conducted in a recent study by Elishakoff and Storch (2005) (see also a recent monograph by Elishakoff (2005)). In these papers, several semi-inverse problems were posed and solved. Namely, it was postulated that the expression in the square parentheses for displacement of the *homogeneous* and *uniform* circular plate under loading q

$$w = \left(\frac{q}{64D}\right)\left[(R^2 - r^2)\left(\frac{5+v}{1+v}R^2 - r^2\right)\right] \qquad (9.1)$$

can serve as a mode shape of an *inhomogeneous* isotropic plate.

In this chapter, Elishakoff (2005) is generated to the simply supported polar orthotropic circular plate. Several closed-form solutions are found, apparently for the first time in the literature.

9.2 Analysis

The governing differential equation of the polar orthotropic circular plate of varying flexural rigidity is derived in Chapter 8, given by Eq. (8.9).

Our analysis is heavily based on the application of the classic formula by Lekhnitskii (1968)

$$w(r) = \frac{q_0 a^4}{8(9-k^2)D_r}\left[\frac{(3-k)(4+k+v_\theta)}{(1+k)(k+v_\theta)}\right.$$
$$\left. - \frac{4(3+v_\theta)}{(1+k)(k+v_\theta)}\left(\frac{r}{R}\right)^{k+1} + \left(\frac{r}{R}\right)^4\right] \qquad (9.2)$$

for the deflection of the circular simply supported polar-orthotropic plate under uniformly distributed load q_0. In Eq. (9.2) $k^2 = D_\theta/D_r$ is the rigidity ratio. We would like to design a circular plate such that is fundamental mode shape, $W(r)$ in Eq. (8.9) would coincide with the expression in brackets in Eq. (9.2), namely,

$$W(r) = \frac{(3-k)(4+k+\nu_\theta)}{(1+k)(k+\nu_\theta)} - \frac{4(3+\nu_\theta)}{(1+k)(k+\nu_\theta)}\left(\frac{r}{R}\right)^{k+1} + \left(\frac{r}{R}\right)^4 \tag{9.3}$$

As the literature shows, the postulate in Eq. (9.2) cannot be realized for the homogeneous plates, with $D_r(r) = $ const, $D_\theta(r) = $ const and $\rho(r) = $ const. However, if previous experience for isotropic case is any guide, Elishakoff (2005), Leknhnitskii (1968), Ari-Gur and Stavsky (1981), Caliò and Elishakoff (2002), we can anticipate finding *heterogeneous* plates that possess expression in Eq. (9.2) as the mode shape.

As follows we specify the value k in Eq. (8.9) as equal some integer m and investigate if there exists a non-negative-valued distributions of $D_r(r)$ and $D_\theta(r)$ that taken together with mode shape

$$W(r) = \frac{1}{(9-m^2)}\frac{(3-m)(4+m+\nu_\theta)}{(1+m)(m+\nu_\theta)}$$
$$- \frac{4(3+\nu_\theta)}{(1+m)(m+\nu_\theta)}\left(\frac{r}{R}\right)^{m+1} + \left(\frac{r}{R}\right)^4 \tag{9.4}$$

satisfy Eq. (8.9).

9.2.1 Semi-inverse method of solution associated with $m = 1$

For $m = 1$, the mode shape in Eq. (9.4) reads

$$W(r) = \frac{1}{8}\left[\frac{10+2\nu_\theta}{2+2\nu_\theta} - \frac{(12+4\nu_\theta)}{(2+2\nu_\theta)}\left(\frac{r}{R}\right)^2 + \left(\frac{r}{R}\right)^4\right] \tag{9.5}$$

The flexural rigidities are sought as polynomials of the fourth-order

$$D_r(r) = b_0 + b_1 r + b_2 r^2 + b_3 r^3 + b_4 r^4 \tag{9.6}$$

$$D_\theta(r) = k^2 D_r(r) \tag{9.7}$$

where k is taken as constant. The substitution of Eqs. (9.5)–(9.7) into Eq. (9.2) leads to the following polynomial equation

$$G_0 + G_1 r + G_2 r^2 + G_3 r^3 + G_4 r^4 + G_5 r^5 = 0 \tag{9.8}$$

where

$$G_0 = -8v_\theta^2 R^2 b_1 + 12k^2 R^2 b_1 - 32v_\theta R^2 b_1 + 4v_\theta v_r R^2 k^2 b_1 + 4k^2 v_\theta R^2 b_1$$
$$+ 12k^2 R^2 v_r b_1 - 24R^2 b_1 \tag{9.9}$$

$$G_1 = 24k^2 v_r R^2 b_2 + 24k^2 R^2 b_2 - 24k^2 v_r v_\theta b_0 + 8k^2 R^2 v_\theta b_2 - 8k^2 b_0$$
$$- 72R^2 b_2 - 24k^2 v_r b_0 + 96v_\theta b_0 + 24v_\theta^2 b_0 + 72b_0 - 24R^2 v_\theta^2 b_2$$
$$- 8k^2 v_\theta b_0 + 8k^2 v_\theta v_r R^2 b_2 - 96R^2 v_\theta b_2$$
$$- \rho h\omega^2 R^4 v_\theta - 5\rho h\omega^2 R^4 \tag{9.10}$$

$$G_2 = -36k^2 v_r v_\theta b_1 + 48v_\theta^2 b_1 - 144R^2 b_3 + 12k^2 R^2 v_\theta b_3 + 12R^2 k^2 v_r v_\theta b_3$$
$$+ 192v_\theta b_1 - 12k^2 b_1 + 144b_1 - 192v_\theta R^2 b_3 - 36k^2 v_r b_1$$
$$- 48R^2 v_\theta^2 b_3 - 12k^2 v_\theta b_1 + 36k^2 R^2 b_3 + 36k^2 R^2 v_r b_3 \tag{9.11}$$

$$G_3 = -48k^2 v_r v_\theta b_2 - 240R^2 b_4 - 80v_\theta^2 R^2 b_4 + 48k^2 R^2 b_4 - 16k^2 b_2$$
$$+ 320v_\theta b_2 - 48v_r k^2 b_2 - 16v_\theta k^2 b_2 + 80v_\theta^2 b_2 - 320R^2 v_\theta b_4$$
$$+ 240b_2 + 16k^2 R^2 v_r v_\theta b_4 + 16k^2 R^2 v_\theta b_4 + 48k^2 R^2 v_r b_4$$
$$+ 2\rho h\omega^2 R^2 v_\theta + 6\rho h\omega^2 R^2 \tag{9.12}$$

$$G_4 = 120v_\theta^2 b_3 + 360b_3 - 60k^2 v_r v_\theta b_3 - 60k^2 v_r b_3 - 20k^2 v_\theta b_3$$
$$+ 480v_\theta b_3 - 20k^2 b_3 \tag{9.13}$$

$$G_5 = 504b_4 - 72k^2 v_r v_\theta b_4 + 672v_\theta b_4 + 168v_\theta^2 b_4 - 24k^2 b_4$$
$$- 24k^2 v_\theta b_4 - \rho h\omega^2 v_\theta - \rho h\omega^2 \tag{9.14}$$

From Eq. (9.14) we get the relationship between the natural frequency squared ω^2 and the coefficient b_4:

$$\omega^2 = 24(21 + 7v_\theta - 3v_r k^2 - 21k^2)b_4/\rho h \tag{9.15}$$

Equation (9.13) shows that b_3 vanishes identically. The equation resulting from substitution of Eq. (9.15) into Eq. (9.12), results in the formula for b_2, as related to b_4

$$b_2 = -2\frac{(3k^2 + 12k^2 v_r + 4k^2 v_r v_\theta - 8v_\theta^2 + k^2 v_\theta - 53v_\theta - 87)}{(1 + v_\theta)(-5v_\theta + k^2 + 3k^2 v_r - 15)}R^2 b_4 \tag{9.16}$$

Equation (9.11) leads to the conclusion that $b_1 = 0$. The equation resulting from substitution of Eqs. (9.15) and (9.16) into Eq. (9.10) leads to the expression for b_0:

$$b_0 = -\frac{b_4}{(1 + v_\theta)(k^2 - 5v_\theta + 3k^2 v_r - 15)(k^2 - 3v_\theta + 3k^2 v_r - 9)}$$
$$\times (-882k^2 v_r - 1632k^2 v_r v_\theta - 246k^2 v_\theta + 5778v_\theta + 36k^2$$
$$- 650v_\theta^2 k^2 v_r + 3312v_\theta^2 - 68v_\theta^3 k^2 v_r + 48k^4 v_\theta v_r + 144k^4 v_r^2 v_\theta$$
$$+ 19k^4 v_r^2 v_\theta^2 + 8k^4 v_r v_\theta^2 - 128v_\theta^2 k^2 - 12v_\theta^3 k^2 + 6k^4 v_\theta + 63k^4 v_r^2$$
$$+ k^4 v_\theta^2 + 750v_\theta^3 + 57v_\theta^4 - 3k^4 + 3159) \tag{9.17}$$

In the view of the relationship

$$v_\theta = k^2 v_r \tag{9.18}$$

we get the following final expression for the flexural rigidity:

$$D_r(r) = \frac{b_4}{(1 + k^2 v_r)^2 (2k^2 v_r - k^2 + 15)(k^2 - 9)} \times [(-135 + 50k^4 v_r$$
$$+ 28k^6 v_r^2 - k^4 + 24k^2 - 171k^4 v_r^2 - 288k^2 v_r - 18k^6 v_r^3 + 2k^8 v_r^3$$
$$- k^8 v_r^2 - 2k^6 v_r) + (6k^4 - 108k^6 v_r^2 + 2304k^8 v_r^2 + 72k^6 v_r^3 - 8k^8 v_r^3$$
$$+ 1566 + 8k^6 v_r + 810k^4 v_r^2 - 328k^4 v_r - 228k^2) R^2 r^2 + (-k^8 v_r^2$$
$$- 4896k^2 v_r - 1734k^4 v_r^2 - 3159 + 246k^4 v_r - 8k^8 v_r^4 - 214k^6 v_r^3$$
$$+ 80k^6 v_r^2 + 2k^4 - 36k^2 + 6k^8 v_r^3 - 6k^6 v_r) R^4 r^4] \tag{9.19}$$

Figure 9.1 depicts the variation of $D_r(r)$ as a function of r, for various values of k, for v_r fixed at 0.35.

9.2.2 Semi-inverse method of solution associated with $m = 2$

For $m = 2$, the mode shape in Eq. (9.4) reads

$$W(r) = \frac{1}{5}\left[\frac{6 + v_\theta}{6 + 3v_\theta} - \frac{(12 + 4v_\theta)}{(6 + 3v_\theta)}\left(\frac{r}{R}\right)^3 + \left(\frac{r}{R}\right)^4\right] \tag{9.20}$$

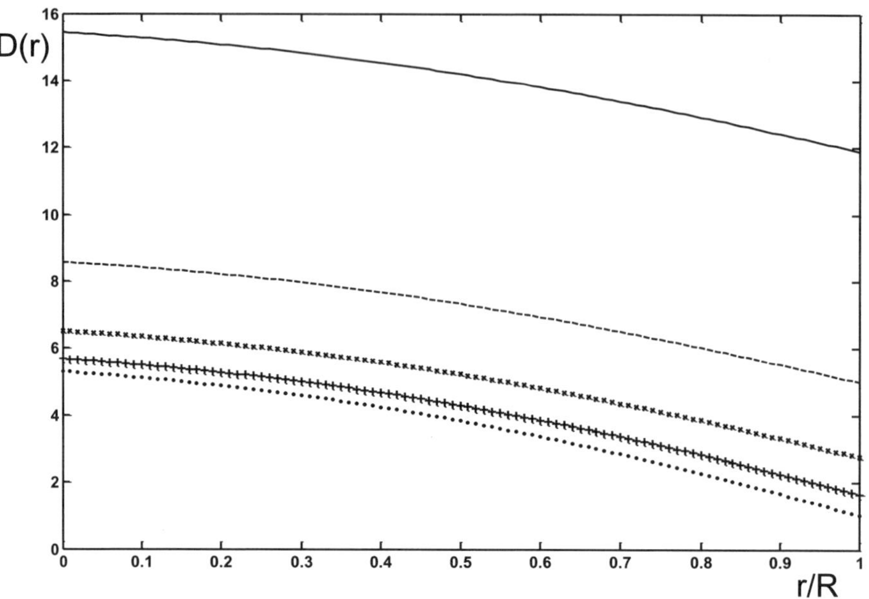

Figure 9.1. Variation of $D(r)$ versus non-dimensional radial coordinate r/R for various values of k, $v_r = 0.35(-k = 2.5; --k = 2; \times \times k = 1.5; ++k = 1; \ldots k = 0.5)$ (here and hereinafter $D_r(r)$ is denoted $D(r)$)

The result of substitution of Eq. (9.20) in coincidence of Eqs. (9.6) and (9.7) yields the equation

$$H_0 + H_1 r + H_2 r^2 + H_3 r^3 + H_4 r^4 + H_5 r^5 = 0 \tag{9.21}$$

where

$$H_0 = -24v_\theta^2 Rb_0 + 36k^2 Rb_0 - 144Rb_0 + 24v_r v_\theta Rk^2 b_1 + 72k^2 v_r Rb_0$$
$$+ 12k^2 Rv_\theta b_0 - 120Rv_\theta b_0 \tag{9.22}$$

$$H_1 = -48k^2 b_0 - 144k^2 v_r b_0 + 72v_\theta^2 b_0 - 432Rb_1 + 432b_0 - 72k^2 v_r v_\theta b_0$$
$$+ 72k^2 Rb_1 - 72R^2 v_\theta^2 b_1 + 48k^2 v_r v_\theta Rb_1 - 360Rv_\theta b_1 + 360v_\theta b_0$$
$$- 24k^2 v_\theta b_0 + 144k^2 v_r Rb_1 + 24k^2 v_\theta R - \rho h \omega^2 R^4 v_\theta - 6\rho h \omega^2 R^4 \tag{9.23}$$

$$H_2 = -720Rv_\theta b_2 + 108k^2 Rb_2 - 108v_r v_\theta k^2 b_1 - 72k^2 b_1 + 36Rk^2 v_\theta b_2$$
$$+ 864b_1 - 144v_\theta^2 Rb_2 + 720v_\theta b_1 + 72k^2 v_r v_\theta Rb_2 + 216k^2 v_r Rb_2$$
$$- 36k^2 v_\theta b_1 - 864Rb_2 + 144v_\theta^2 b_1 - 216k^2 v_2 b_1 \tag{9.24}$$

$$H_3 = 48k^2 Rv_\theta b_3 - 1200v_\theta Rb_3 + 96v_r v_\theta Rk^2 b_3 + 144k^2 Rb_3 + 1440b_2$$
$$- 288k^2 v_r b_2 - 144v_r v_\theta k^2 b_2 + 288v_r k^2 Rb_3 - 48k^2 v_\theta b_2 - 96k^2 b_2$$
$$+ 1200v_\theta b_2 - 1440Rb_3 + 240v_\theta^2 b_2 - 240Rv_\theta^2 b_3 \tag{9.25}$$

$$H_4 = -60k^2 v_\theta b_3 - 120k^2 b_3 + 180Rk^2 b_4 - 216Rb_4 + 2160b_3$$
$$+ 60k^2 Rv_\theta b_4 - 180k^2 v_r v_\theta b_3 - 360Rv_\theta^2 b_4 + 120k^2 Rv_r v_\theta b_4$$
$$+ 360k^2 Rv_r b_4 - 1800Rv_\theta b_4 + 1800v_\theta b_3 + 360v_\theta^2 b_3$$
$$- 360k^2 v_r b_3 + 12\rho h\omega^2 R + 4\rho h\omega^2 Rv_\theta \tag{9.26}$$

$$H_5 = 3024b_4 + 504v_\theta^2 b_4 - 72v_\theta k^2 b_4 - 144k^2 b_4 - 216k^2 v_r v_\theta b_4$$
$$- 432k^2 v_r b_4 + 2520v_\theta b_4 - 3\rho h\omega^2 v_\theta - 6\rho h\omega^2 \tag{9.27}$$

From Eq. (9.27) we get the expression for natural frequency squared ω^2:

$$\omega^2 = 24(21 + 7v_\theta - 3v_r k^2 - 21k^2)b_4/\rho h \tag{9.28}$$

It is remarkable, that the expression (9.28) coincides with Eq. (9.15) for the squared natural frequency, although it is derived for another mode shape. Substitution of Eq. (9.28) into Eq. (9.26) leads to

$$b_3 = -\frac{1}{5}\frac{(9k^2 + 42k^2 v_r + 14k^2 v_r v_\theta - 26v_\theta^2 + 3k^2 v_\theta - 186v_\theta - 324)}{(2 + v_\theta)(-6v_\theta + k^2 + 3k^2 v_r - 18)}Rb_4 \tag{9.29}$$

Substituting Eq. (9.29) into Eq. (9.25) we get

$$b_2 = -\frac{1}{5(2 + v_\theta)(k^2 - 6v_\theta + 3k^2 v_r - 18)(k^2 - 5v_\theta + 3k^2 v_r - 15)}$$
$$\times (-186v_\theta + 9k^2 - 324 + 42k^2 v_r + 14k^2 v_r v_\theta - 26v_\theta^2 + 3k^2 v_\theta)$$
$$\times (-25v_\theta + k^2 v_\theta + 6k^2 v_r + 2k^2 v_r v_\theta - 5v_\theta^2 + 3k^2 - 30) \tag{9.30}$$

Equation (9.24) leads to the coefficient b_1:

$$b_1 = -\frac{b_4 R^2}{5(2 + v_\theta)^2(-6v_\theta + k^2 + 3k^2 v_r - 18)(-5v_\theta + k^2 + 3k^2 v_r - 15)}$$
$$\times (-186v_\theta + 9k^2 - 324 + 14k^2 v_r v_\theta - 26v_\theta^2 + 3k^2 v_\theta)(-30 + 3k^2$$
$$+ k^2 v_\theta + 6k^2 v_r + 2k^2 v_r v_\theta - 25v_\theta - v_\theta^2) \tag{9.31}$$

Note that from Eq. (9.22), we conclude that $b_0 = 0$. Equation (9.23) yields another expression for b_1:

$$b_1 = -\frac{b_4(-7v_\theta + k^2 + 3k^2 v_r - 21)(6 + v_\theta)R^3}{-18 + k^2 v_\theta + 2k^2 v_r v_\theta + 6k^2 v_r + 3k^2 - 15v_\theta - 3v_\theta^2} \tag{9.32}$$

If the expression in Eqs. (9.31) and (9.32) differ from each other, the problem contains a contradiction. In order to have a consistent set, we demand the expression for b_1, in Eq. (9.31) and in Eq. (9.32) to be equal. This results in the following polynomial equation:

$$\sum_{j=0}^{8} I_j v_r^j = 0 \tag{9.33}$$

where

$$I_0 = -3k^8 - 1260k^6 + 131004k^4 - 2372976k^2 + 12130560 \tag{9.34}$$

$$I_1 = 76k^{10} - 6930k^8 + 320580k^6 - 5285304k^4 + 27511488k^2 \tag{9.35}$$

$$I_2 = 78k^{12} - 6880k^{10} + 277215k^8 - 4609980k^6 + 25441776k^4 \tag{9.36}$$

$$I_3 = 24k^{14} - 2580k^{12} + 111200k^{10} - 2031330k^8 + 12453480k^6 \tag{9.37}$$

$$I_4 = 2k^{16} - 390k^{14} + 21885k^{12} - 485570k^{10} + 3511368k^8 \tag{9.38}$$

$$I_5 = -20k^{16} + 2028k^{14} - 62892k^{12} + 582054k^{10} \tag{9.39}$$

$$I_6 = 70k^{16} - 4068k^{14} - 54993k^{12} \tag{9.40}$$

$$I_7 = -100k^{16} + 2658k^{14} \tag{9.41}$$

$$I_8 = 48k^6 \tag{9.42}$$

Numerical evaluation of the roots of Eq. (9.33) reveals the following: For $k = 1$, $k = 2$ or $k = 3$ it does not possess real positive roots for v_r.

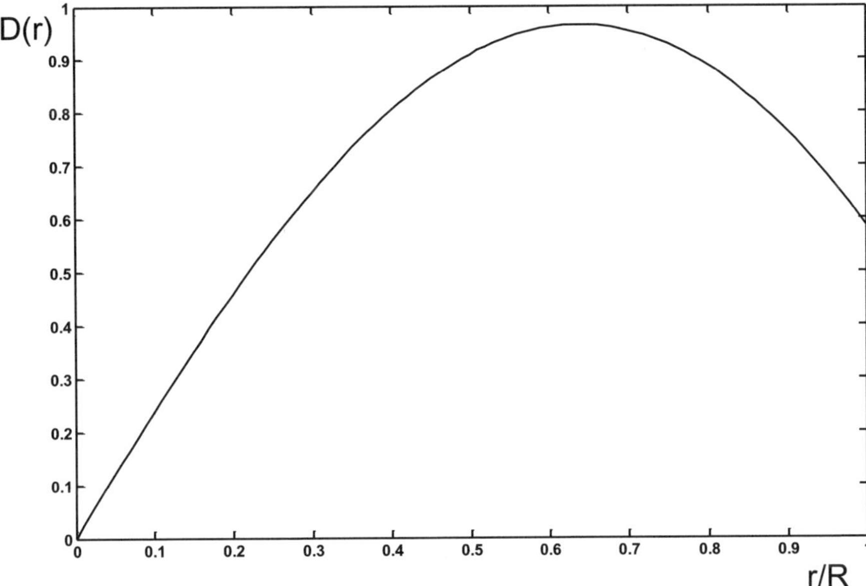

Figure 9.2. Variation of $D(r)$ versus non-dimensional radial coordinate r/R for $k = 4$ and $\nu_r = 0.11086$

For $k = 4$, the only positive root of Eq. (9.33) is $\nu_r = 0.11086$. Figure 9.2 depicts the appropriate variation of the radial flexural rigidity $D_r(r)$. The values of the circumferential rigidity are obtained by dividing $D_r(r)$ by the appropriate value of k^2. For $k = 5$, Eq. (9.33) yields $\nu_r = 0.44134$ (Fig. 9.3); $k = 6$ corresponds to $\nu_r = 0.561495$ (Fig. 9.4); Eq. (9.33) has two positive roots $\nu_r = 0.295815$ and $\nu_r = 0.6490527$ for $k = 7$. Two resulting curves for $D_r(r)$ are shown in Fig. 9.5. Likewise, two roots $\nu_r = 0.3256135$ and $\nu_r = 0.71579117$ correspond to $k = 8$ (Fig. 9.6). The value $k = 9$ is associated with roots $\nu_r = 0.3507312$ and $\nu_r = 0.7670678$ (Fig. 9.7), whereas $k = 10$ corresponds to $\nu_r = 0.37285365$ and $\nu_r = 0.80664837$ (Fig. 9.8).

9.2.3 Semi-inverse method of solution associated with $m = 3$

As is seen from Eq. (9.2) when m tends to three the denominator in front of the square parenthesis approaches zero. The first term in the square

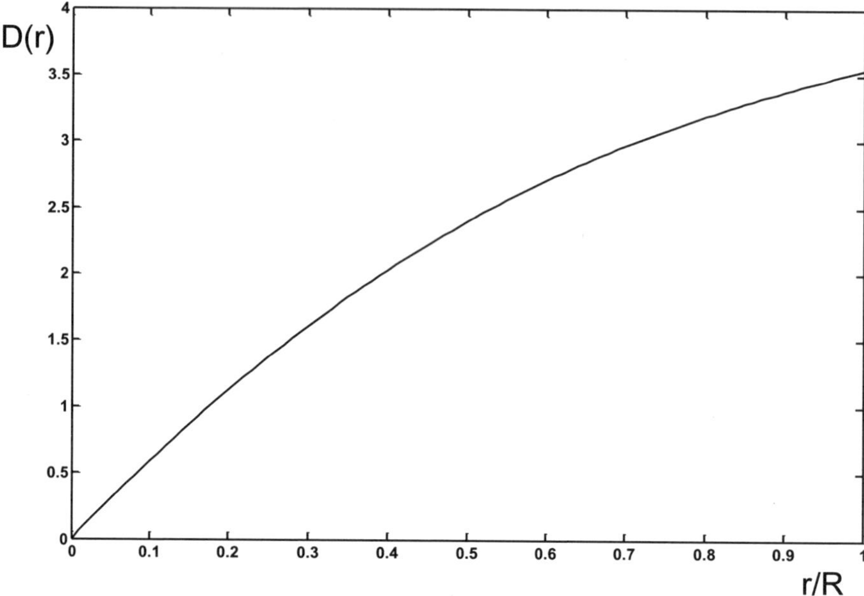

Figure 9.3. Variation of $D(r)$ versus non-dimensional radial coordinate r/R for $k = 5$ and $\nu_r = 0.44134$

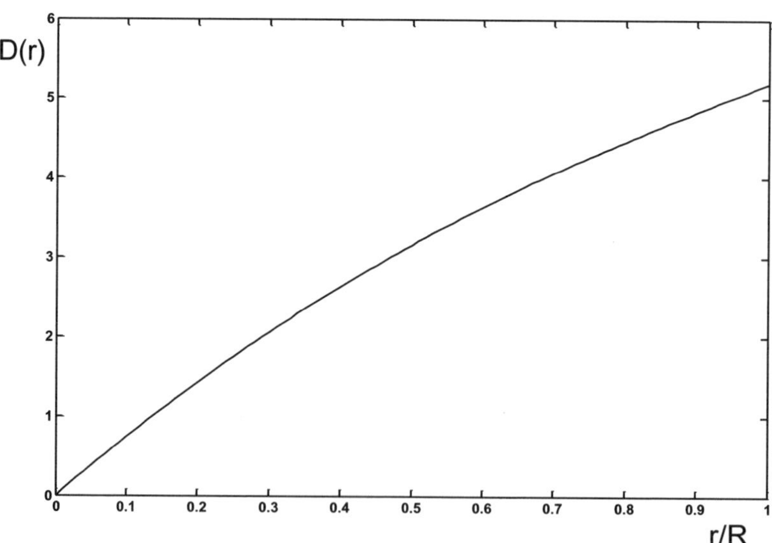

Figure 9.4. Variation of $D(r)$ versus non-dimensional radial coordinate r/R for $k = 6$ and $\nu_r = 0.561495$

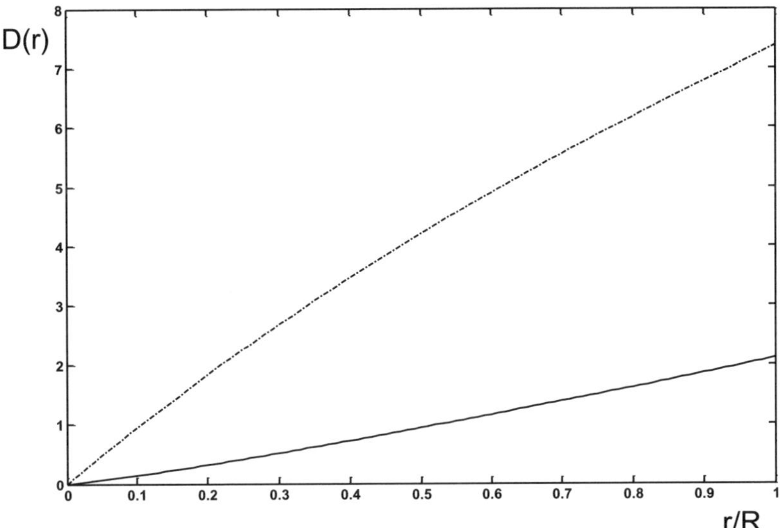

Figure 9.5. Variation of $D(r)$ versus non-dimensional radial coordinate r/R for $k = 7(-v_r = 0.295815; - \cdot -v_r = 0.3507312)$

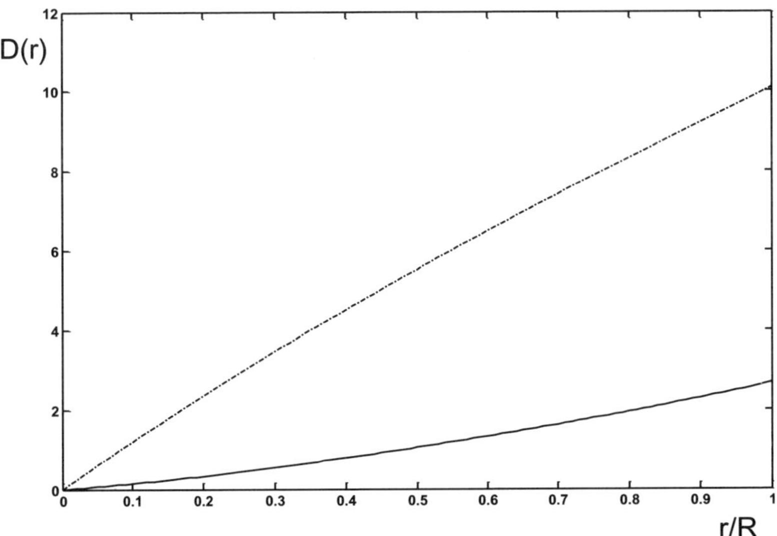

Figure 9.6. Variation of $D(r)$ versus non-dimensional radial coordinate r/R for $k = 8(-v_r = 0.3256135; - \cdot -v_r = 0.71579117)$

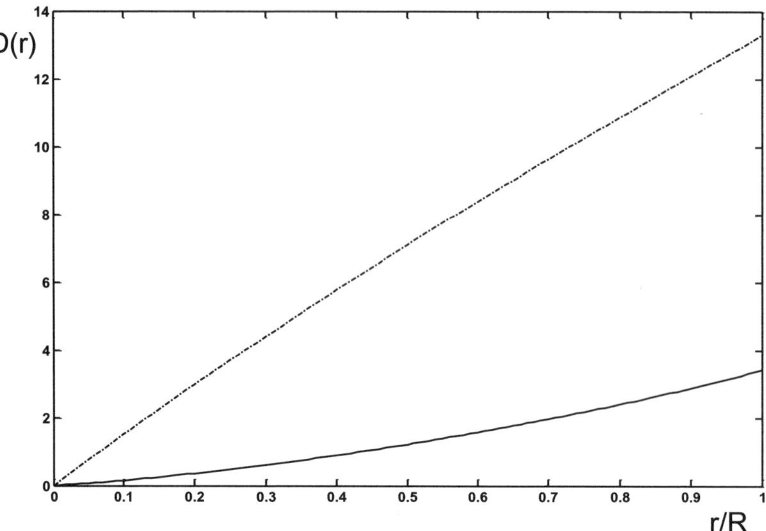

Figure 9.7. Variation of $D(r)$ versus non-dimensional radial coordinate r/R for $k = 9(-v_r = 0.3507312; - \cdot -v_r = 0.7670678)$

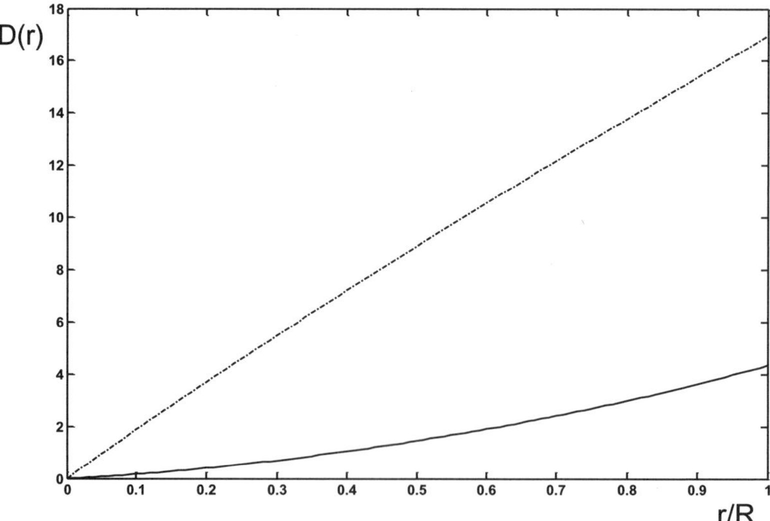

Figure 9.8. Variation of $D(r)$ versus non-dimensional radial coordinate r/R for $k = 10(-v_r = 0.37285365; - \cdot -v_r = 0.80664837)$

parentheses tends to zero, but the product

$$\lim_{m \to 3} \frac{1}{(9 - m^2)} \frac{(3 - m)(4 + m + v_\theta)}{(1 + m)(m + v_\theta)} = \frac{1}{54 + 10v_\theta} \tag{9.43}$$

turns out to approach a finite value $(54 + 10v_\theta)^{-1}$.

The sum of the second and the third terms in the square parenthesis in Eq. (9.4) also tends to zero. As Lekhnitskii (1968) stresses, "formulas for the case when $m = 3$ will be found by going to the limit". Indeed,

$$\lim_{m \to 3} \frac{1}{(9 - m^2)} \left[\left(\frac{r}{R} \right)^4 - \frac{4(3 + v_\theta)}{(1 + m)(m + v_\theta)} \left(\frac{r}{R} \right)^{m+1} \right]$$

$$= \frac{r^4 [-7 - v_\theta - 12 \log(R) - 4 \log(R) v_\theta + 12 \log(r) + 4 \log(r) v_\theta]}{24 R^4 (3 + v_\theta)}$$

$$\tag{9.44}$$

using the L'Hopital's Rule.

Then, the candidate mode shape for $m = 3$ becomes:

$$W(r) = \frac{1}{(54 + 10v_\theta)} + \frac{r^4 [-7 - v_\theta - (12 + 4v_\theta) \ln(r/R)]}{24 R^4 (3 + v_\theta)} \tag{9.45}$$

Note that Ari-Gur and Stavsky (1981) also report displacements including logarithmic function of the radial coordinate for $k = 3$. The mode shape (9.45) does not constitute a polynomial function; since in this study we limit ourselves with polynomial mode shapes, the reconstruction of flexural rigidities for the case $m = 3$ will not be pursued. This does not mean that there are no non-polynomial mode shapes for heterogeneous structures; for non-polynomial mode shapes in the semi-inverse problems one can consult Elishakoff's (2005) monograph, as well as by Caliò and Elishakoff (2002). In the former monograph by Elishakoff (2005), the rational functions are obtained as mode shapes, whereas in the latter, Caliò and Elishakoff (2002), trigonometric mode shapes are utilized.

9.2.4　Semi-inverse method of solution associated with $m = 4$

For $m = 4$, the mode shape in Eq. (9.4) becomes

$$W(r) = \frac{(-8 - v_\theta)}{(20 + 5v_\theta)} + \left(\frac{r}{R} \right)^4 - \frac{(12 + 4v_\theta)}{(20 + 5v_\theta)} \left(\frac{r}{R} \right)^5 \tag{9.46}$$

The result of substitution of Eq. (9.46) in coincidence of Eqs. (9.6), and (9.7) yields the equation

$$J_0 + J_1 r + J_2 r^2 + J_3 r^3 + J_4 r^4 + J_5 r^5 + J_6 r^6 = 0 \tag{9.47}$$

where

$$\begin{aligned}
J_0 = {}& 8\rho h\omega^2 R^5 + 1440 b_0 R + \rho h\omega^2 R^5 v_\theta + 840 b_0 R v_\theta \\
& + 120 b_0 v_\theta^2 R - 40 k^2 b_0 R v_\theta - 160 k^2 R b_0 - 480 k^2 b_0 R v_r \\
& - 120 b_0 R k^2 v_r v_\theta \tag{9.48}
\end{aligned}$$

$$\begin{aligned}
J_1 = {}& 2880 b_1 R + 60 b_0 k^2 v_\theta + 1680 R b_1 v_\theta - 240 b_0 v_\theta^2 - 720 k^2 b_1 v_r R \\
& - 1680 b_0 v_\theta - 60 k^2 R v_\theta b_1 + 180 k^2 b_0 + 240 b_0 k^2 v_r v_\theta - 240 k^2 b_1 R \\
& - 180 b_1 k^2 v_r v_\theta R + 240 b_1 R v_\theta^2 - 2880 b_0 + 720 k^2 b_0 v_r \tag{9.49}
\end{aligned}$$

$$\begin{aligned}
J_2 = {}& 960 k^2 b_1 v_r - 2800 b_1 v_\theta + 80 k^2 b_1 v_\theta + 240 k^2 b_1 - 4800 b_1 \\
& - 960 k^2 b_2 v_r R + 2800 R v_\theta b_2 - 240 k^2 b_2 R v_r v_\theta + 400 b_2 R v_\theta^2 \\
& - 80 k^2 b_2 R v_\theta + 4800 b_2 R - 400 b_1 v_\theta^2 - 320 k^2 b_2 R \\
& + 320 k^2 b_1 v_r v_\theta \tag{9.50}
\end{aligned}$$

$$\begin{aligned}
J_3 = {}& 7200 b_3 R + 4200 b_3 R v_\theta + 400 k^2 b_2 v_r v_\theta - 4200 b_2 v_\theta - 400 k^2 b_3 R \\
& - 600 b_2 v_\theta^2 + 300 k^2 b_2 - 7200 b_2 - 100 k^2 b_3 R v_\theta \\
& + 1200 k^2 b_2 v_r - 1200 k^2 b_3 v_r R + 600 b_3 R v_\theta^2 - 300 k^2 b_3 R v_r v_\theta \\
& + 100 k^2 b_2 v_\theta \tag{9.51}
\end{aligned}$$

$$\begin{aligned}
J_4 = {}& 10080 b_4 R - 20 \rho h\omega^2 R - 840 b_3 v_\theta^2 + 840 b_4 R v_\theta^2 - 5880 b_3 v_\theta \\
& + 5880 b_4 R v_\theta - 10080 b_3 - 5 \rho h\omega^2 R v_\theta + 360 k^2 b_3 - 1440 k^2 b_4 v_r R \\
& + 480 k^2 b_3 v_r v_\theta + 120 k^2 b_3 v_\theta - 120 k^2 b_4 R v_\theta + 1440 k^2 b_3 v_r \\
& - 360 k^2 b_4 v_r R v_\theta - 480 k^2 b_4 R \tag{9.52}
\end{aligned}$$

$$\begin{aligned}
J_5 = {}& 560 k^2 b_4 v_r v_\theta + 1680 k^2 b_4 v_r - 20 b_4 v_\theta^2 + 140 k^2 b_4 v_\theta + 12 \rho h\omega^2 \\
& - 13440 b_4 + 4 \rho h\omega^2 v_\theta + 420 k^2 b_4 - 7840 b_4 v_\theta \tag{9.53}
\end{aligned}$$

From Eq. (9.53) we get the expression for natural frequency squared ω^2:

$$\omega^2 = 35(32 + 8 v_\theta - 4 v_r k^2 - k^2) b_4 / \rho h \tag{9.54}$$

Substitution of Eq. (9.54) into Eq. (9.52) leads to

$$b_3 = -\frac{1}{24}\frac{(11k^2 + 68k^2v_r - 112v_\theta - 616)(4 + v_\theta)}{(-49v_\theta + 3k^2 + 12k^2v_r + k^2v_\theta + 4k^2v_rv_\theta - 7v_\theta^2 - 84)}Rb_4$$

(9.55)

Substituting Eq. (9.55) into Eq. (9.51) we get

$$b_2 = K/L \tag{9.56}$$

$$K = -(-6v_\theta^2 + 12k^2v_r - 42v_\theta + 3k^2v_rv_\theta + 4k^2 + k^2v_\theta - 72)$$
$$\times (-112v_\theta - 616 + 11k^2 + 68k^2v_r)(4 + v_\theta)^2 R^2 b_4 \tag{9.57}$$
$$L = 24(-7v_\theta^2 + 12k^2v_r - 49v_\theta + 4k^2v_rv_\theta + 3k^2 + k^2v_\theta - 84)$$
$$\times (-6v_\theta^2 + 4k^2v_rv_\theta - 42v_\theta + 12k^2v_r + 3k^2 + k^2v_\theta - 72) \tag{9.58}$$

Equation (9.50) leads to the coefficient b_1:

$$b_1 = M/N \tag{9.59}$$

where

$$M = -(-5v_\theta^2 + 12k^2v_r - 35v_\theta + 3k^2v_rv_\theta + 4k^2 + k^2v_\theta - 60)$$
$$\times (-6v_\theta - 18 + k^2 + 3k^2v_r)(-112v_\theta - 616 + 11k^2 + 68k^2v_r)$$
$$\times (4 + v_\theta)^2 R^3 b_4 \tag{9.60}$$
$$N = 24(-7v_\theta^2 + 12k^2v_r - 49v_\theta + 4k^2v_rv_\theta + 3k^2 + k^2v_\theta - 84)$$
$$\times (-6v_\theta^2 + 4k^2v_rv_\theta - 42v_\theta + 12k^2v_r + 3k^2 + k^2v_\theta - 72)$$
$$\times (3k^2 + 12k^2v_r - 35v_\theta + k^2v_\theta^2 + 4k^2v_rv_\theta - 60) \tag{9.61}$$

From Eq. (9.49), we get

$$b_0 = O/P \tag{9.62}$$

where

$$O = -(-5v_\theta^2 + 12k^2v_r - 35v_\theta + 3k^2v_rv_\theta + 4k^2 + k^2v_\theta - 60)$$
$$\times (-6v_\theta - 18 + k^2 + 3k^2v_r)(-112v_\theta - 616 + 11k^2 + 68k^2v_r)$$
$$\times (-5v_\theta + 3k^2v_r + k^2 - 15)(4 + v_\theta)^3 R^4 b_4 \tag{9.63}$$

$$P = 24(-7v_\theta^2 + 12k^2v_r - 49v_\theta + 4k^2v_rv_\theta + 3k^2 + k^2v_\theta - 84)(-6v_\theta^2$$
$$+ 4k^2v_rv_\theta - 42v_\theta + 12k^2v_r + 3k^2 + k^2v_\theta - 72)(3k^2 + 12k^2v_r$$
$$- 35v_\theta + k^2v_\theta^2 + 4k^2v_rv_\theta - 60)(k^1v_\theta - 4v_\theta^2 + 4k^2v_rv_\theta - 28v_\theta$$
$$+ 3k^2 + 12k^2v_r - 48) \tag{9.64}$$

Equation (9.48) yields another expression for b_0:

$$b_0 = -\frac{7(-8v_\theta + k^2 + 4k^2v_r - 32)(8 + v_\theta)R^4b_4}{(-3v_\theta + 3k^2v_r + k^2 - 9)(4 + v_\theta)} \tag{9.65}$$

We demand the expression for b_0, in Eq. (9.62) and in Eq. (9.65) to be equal. This results in the following polynomial equation:

$$\sum_{j=0}^{9} R_j v_r^j = 0 \tag{9.66}$$

where

$$R_0 = -393920k^8 + 31167360k^6 - 1023206400k^4$$
$$+ 14553686016k^2 - 75246796800 \tag{9.67}$$

$$R_1 = 5765k^{12} - 856040k^{10} + 58795440k^8 - 1855455360k^6$$
$$+ 26477328384k^4 - 13942554620k^2 \tag{9.68}$$

$$R_2 = 4300k^{14} - 644050k^{12} + 42939440k^{10} - 1365596320k^8$$
$$+ 20040795648k^6 - 109161271296k^4 \tag{9.69}$$

$$R_3 = 1390k^{16} - 227400k^{14} + 15828175k^{12} - 529797400k^{10}$$
$$+ 8218352640k^8 - 47164667904k^6 \tag{9.70}$$

$$R_4 = 200k^{18} - 40100k^{16} + 3160400k^{14} - 117354050k^{12}$$
$$+ 1992532896k^{10} - 12332498688k^8 \tag{9.71}$$

$$R_5 = 10k^{20} - 3310k^{18} + 337685k^{16} - 15019770k^{14}$$
$$- 292908120k^{12} - 2015921280k^{10} \tag{9.72}$$

$$R_6 = -100k^{20} + 17740k^{18} - 1072240k^{16} + 25774302k^{14}$$
$$- 205541136k^{12} \tag{9.73}$$

$$R_7 = 350k^{20} - 38200k^{18} + 1281213k^{16} - 12579120k^{14} \qquad (9.74)$$

$$R_8 = -500k^{20} + 31086k^{18} - 417732k^{16} \qquad (9.75)$$

$$R_9 = 240k^{20} - 5688k^{18} \qquad (9.76)$$

Numerical evaluation of the roots of Eq. (9.66) reveals the following: For $k = 1$, $k = 2$ or $k = 3$, Eq. (9.66) does not possess real positive roots for v_r. For $k = 4$, the positive roots of Eq. (9.66) are $v_r = 1/32$ and $v_r = 1/4$; however, the flexural rigidities associated with either of the value turns out to be negative. Thus, for $k = 4$ no physically realizable solution is found. For $k = 5$, Eq. (9.66) yields $v_r = 0.086627$ (Fig. 9.9) and $v_r = 0.199999$, the latter value not yielding the positive-valued flexural rigidity. The value $k = 6$ corresponds to $v_r = 1/6$ and $v_r = 0.362716$ (Fig. 9.10), the former value not corresponding to physically realizable flexural rigidity. Equation (9.66) has two positive roots $v_r = 0.222404$ and $v_r = 0.482459$ for $k = 7$ (Fig. 9.11). Likewise, two roots $v_r = 0.262989$ and $v_r = 0.578958$ correspond to $k = 8$ (Fig. 9.12). The value $k = 9$ (Fig. 9.13) is associated with roots $v_r = 0.295275$, and $v_r = 0.656521$, whereas $k = 10$ (Fig. 9.14) is associated with $v_r = 0.323355$ and $v_r = 0.717262$.

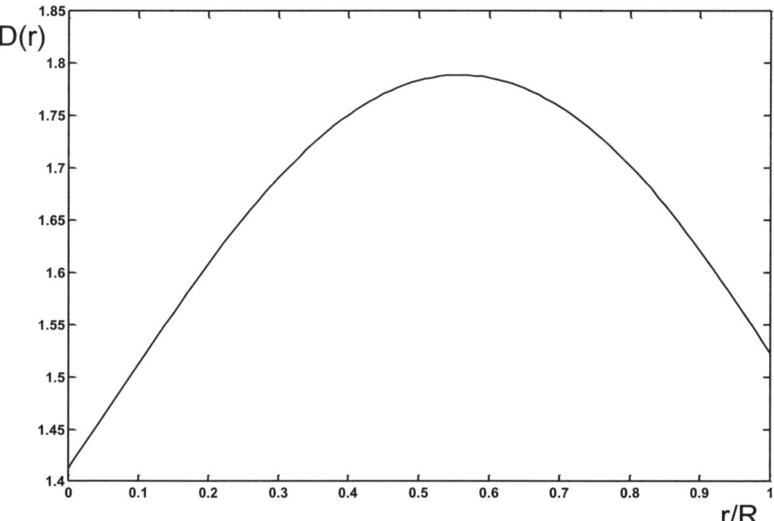

Figure 9.9. Variation of $D(r)$ versus non-dimensional radial coordinate r/R for $k = 5$ and $v_r = 0.086627$

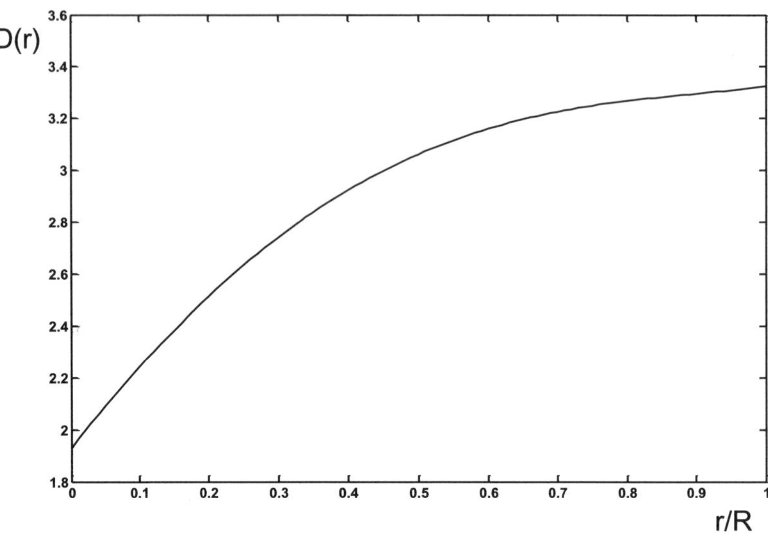

Figure 9.10. Variation of $D(r)$ versus non-dimensional radial coordinate r/R for $k = 6$ and $v_r = 0.362716$

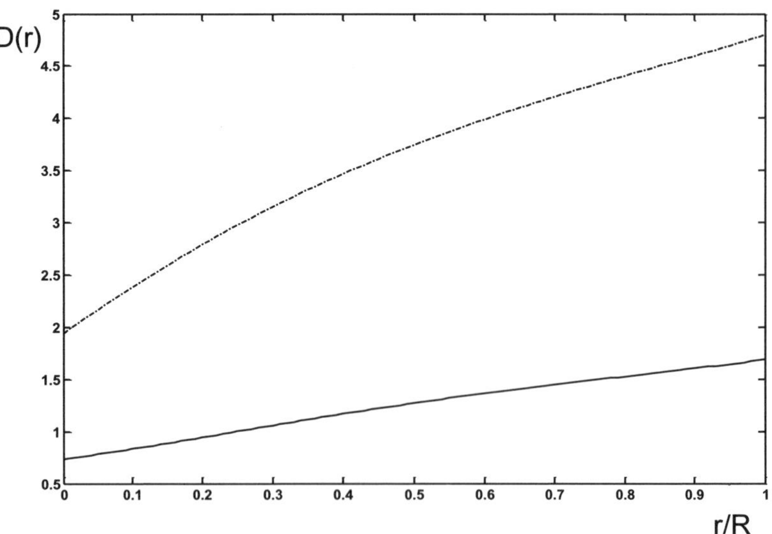

Figure 9.11. Variation of $D(r)$ versus non-dimensional radial coordinate r/R for $k = 7(-v_r = 0.222404; -\cdot-v_r = 482459)$

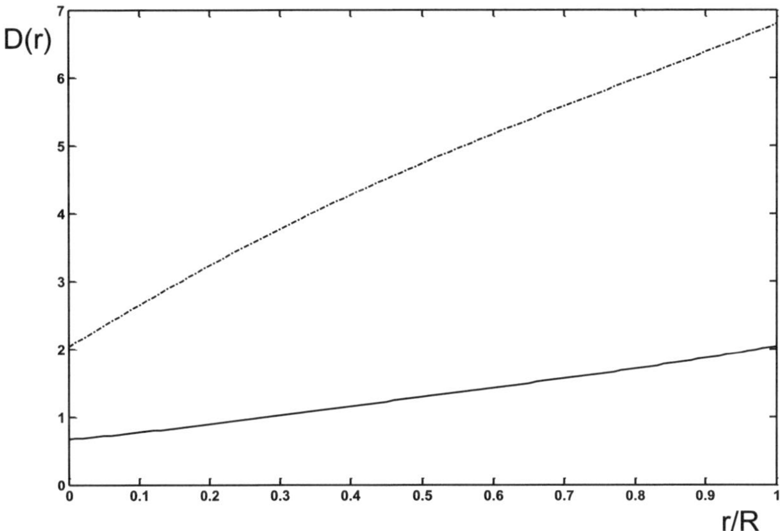

Figure 9.12. Variation of $D(r)$ versus non-dimensional radial coordinate r/R for $k = 8(-\nu_r = 0.262989; -\cdot-\nu_r = 0.578958)$

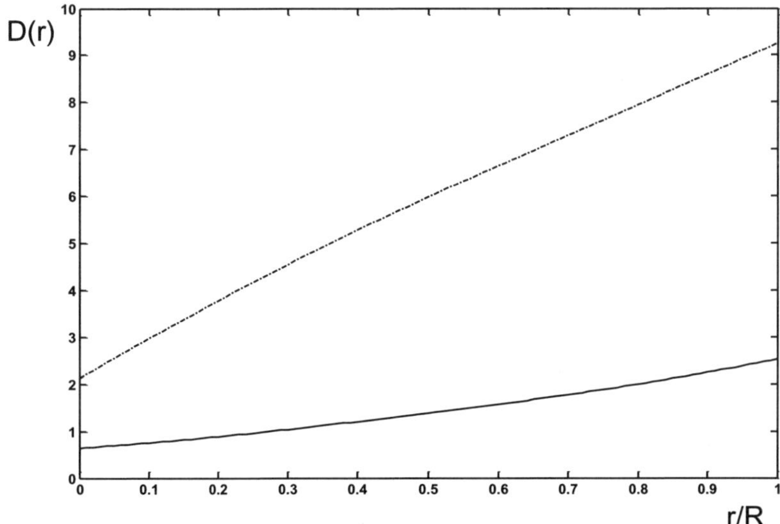

Figure 9.13. Variation of $D(r)$ versus non-dimensional radial coordinate r/R for $k = 9(-\nu_r = 0.295275; -\cdot-\nu_r = 0.656521)$

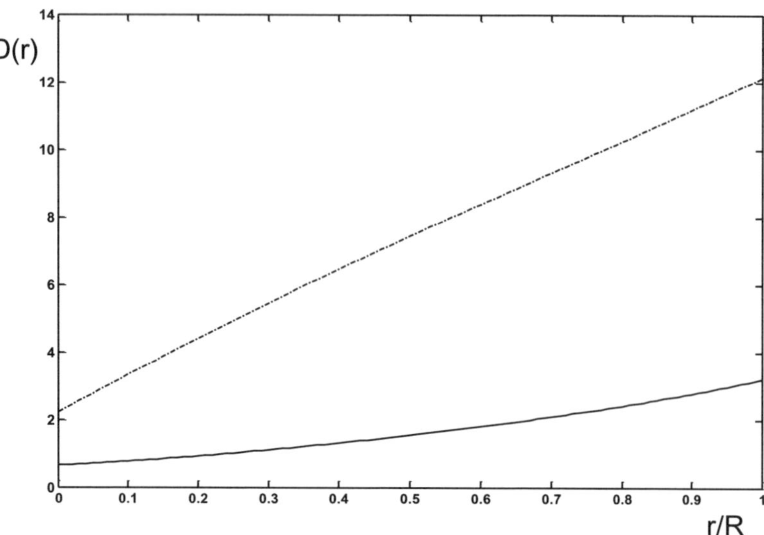

Figure 9.14. Variation of $D(r)$ versus non-dimensional radial coordinate r/R for $k = 10(-v_r = 0.323355; - \cdot -v_r = 0.717262)$

9.3 Vibration tailoring: Numerical example

Equations (9.15), (9.28), and (9.54) all are in analogous form

$$\omega^2 = \varphi(v_r, v_\theta, k)b_4/\rho h \tag{9.77}$$

where the form of the function ϕ depends on the postulated mode shape or, more specifically, for various values of m we have

$$\varphi(v_r, v_\theta, k) = \begin{cases} 24(21 + 7v_\theta - 3v_r k^2 - 21k^2), & \text{for } m = 1 \\ 24(21 + 7v_\theta - 3v_r k^2 - 21k^2), & \text{for } m = 2 \\ 35(32 + 8v_\theta - 3v_r k^2 - 21k^2), & \text{for } m = 3 \end{cases} \tag{9.78}$$

The coefficient b_4 was up to now treated as an arbitrary coefficient. We now demand that the circular plate's fundamental frequency must be equal pre-specified value Ω, i.e.

$$\omega = \Omega \tag{9.79}$$

This demand allows specification of the coefficient b_4:

$$b_4 = \Omega^2 \rho h/\varphi(v_r, v_\theta, k) \tag{9.80}$$

Once coefficient b_4 is determined, the associated flexural rigidities $D_r(r)$ and $D_\theta(r)$ can be evaluated. Let for example, $R = 0.2\,\text{m}$, $\rho = 100\,\text{kg/m}^3$, $h = 0.06\,\text{m}$ and the demanded natural frequency value $\Omega = 90\,\text{Hz}$. The tailoring can be accomplished by choosing $m = 1(k = 1, v_r = 0.35)$, leading to $b_4 = 1446.428571$. The appropriate flexural rigidity equals

$$D_r(r) = 5.6783 - 1.6257r - 1.681r^2 - 1.7248r^3 + r^4 \qquad (9.81)$$

Chapter 10

Vibration Tailoring of Clamped–Clamped Polar Orthotropic Inhomogeneous Circular Plates

In this case, the same method of analysis is used for the semi-inverse problem of free vibration of clamped–clamped inhomogeneous polar orthotropic plates, as in Chapter 9 for the semi–inverse problem of free vibration of simply-supported inhomogeneous polar orthotropic plates. Several closed-form solutions are found by using an exact mode shape derived by Lekhnitskii (1968) for the clamped edge of uniform polar orthotropic circular plate under uniform loading, as shown in Fig. 10.1.

10.1 Analysis

Consider a clamped–clamped orthotropic plate under statically applied uniformly distributed load of intensity q_0. The deflection is given by Lecknitskii's classic formula

$$w(r) = \frac{q_0 a^4}{8(9 - k^2)(1 + k) D_r} \left[3 - k - 4 \left(\frac{r}{R} \right)^{k+1} + (1 + k) \left(\frac{r}{R} \right)^4 \right]$$

$$(10.1)$$

In this design of a circular plate, we demand the fundamental mode shape, $W(r)$ in the governing differential equation (8.9) to coincide with the expression in brackets in Eq. (10.1), that is

$$W(r) = 3 - k - 4 \left(\frac{r}{R} \right)^{k+1} + (1 + k) \left(\frac{r}{R} \right)^4 \qquad (10.2)$$

177

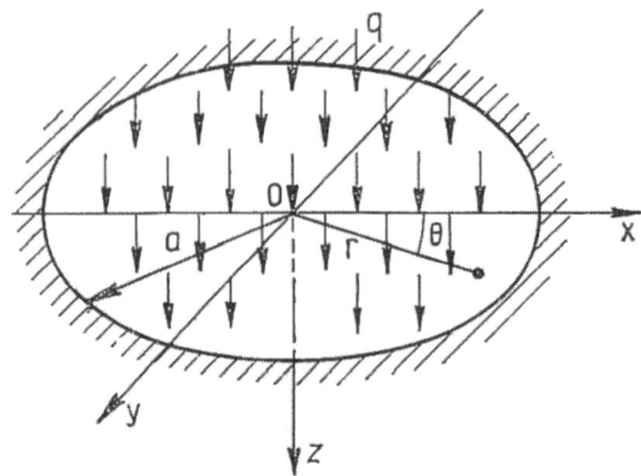

Figure 10.1.　Uniform polar orthotropic plate under uniform loading

Similarly, as in Chapter 9, we denote the value k in Eq. (10.2) as some equal integer m

$$W(r) = 3 - m - 4\left(\frac{r}{R}\right)^{m+1} + (1+m)\left(\frac{r}{R}\right)^{4} \qquad (10.3)$$

and examine if there exists a non-negative-valued variations of $D_r(r)$ and $D_\theta(r)$ that taken together with mode shape satisfy the governing differential equation.

It should be noted that the proposed method constitutes the *semi-inverse* method. This is due to the fact that the fundamental mode shape, that ordinarily is obtained as a result of the direct vibration analysis, is being assumed here to be given. One may ask why should one postulate that the mode shape equals the static displacement. The reply of this question is this: Any polynomial function that satisfies the boundary conditions can be postulated as a *candidate* mode shape; it can be anticipated that some of the candidate mode shapes will result in the physically realizable variations of the flexural rigidities.

10.1.1　Semi-inverse method of solution associated with $m = 1$

For $m = 1$, the mode shape in Eq. (10.3) reads

$$W(r) = 2 - 4\left(\frac{r}{R}\right)^{2} + 2\left(\frac{r}{R}\right)^{4} \qquad (10.4)$$

The substitution of Eqs. (9.5), (9.6) and (10.4) into Eq. (9.8) leads to the following polynomial equation

$$S_0 + S_1 r + S_2 r^2 + S_3 r^3 + S_4 r^4 + S_5 r^5 = 0 \qquad (10.5)$$

where

$$S_0 = -4k^2 R^2 b_1 + 8R^2 b_1 + 8v_\theta R^2 b_1 - 4v_r R^2 k^2 b_1 \qquad (10.6)$$

$$S_1 = 24R^2 b_2 - 24v_\theta b_0 + 8k^2 b_0 - 72b_0 - 8k^2 b_0 + 24v_\theta R^2 b_2$$
$$- 8k^2 v_r R^2 b_2 + 24k^2 v_r b_0 - 8k^2 R^2 b_2 + \rho h \omega^2 R^4 \qquad (10.7)$$

$$S_2 = 12k^2 b_1 + 48v_\theta R^2 b_3 - 12k^2 v_r R^2 b_3 + 36k^2 v_r b_1 - 144b_1$$
$$+ 48R^2 b_3 - 12k^2 R^2 b_3 - 48v_\theta b_1 \qquad (10.8)$$

$$S_3 = 16k^2 b_2 + 48k^2 v_r b_2 - 16k^2 R^2 b_4 + 80v_\theta R^2 b_4 + 80R^2 b_4$$
$$- 80v_\theta b_2 - 16v_r k^2 R^2 b_4 - 240b_2 - 2\rho h \omega^2 R^2 \qquad (10.9)$$

$$S_4 = 20k^2 b_3 + 60k^2 v_r b_3 - 120v_\theta b_3 + 360b_3 \qquad (10.10)$$

$$S_5 = -504b_4 - 168v_\theta b_4 + 24k^2 b_4 - 72k^2 v_r b_4 + \rho h \omega^2 \qquad (10.11)$$

From Eq. (10.11), we get the relationship between the natural frequency squared ω^2 and the coefficient b_4:

$$\omega^2 = 24(21 + 7v_\theta - 3v_r k^2 - 21k^2)b_4/\rho h \qquad (10.12)$$

Equation (10.10) shows that b_3 vanishes identically. The equation resulting from substitution of Eq. (10.12) into Eq. (10.9), results in the formula for b_2, as related to b_4

$$b_2 = -2\frac{(k^2 + 4k^2 v_r - 8v_\theta - 29)}{(-5v_\theta + k^2 + 3k^2 v_r - 15)} R^2 b_4 \qquad (10.13)$$

Equation (10.8) leads to the conclusion that $b_1 = 0$. The equation resulting for substitution of Eqs. (10.12) and (10.13) into Eq. (10.7) leads to the expression for b_0:

$$b_0 = -\frac{b_4}{(k^2 - 5v_\theta + 3k^2 v_r - 15)(k^2 - 3v_\theta + 3k^2 v_r - 9)}$$
$$\times (-242k^2 v_r - 44k^2 + 408v_\theta - 68k^2 v_\theta v_r - 14k^2 + 57v_\theta^2$$
$$+ 8k^4 v_r + 19k^4 v_r + k^4 + 771) \qquad (10.14)$$

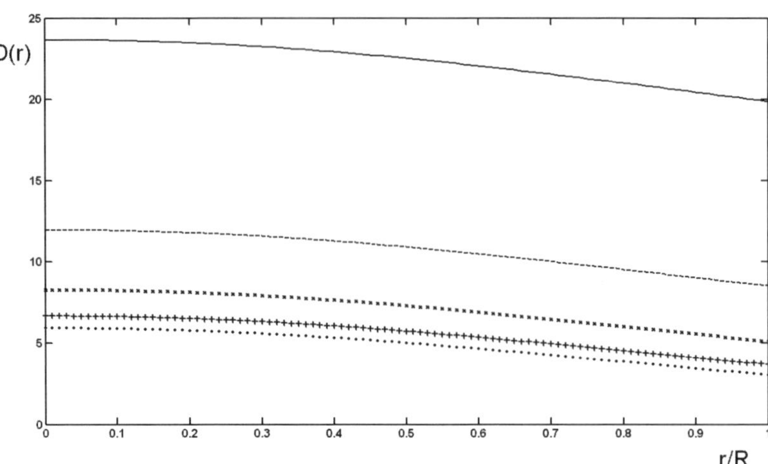

Figure 10.2. Variation of $D(r)$ versus non-dimensional radial coordinate r/R for various values of k, and $v_r = 0.35(-k = 2.5; -- k = 2; \times \times k = 1.5; ++ k = 1; \ldots k = 0.5)$ (here and hereinafter $D_r(r)$ is denoted as $D(r)$)

In the view of the relationship $v_\theta = k^2 v_r$, we get the following final expression for the flexural rigidity:

$$
\begin{aligned}
D_r(r) = {} & \frac{b_4}{(2k^2 v_r - k^2 + 15)(k^2 - 9)} \times [(-18k^2 v_r + 2k^4 v_r + 24k^2 \\
& - k^4 - 135)r^4 + (72k^2 v_r + 522 + 2k^4 - 8k^4 - 8k^4 v_r \\
& - 76k^2)R^2 r^2 + (-8k^4 v_r^2 - k^4 - 771 v_r^4 + 44k^2 + 6k^4 v_r \\
& - 166k^2 v_r)R^4 r^4]
\end{aligned}
\tag{10.15}
$$

Figure 10.2 depicts the variation of $D_r(r)/b_4 R^4$ as a function of r, for various values of k, for v_r fixed at 0.35.

10.1.2 Semi-inverse method of solution associated with $m = 2$

For $m = 2$, the mode shape in Eq. (10.3) reads

$$
W(r) = 1 - 4\left(\frac{r}{R}\right)^3 + 3\left(\frac{r}{R}\right)^4
\tag{10.16}
$$

The result of substitution of Eq. (10.16) in coincidence of Eqs. (9.7), and (9.8) yields the equation

$$T_0 + T_1 r + T_2 r^2 + T_3 r^3 + T_4 r^4 + T_5 r^5 = 0 \qquad (10.17)$$

where

$$T_0 = 24 v_\theta R b_0 - 24 k^2 v_r R b_0 + 48 R b_0 - 12 k^2 R b_0 \qquad (10.18)$$

$$T_1 = 72 k^2 v_r b_0 - 216 b_0 + 24 k^2 b_0 - 24 k^2 R b_1 - 48 k^2 R v_r b_1$$
$$+ 72 v_\theta R b_1 - 72 v_\theta b_0 + 144 R b_1 + \rho h \omega^2 R^4 \qquad (10.19)$$

$$T_2 = -432 b_1 - 144 v_\theta b_1 + 108 v_r k^2 b_1 + 144 v_\theta R b_2 - 72 R k^2 v_r b_2$$
$$+ 288 b_2 R + 36 k^2 b_1 - 36 k^2 R b_2 \qquad (10.20)$$

$$T_3 = 144 k^2 v_r b_2 - 720 b_2 - 240 v_\theta b_2 + 240 v_\theta R b_3 - 48 k^2 R b_3$$
$$- 96 k^2 R v_r b_3 + 480 R b_3 + 48 k^2 b_2 \qquad (10.21)$$

$$T_4 = 180 k v_r b_3 - 360 v_\theta b_3 + 360 R v_\theta b_4 - 1080 b_3 + 720 R b_4$$
$$- 120 k^2 R v_r b_4 + 60 k^2 b_3 - 60 R k^2 b_4 - 4 \rho h \omega^2 R \qquad (10.22)$$

$$T_5 = 72 k^2 b_4 + 216 k^2 v_r b_4 - 504 v_\theta b_4 - 1512 b_4 + 3 \rho h \omega^2 \qquad (10.23)$$

From Eq. (10.23), we get the expression for natural frequency squared ω^2:

$$\omega^2 = 24(21 + 7 v_\theta - 3 v_r k^2 - 21 k^2) b_4 / \rho h \qquad (10.24)$$

It is significant, that the expression (10.24) coincides with Eq. (10.12) for the squared natural frequency. Also, the expressions (10.24) and (10.12) match the expressions (9.15) and (9.28) for the simply supported case. Substitution of Eq. (10.24) into Eq. (10.22) yields

$$b_3 = -\frac{1}{5} \frac{(3k^2 + 14 k^2 v_r - 26 v_\theta - 21)}{(-6 v_\theta + k^2 + 3 k^2 v_r - 18)} R b_4 \qquad (10.25)$$

Substituting Eq. (10.25) into Eq. (10.21) we get

$$b_2 = -\frac{b_4 R^2}{5(k^2 - 6 v_\theta + 3 k^2 v_r - 18)(k^2 - 5 v_\theta + 3 k^2 v_r - 15)}$$
$$\times (-26 v_\theta + 3 k^2 - 107 + 14 k^2 v_r)(-5 v_\theta + 2 k^2 v_r - 10 + k^2)$$

$$(10.26)$$

Equation (10.20) leads to the coefficient b_1:

$$b_1 = -\frac{b_4 R^3}{5(3k^2 v_r - 5v_\theta - 15 + k^2)(-6v_\theta + k^2 + 3k^2 v_r - 18)}$$
$$(-5v_\theta + k^2 + 3k^2 v_r - 15)$$

$$\times (3k^2 - 108 + 14k^2 v_r - 26v_\theta)(-10 + k^2 - 5v_\theta + 2k^2 v_r)$$

$$\times (-4v_\theta + k^2 - 8 + 2k^2 v_r) \tag{10.27}$$

From Eq. (10.18), we conclude that $b_0 = 0$. Eq. (10.19) yields another expression for b_1:

$$b_1 = -\frac{b_4(-7v_\theta + k^2 + 3k^2 v_r - 21)(6 + v_\theta)R^3}{-6 + 2k^2 v_r + k^2 - 3v_\theta} \tag{10.28}$$

In order to resolve the contradiction of Eqs. (10.27) and (10.28), we order these two equations to be equal. This results in the following polynomial equation:

$$\sum_{j=0}^{4} U_j v_r^j = 0 \tag{10.29}$$

where

$$U_0 = 2k^8 - 150k^6 + 4899k^4 - 64386k^2 + 288360 \tag{10.30}$$

$$U_1 = -20k^8 + 1188k^6 - 24576k^4 + 152298k^2 \tag{10.31}$$

$$U_2 = 70k^8 - 2868k^6 + 28137k^4 \tag{10.32}$$

$$U_3 = -100k^8 + 2082k^6 \tag{10.33}$$

$$U_4 = 48k^8 \tag{10.34}$$

Numerical evaluation of the roots of Eq. (10.29) reveals the following: For $k = 1$, $k = 2$ or $k = 3$ it does not possess real positive roots for v_r. For $k = 4$, the only positive root of Eq. (10.29) is $v_r = 0.1080$. Figure 10.3 represents the appropriate variation of the radial flexural rigidity $D_r(r)$. For $k = 5$, Eq. (10.29) yields a positive root, $v_r = 0.4431$ (Fig. 10.4); $k = 6$ corresponds to two positive roots $v_r = 0.2574$ and $v_r = 0.5628$ (Fig. 10.5); Eq. (10.29) has two positive roots $v_r = 0.2964$ and $v_r = 0.6498$ for $k = 7$. Figure 10.6 shows the two resulting curves for $D_r(r)$. Similarly, two roots

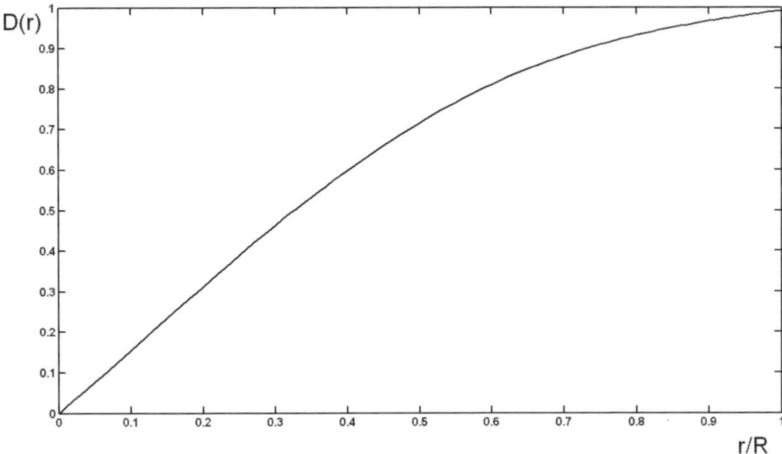

Figure 10.3. Variation of $D(r)$ versus non-dimensional radial coordinate r/R for $k = 4$ and $v_r = 0.1080$

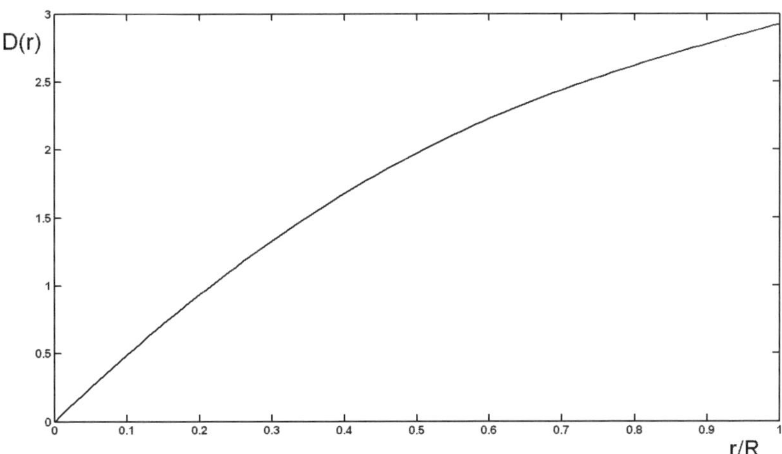

Figure 10.4. Variation of $D(r)$ versus non-dimensional radial coordinate r/R for $k = 5$ and $v_r = 0.4431$

$v_r = 0.3262$ and $v_r = 0.7162$ match to $k = 8$ (Fig. 10.7). In addition, the value $k = 9$ is represented with roots $v_r = 0.3512$ and $v_r = 0.7673$ (Fig. 10.8), whereas $k = 10$ corresponds to $v_r = 0.3732$ and $v_r = 0.8068$ (Fig. 10.9).

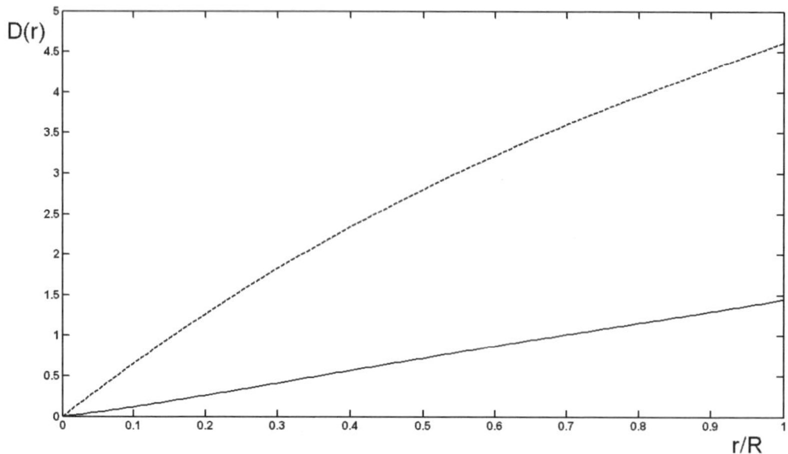

Figure 10.5. Variation of $D(r)$ versus non-dimensional radial coordinate r/R for $k = 6$ (—$\nu_r = 0.2574$; - - -$\nu_r = 0.5628$)

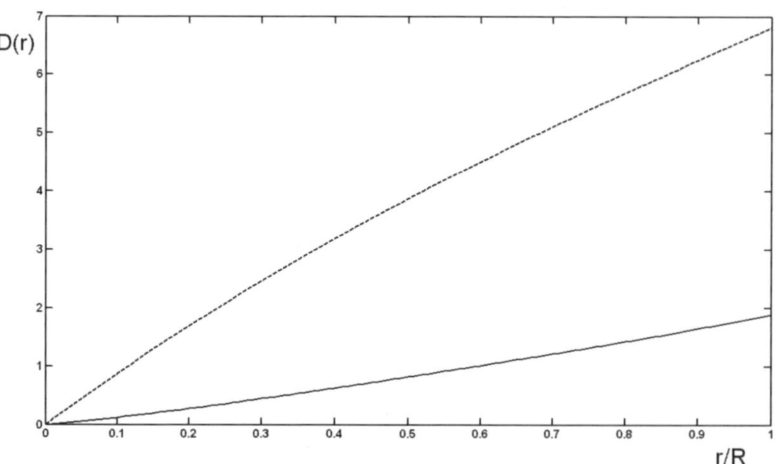

Figure 10.6. Variation of $D(r)$ versus non-dimensional radial coordinate r/R for $k = 7$ (—$\nu_r = 0.2964$; - - -$\nu_r = 0.6498$)

10.1.3   Semi-inverse method of solution associated with $m = 3$

Likewise, as in Chapter 9, when m tends to three the denominator of Eq. (10.3) in front of the square parenthesis approaches zero. The first term

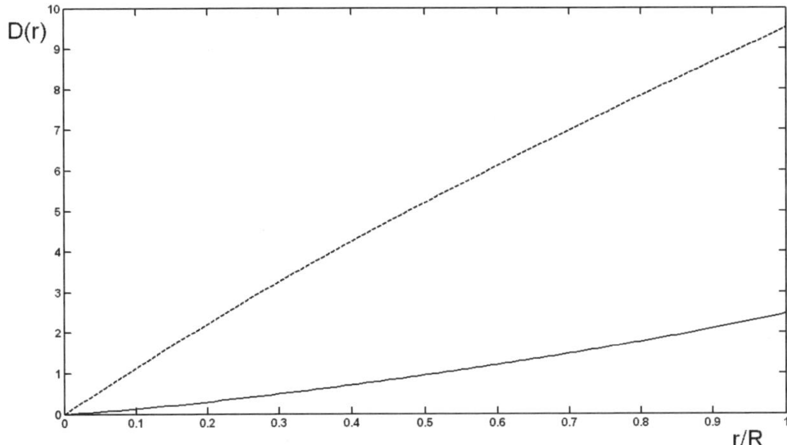

Figure 10.7. Variation of $D(r)$ versus non-dimensional radial coordinate r/R for $k = 8$ ($—v_r = 0.3262; ---v_r = 0.7162$)

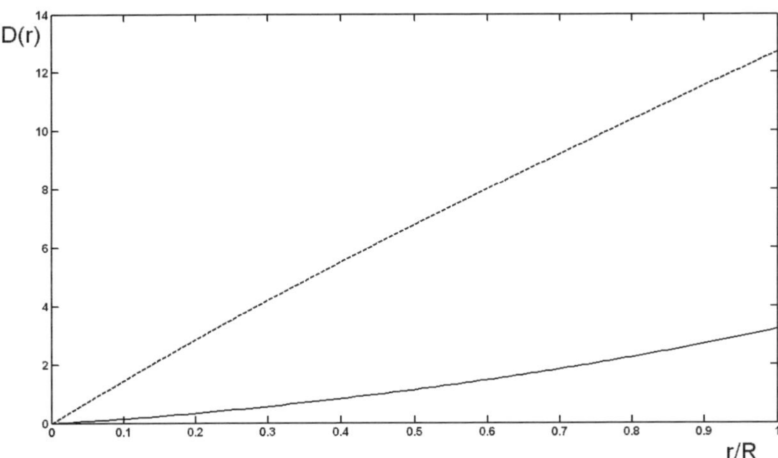

Figure 10.8. Variation of $D(r)$ versus non-dimensional radial coordinate r/R for $k = 9$ ($—v_r = 0.3512; ---v_r = 0.7673$)

in the square parentheses tends to zero, but the product

$$\lim_{m \to 3} \frac{1}{(9 - m^2)} \frac{(3 - m)}{(1 + m)} = \frac{1}{24} \tag{10.35}$$

turns out to approach a value equals to $1/24$.

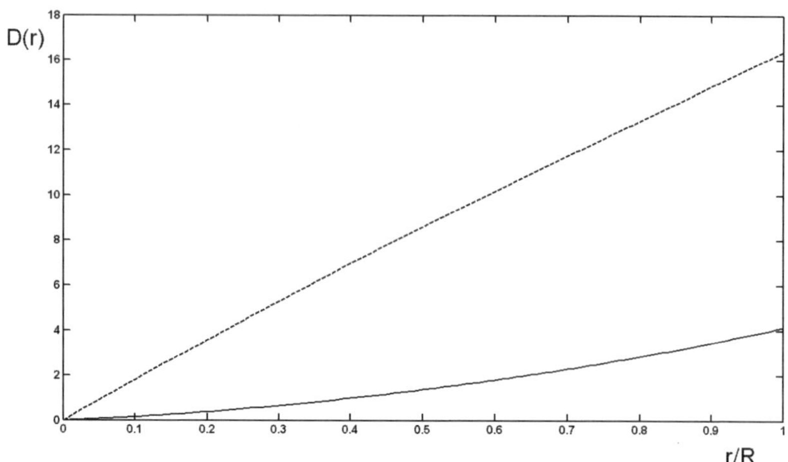

Figure 10.9. Variation of $D(r)$ versus non-dimensional radial coordinate r/R for $k = 10$ ($—\nu_r = 0.3732$; $- - -\nu_r = 0.8068$)

Also, the sum of the second and the third terms in the square parenthesis in Eq. (10.3) also tends to zero. By applying the L'Hopital's Rule, we found that

$$\lim_{m \to 3} \frac{1}{(9 - m^2)(1 + m)} \left[-4 \left(\frac{r}{R} \right)^{m+1} + (1 + m) \left(\frac{r}{R} \right)^4 \right]$$

$$= -\frac{1}{24} r^4 [1 - \log(r)] \tag{10.36}$$

Thus, the candidate mode shape for $m = 3$ becomes:

$$W(r) = \frac{1}{24}[1 - r^4[1 - \log(r)]] \tag{10.37}$$

It is obvious that the mode shape (10.37) does not form a polynomial function. As we mentioned in Chapter 9, in this study we are dealing only with polynomial mode shapes. Consequently, the reconstruction of flexural rigidities for the case $m = 3$ will not be pursued.

10.1.4 Semi-inverse method of solution associated with $m = 4$

For $m = 4$, the mode shape in Eq. (10.3) becomes

$$W(r) = -1 + 5 \left(\frac{r}{R} \right)^4 - 4 \left(\frac{r}{R} \right)^5 \tag{10.38}$$

The result of substitution of Eq. (10.38) in coincidence of Eqs. (9.6), and (9.7) yields the equation

$$V_0 + V_1 r + V_2 r^2 + V_3 r^3 + V_4 r^4 + V_5 r^5 + V_6 r^6 = 0 \tag{10.39}$$

where

$$V_0 = 0 \tag{10.40}$$

$$V_1 = \rho h \omega^2 R^5 + 360 b_0 R - 120 k^2 R v_r + 360 b_0 R$$
$$+ 120 b_0 v_\theta R - 40 k^2 b_0 R \tag{10.41}$$

$$V_2 = -180 k^2 b_1 v_r R + 720 b_1 R + 240 k^2 b_0 v_r - 60 k^2 b_0 v_r$$
$$- 240 b_0 v_\theta - 960 b_0 + 240 R v_\theta b_1 + 60 k^2 b_0 \tag{10.42}$$

$$V_3 = -80 k^2 b_2 R - 400 b_1 v_\theta + 320 k^2 b_1 v_r - 240 k^2 b_2 v_r R$$
$$+ 400 b_2 v_\theta R + 1200 R b_2 - 1600 b_1 - 80 k^2 b_1 \tag{10.43}$$

$$V_4 = 400 k^2 b_2 v_r + 600 b_3 v_\theta R + 100 k^2 b_2 - 2400 b_2 - 600 b_2 v_\theta$$
$$- 300 k^2 b_3 R v_r + 1800 b_3 R - 100 k^2 b_3 R \tag{10.44}$$

$$V_5 = 2520 b_4 R - 120 k^2 b_4 R - 2260 b_3 - 360 k^2 b_4 v_r R - 5 \rho h \omega^2 - 3360 b_3$$
$$- 360 k^2 v_r R b_4 + 480 k^2 b_3 v_r + 840 b_4 v_\theta R + 120 k^2 b_3 - 840 b_3 v_\theta \tag{10.45}$$

$$V_6 = 560 k^2 b_4 v_r - 4480 b_4 - 1120 b_4 v_\theta + 140 k^2 b_4 + 4 \rho h \omega^2 \tag{10.46}$$

From Eq. (10.46), we get the expression for natural frequency squared ω^2:

$$\omega^2 = 35(32 + 8 v_\theta - 4 v_r k^2 - k^2) b_4 / \rho h \tag{10.47}$$

Substitution of Eq. (10.47) into Eq. (10.45) leads to

$$b_3 = -\frac{1}{24} \frac{(11 k^2 + 68 k^2 v_r - 112 v_\theta - 616)}{(-7 v_\theta + k^2 + 4 k^2 v_r - 28)} R b_4 \tag{10.48}$$

Substituting Eq. (10.48) into Eq. (10.44) we get

$$b_2 = -\frac{1}{24}$$

$$\times \frac{(68 k^2 v_r - 616 - 112 v_\theta + 11 k^2)(-18 - 6 v_\theta + 3 k^2 v_r + k^2)}{(-7 v_\theta + k^2 - 28 + 4 k^2 v_r)(4 k^2 v_r - 6 v_\theta - 24 + k^2)} R^2 b_4 \tag{10.49}$$

Equation (10.43) leads to the coefficient b_1:

$$b_1 = W/X \tag{10.50}$$

where

$$W = -(-5\nu_\theta + k^2 - 15 + 3k^2\nu_r)(-6\nu_\theta - 18 + k^2 + 3k^2\nu_r)$$

$$\times (-112\nu_\theta - 616 + 11k^2 + 68k^2\nu_r)R^3 b_4 \tag{10.51}$$

$$X = 24(-7\nu_\theta + k^2 + 4k^2\nu_r + k^2)(-6\nu_\theta + 4k^2\nu_r + k^2 - 24)$$

$$\times (k^2 + 4k^2\nu_r - 5\nu_\theta - 20) \tag{10.52}$$

From Eq. (10.41), we get

$$b_0 = -\frac{7}{8}\frac{(4k^2\nu_r - 32 - 8\nu_\theta + k^2)}{(-3\nu_\theta + 3k^2\nu_r - 9 + k^2)}R^4 b_4 \tag{10.53}$$

Equation (10.42) yields another expression for b_1:

$$b_1 = Y/Z \tag{10.54}$$

where

$$Y = -7(8k^4\nu_r + 512 + 256\nu_\theta + k^4 - 12\nu_\theta k^2 - 48k^2 - 48k^2\nu_r\nu_\theta$$

$$+ 32\nu_\theta^2 - 192k^2\nu_r + 16k^4\nu_r^2) \tag{10.55}$$

$$Z = 8(-3\nu_\theta + k^2 - 9 + 3k^2\nu_r)(3k^2\nu_r - 4\nu_\theta - 12 + k^2) \tag{10.56}$$

We demand the expression for b_1, in Eqs. (10.50) and (10.54) to be equal. This results in the following polynomial equation:

$$\sum_{j=0}^{4} \Phi_j \nu_r^j = 0 \tag{10.57}$$

where

$$\Phi_0 = 10k^{10} - 1310k^8 + 74235k^6 - 2020770k^4$$

$$+ 25946040k^2 - 126544320 \tag{10.58}$$

$$\Phi_1 = -100k^{10} + 10740k^8 - 10740k^8 - 436740k^6$$

$$+ 7439322k^4 - 44645136k^2 \tag{10.59}$$

$$\Phi_2 = 350k^{10} - 28300k^8 + 722133k^6 - 5670240k^4 \tag{10.60}$$

$$\Phi_3 = -500k^{10} - 26286k^8 - 303972k^6 \tag{10.61}$$

$$\Phi_4 = -5688k^8 + 240k^{10} \tag{10.62}$$

Numerical evaluation of the roots of Eq. (10.57) reveals the following: For $k = 1$, $k = 2$ or $k = 3$ Eq. (10.57) does not possess real positive roots for v_r. For $k = 4$, the only positive roots of Eq. (10.57) is $v_r = 1/32$. Figure 10.10 shows the variation of $D(r)$ versus non-dimensional radial coordinate r/R. For $k = 5$, Eq. (10.57) does not have real positive roots. The value $k = 6$ corresponds to a positive root $v_r = 0.3644$ (Fig. 10.11). Equation (10.57) has two positive roots $v_r = 0.2228$ and $v_r = 0.4839$ for $k = 7$ (Fig. 10.12). Also, two roots $v_r = 0.2635$ and $v_r = 0.5799$ correspond to $k = 8$ (Fig. 10.13). The value $k = 9$ (Fig. 10.14) is associated with roots $v_r = 0.2958$ and $v_r = 0.6570$, whereas $k = 10$ (Fig. 10.15) is associated with $v_r = 0.3239$ and $v_r = 0.7175$.

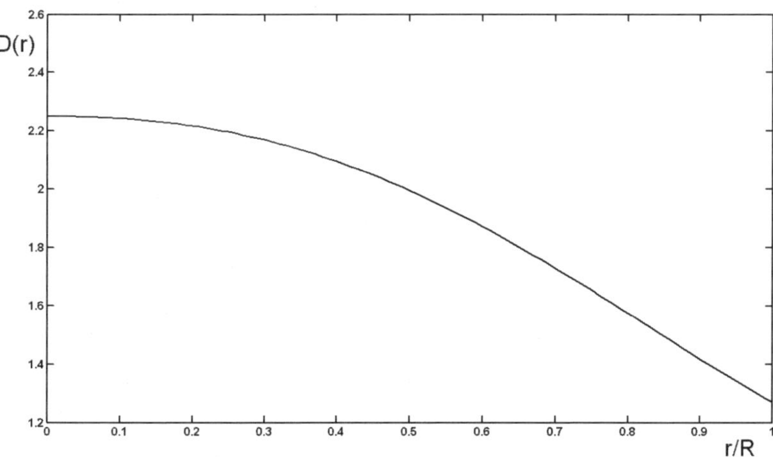

Figure 10.10. Variation of $D(r)$ versus non-dimensional radial coordinate r/R for $k = 4$ and $v_r = 1/32$

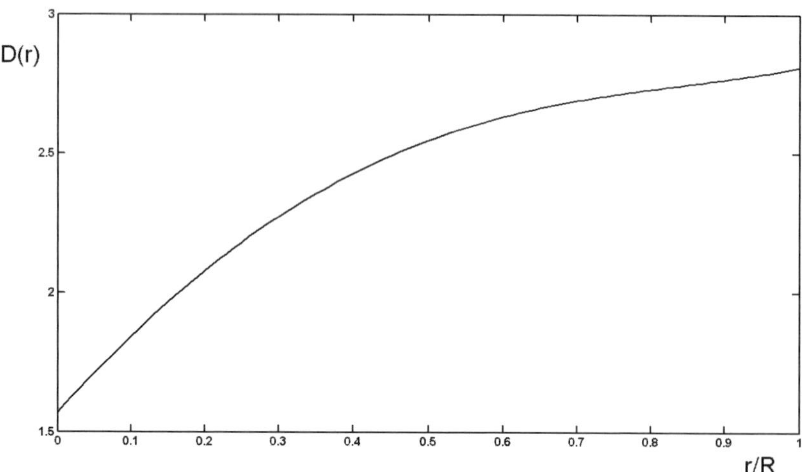

Figure 10.11. Variation of $D(r)$ versus non-dimensional radial coordinate r/R for $k = 6$ and $\nu_r = 0.3644$

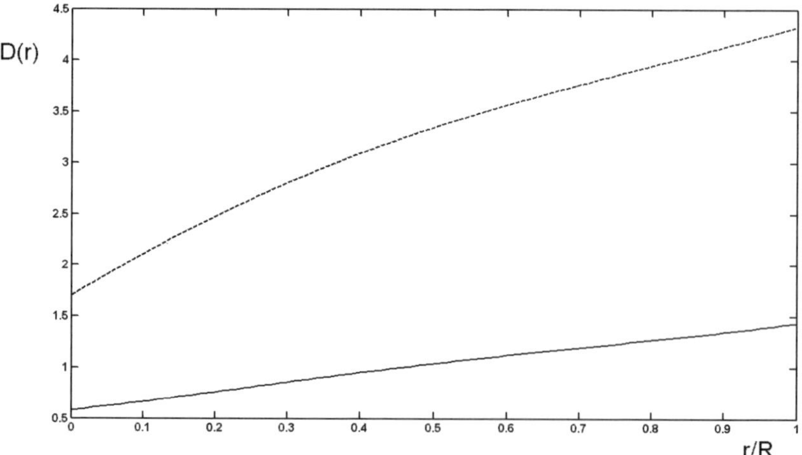

Figure 10.12. Variation of $D(r)$ versus non-dimensional radial coordinate r/R for $k = 7$ ($—\nu_r = 0.2228$; $\text{- - -}\nu_r = 4839$)

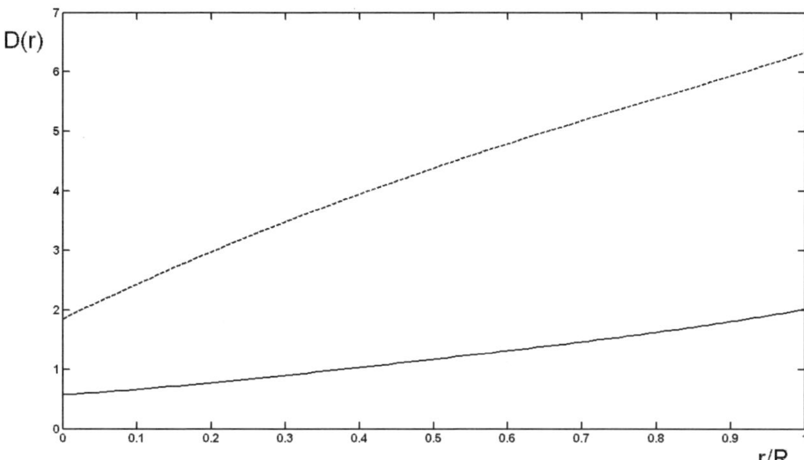

Figure 10.13. Variation of $D(r)$ versus non-dimensional radial coordinate r/R for $k = 8$ ($—\nu_r = 0.2635$; $---\nu_r = 0.5799$)

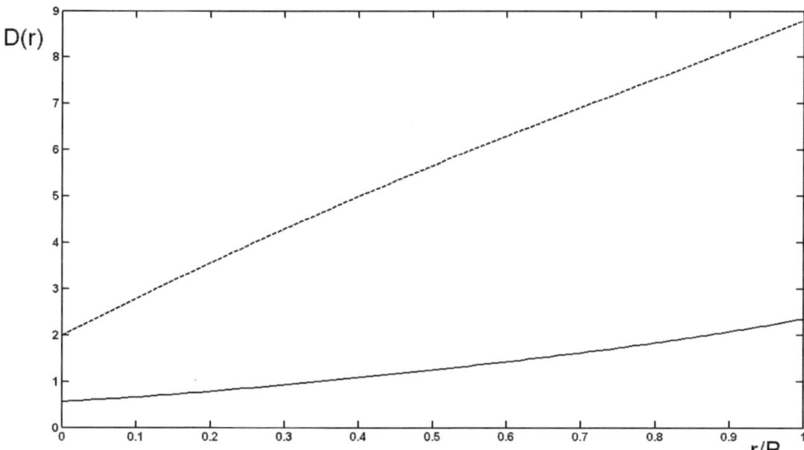

Figure 10.14. Variation of $D(r)$ versus non-dimensional radial coordinate r/R for $k = 9$ ($—\nu_r = 0.2958$; $---\nu_r = 0.6570$)

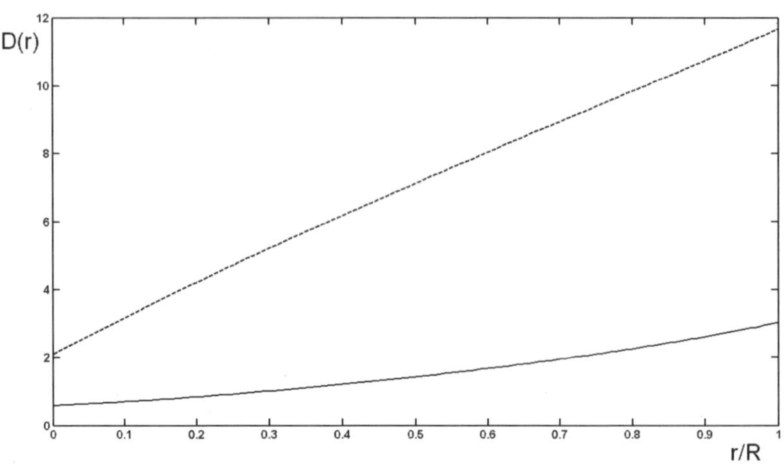

Figure 10.15. Variation of $D(r)$ versus non-dimensional radial coordinate r/R for $k = 10$ (—$v_r = 0.7175$; - - -$v_r = 0.3239$)

10.2 Vibration tailoring: Numerical example

Equations (10.12), (10.24), and (10.47) are in the similar form as shown in Eq. (9.77). For various values of m, we have

$$\phi(v_r, v_\theta, k) = \begin{cases} 24(21 + 7v_\theta - 3v_r k^2 - 21k^2), & \text{for } m = 1 \\ 24(21 + 7v_\theta - 3v_r k^2 - 21k^2), & \text{for } m = 2 \\ 35(32 + 8v_\theta - 3v_r k^2 - 21k^2), & \text{for } m = 3 \end{cases} \quad (10.63)$$

In this case, we need to examine the behavior of this clamped–clamped polar orthotropic inhomogeneous circular plate by specifying the following properties: $R = 0.15\,\text{m}$, $\rho = 100\,\text{kg/m}^3$, $h = 0.08\,\text{m}$ and $\Omega = 95\,\text{Hz}$. Also, by choosing $m = 1$ ($k = 2$, $v_r = 0.35$) leads to $b_4 = -52.40998839$. Then, the flexural rigidity is

$$D_r(r) = 52.41r^4 - 5.1886r^2 + 0.3279 \quad (10.64)$$

However, if we choose $m = 2$ ($k = 4$, $v_r = 0.1080$) we get $b_4 = -9.764526153$. The corresponding flexural rigidity reads

$$D_r(r) = 9.7645r^4 - 3.2921r^3 + 0.1641r^2 + 0.0492r \quad (10.65)$$

As is seen, the vibration tailoring can be accomplished in several ways.

Chapter 11

Vibration Tailoring of a Polar Orthotropic Circular Plate with Translational Spring

In this chapter the results of the paper by Elishakoff and Meyer (2005f) will be generalized for the polar orthotropic circular plate with translational spring. By using the semi-inverse method and postulating the mode shape as a polynomial, we derive closed-form solutions.

11.1 Analysis

The governing differential equation of the polar orthotropic circular plate of varying flexural rigidity is given by Eq. (8.9).

We postulated the following mode shape

$$W(r) = \alpha_0 + \alpha_2 r^2 + r^4 \tag{11.1}$$

where α_0 and α_2 are unknown coefficients. The coefficients α_0 and α_2 will be determined at a later stage.

The flexural rigidities are sought as polynomials of the fourth-order

$$D_r(r) = b_0 + b_1(r - R) + b_2(r - R)^2 + b_3(r - R)^3$$

$$+ b_4(r - R)^4 \tag{11.2}$$

$$D_\theta(r) = k^2 D_r(r) \tag{11.3}$$

where k is taken as constant.

193

11.1.1 Boundary conditions

We study a circular plate with a translational spring that the boundary conditions are

$$M_r(R) = 0 \tag{11.4}$$

and

$$Q_r(R) + k_w W(R) = 0 \tag{11.5}$$

where

$$M_r(r) = -D_r \left(W'' + \frac{v_\theta}{r} W' \right) \tag{11.6}$$

$$Q_r = \frac{M_r}{r} + \frac{dM_r}{dr} - \frac{M_\theta}{r} \tag{11.7}$$

$$M_\theta(r) = -D_\theta \left(v_r W'' + \frac{1}{r} W' \right) \tag{11.8}$$

k_w denotes the stiffness per unit of length of the translational spring.

11.1.2 Method of solution

First of all, we need to calculate the coefficients α_0 and α_2 of the mode shape given in Eqs. (11.1). These coefficients can be found by applying the boundary conditions given in Eqs. (11.4) and (11.5), which read

$$2\alpha_2 + 12R^2 + k^2 v_r (2\alpha_2 + 4R^2) = 0 \tag{11.9}$$

$$24b_0 R + 32b_0 R k^2 v_r + 8b_0 R k^2 + k_w \alpha_0 + k_w \alpha_0 k^2 v_r$$
$$- 5k_w R^4 + k_w R^4 k^2 v_r = 0 \tag{11.10}$$

Solution of Eqs (11.9) and (11.10) give

$$\alpha_0 = -R[24b_0 + 32b_0 k^2 v_r + 8b_0 k^2$$
$$- \frac{2k_w R^3 (3 + k^2 v_r)}{1 + k^2 v_r} - \frac{2k_w R^3 (3 + k^2 v_r) k^2 v_r}{1 + k^2 v_r}$$
$$+ k_w R^3 + k_w R^3 k^2 v_r]/(1 + k^2 v_r) k_w \tag{11.11}$$

$$\alpha_2 = -2 \frac{R^2 (3 + k^2 v_r)}{1 + k^2 v_r} \tag{11.12}$$

Substituting Eqs. (11.11) and (11.12) into Eq. (11.1), we get the equation of mode shape depending upon the coefficient b_0 and the translational spring k_w:

$$W(r) = -\frac{R}{(1+k^2 v_r)k_w}\left[24b_0 + 32b_0 k^2 v_r + 8b_0 k^2 - \frac{2k_w R^3(3+k^2 v_r)}{1+k^2 v_r}\right.$$

$$\left. - \frac{2k_w R^3(3+k^2 v_r)k^2 v_r}{1+k^2 v_r} + k_w R^3 + k_w R^3 k^2 v_r\right]$$

$$- \frac{2R^2(3+k^2 v_r)}{1+k^2 v_r}r^2 + r^4 \tag{11.13}$$

It appears to be instructive to investigate limiting cases, namely when k_w approaches zero or infinity. For the case $k_w = 0$, Eq. (11.10) reduces to

$$24b_0 R + 32b_0 Rk^2 v_r + 8b_0 Rk^2 = 0 \tag{11.14}$$

Solution of Eq. (11.14) yields $b_0 = 0$. The equation for α_2 remains the same as in Eq. (11.12). Thus, the mode shape becomes

$$W(r) = \alpha_0 - \frac{2R^2(3+k^2 v_r)}{1+k^2 v_r}r^2 + r^4 \tag{11.15}$$

For k_w becoming unbounded, Eq. (11.10) becomes

$$\alpha_0 + \alpha_0 k^2 v_r - 5R^4 + R^4 k^2 v_r = 0 \tag{11.16}$$

For α_0, we obtain:

$$\alpha_0 = -R\left[-\frac{2R^3(3+k^2 v_r)}{1+k^2 v_r} - \frac{2R^3(3+k^2 v_r)k^2 v_r}{1+k^2 v_r}\right.$$

$$\left. + R^3 + R^3 k^2 v_r\right]\Big/(1+k^2 v_r) \tag{11.17}$$

For the coefficient α_2, we get the same expression as given in Eq. (11.12). Hence, the mode shape is expressed as

$$W(r) = -\frac{R}{(1+k^2 v_r)}\left[-\frac{2R^3(3+k^2 v_r)}{1+k^2 v_r} - \frac{2R^3(3+k^2 v_r)k^2 v_r}{1+k^2 v_r}\right.$$

$$\left. + R^3 + R^3 k^2 v_r\right] - \frac{2R^2(3+k^2 v_r)}{1+k^2 v_r}r^2 + r^4 \tag{11.18}$$

It must be noted that Eq. (11.18) can be obtained directly from Eq. (11.13) when k_w is set to approach to infinity. The first three terms in parentheses in Eq. (11.13) are cancelled out whereas the remaining three terms in parenthesis coincide with Eq. (11.18).

Semi-inverse method of solution associated with $k = 1$

The substitution of Eqs. (11.2), (11.3) and (11.13) into the governing differential equation, where $k = 1$, leads to the following polynomial equation

$$\Lambda_0 + \Lambda_1 r + \Lambda_2 r^2 + \Lambda_3 r^3 + \Lambda_4 r^4 + \Lambda_5 r^5 = 0 \tag{11.19}$$

where

$$\begin{aligned}
\Lambda_0 = {} & 32b_2 R^3 v_r + 24b_2 R^3 + 8b_2 R^3 v_r^2 + 48b_4 R^5 - 36b_3 R^4 - 12b_1 R^2 \\
& - 4b_1 R^2 v_r^2 - 48b_3 R^4 v_r - 12b_3 R^4 v_r^2 - 16b_1 R^2 v_r + 64b_4 R^5 v_r \\
& + 16b_4 R^5 v_r^2
\end{aligned} \tag{11.20}$$

$$\begin{aligned}
\Lambda_1 = {} & -5\rho h\omega^2 k_w R^4 + 16k_w b_2 R^2 + 80k_w b_3 R^3 - 64k_w b_1 R v_r \\
& + 32\rho h\omega^2 R b_0 + 48k_w b_3 R^3 v_r^2 - 16k_w b_2 R^2 v_r^2 - 224k_w b_4 R^4 \\
& - 320k_w b_4 R^4 v_r - 96k_w b_4 R^4 v_r^2 - 64k_w b_1 R + 64k_w b_0 \\
& + 128k_w b_3 R^3 v_r + 32\rho h\omega^2 R b_0 v_r - \rho h\omega^2 k_w R^4 v_r \\
& + 64k_w b_0 v_r
\end{aligned} \tag{11.21}$$

$$\begin{aligned}
\Lambda_2 = {} & 96b_4 R^3 v_r^2 + 288b_3 R^2 - 288b_2 R v_r - 96b_4 R^3 + 12b_1 v_r^2 \\
& - 264b_2 R + 144b_1 v_r + 288b_3 R^2 v_r - 24b_2 R v_r^2 + 132b_1
\end{aligned} \tag{11.22}$$

$$\begin{aligned}
\Lambda_3 = {} & -768b_3 R v_r + 2\rho h\omega^2 R^2 v_r - 672b_3 k + 224b_2 + 6\rho h\omega^2 R^2 \\
& + 32b_2 v_r^2 + 1152b_4 R^2 - 96b_3 R v_r^2 + 128b_4 R^2 v_r^2 + 256b_2 v_r \\
& + 1280b_4 R^2 v_r
\end{aligned} \tag{11.23}$$

$$\begin{aligned}
\Lambda_4 = {} & -240b_4 R v_r^2 - 1600b_4 R v_r + 340b_3 - 1360b_4 R \\
& + 60b_3 v_r^2 + 400b_3 v_r
\end{aligned} \tag{11.24}$$

$$\Lambda_5 = 480b_4 + 96b_4 v_r^2 + 576b_4 v_r - \rho h\omega^2 - \rho h\omega^2 v_r \tag{11.25}$$

From Eq. (11.25), we get the relationship between the natural frequency squared ω^2 and the coefficient b_4:

$$\omega^2 = 96(v_r + 5)b_4/\rho h \tag{11.26}$$

The equation resulting from substitution of Eq. (11.26) into Eq. (11.24), results in the formula for b_3, as related to b_4

$$b_3 = 4Rb_4 \tag{11.27}$$

Substitution of Eqs. (11.26) and (11.27) into Eq. (11.23) results in the expression for b_2:

$$b_2 = 2\frac{(v_r - 3)R^2}{1 + v_r}b_4 \tag{11.28}$$

Equation (11.22) gives the expression for b_1:

$$b_1 = -4\frac{(v_r + 5)R^3}{1 + v_r}b_4 \tag{11.29}$$

The equation resulting from substitution of Eqs. (11.26)–(11.29) into Eq. (11.24) yields to the expression for b_0:

$$b_0 = \frac{1}{2}\frac{(v_r^2 + 12v_r + 35)k_w R^4 b_4}{(1 + v_r)(48Rb_4v_r + k_w + 240Rb_4)} \tag{11.30}$$

As a result, the expression for the flexural rigidity becomes:

$$
\begin{aligned}
D_r(r) = \Bigg\{ &\frac{1}{2}\frac{(v_r^2 + 12v_r + 35)R^4 k_w}{(1 + v_r)(48Rb_4v_r + k_w + 240Rb_4)} + 4\frac{R^4(v_r + 5)}{1 + v_r} - 3R^4 \\
&+ 2\frac{(v_r - 3)R^4}{1 + v_r} + \left[-4\frac{(v_r - 3)R^3}{1 + v_r} + 8R^3 - 4\frac{R^3(v_r + 5)}{1 + v_r} \right] r \\
&+ \left[2\frac{(v_r - 3)R^2}{1 + v_r} - 6R^2 \right] r^2 + r^4 \Bigg\} b_4
\end{aligned}
\tag{11.31}
$$

In the view of the relationship

$$\beta = k_w/b_4 R \tag{11.32}$$

which denotes a non-dimensional constant, we get the final expression

$$
D_r(r) = \left\{ \frac{1}{2} \frac{(v_r^2 + 12v_r + 35)R^4\beta}{(1+v_r)(48v_r + \beta + 240)} + 4\frac{R^4(v_r + 5)}{1+v_r} - 3R^4 \right.
$$

$$
+ 2\frac{(v_r - 3)R^4}{1+v_r} + \left[-4\frac{(v_r - 3)R^3}{1+v_r} + 8R^3 - 4\frac{R^3(v_r + 5)}{1+v_r} \right] r
$$

$$
\left. + \left[2\frac{(v_r - 3)R^2}{1+v_r} - 6R^2 \right] r^2 + r^4 \right\} b_4 \tag{11.33}
$$

or

$$
\frac{D_r(r)}{R^4 b_4} = \frac{1}{2} \frac{(v_r^2 + 12v_r + 35)\beta}{(1+v_r)(48v_r + \beta + 240)} + 4\frac{(v_r + 5)}{1+v_r} - 3 + 2\frac{(v_r - 3)}{1+v_r}
$$

$$
+ \left[-4\frac{(v_r - 3)}{1+v_r} + 8 - 4\frac{R^3(v_r + 5)}{1+v_r} \right] \frac{r}{R}
$$

$$
+ \left[2\frac{(v_r - 3)}{1+v_r} - 6 \right] \left(\frac{r}{R}\right)^2 + \left(\frac{r}{R}\right)^4 \tag{11.34}
$$

Also, the mode shape in Eq. (11.13) becomes

$$
W(r) = -32\frac{(1/2v_r^2 + 6v_r + 35/2)}{(1+v_r)(48v_r + \beta + 240)} R^4 + 2\frac{(3 + v_r)}{(1+v_r)} R^4
$$

$$
- R^4 - 2\frac{(3 + v_r)}{1+v_r} R^2 r^2 + r^4 \tag{11.35}
$$

Note that Eqs. (11.33) and (11.35) coincide with respective expression derived in the study by Elishakoff and Meyer (2005). Figures 11.1 and 11.2 depict the variation of flexural rigidity $D_r(r)/R^4 b_4$ versus the radial coordinate r/R for different values of β and Poisson's ratio $v_r = 0.3$.

For the case when k_w approaches to zero, we utilize Eq. (11.15). Equation (11.15) and (11.2) are then substituted in the governing differential equation (8.9) and get expressions for b_j. The expression for b_0 is

$$
b_0 = \frac{1}{2} \frac{(-3a_0 v_r^2 + 2v_r^2 + 18v_r - 18v_r - 15a_0 + 40)R^4}{(1+v_r)} \tag{11.36}
$$

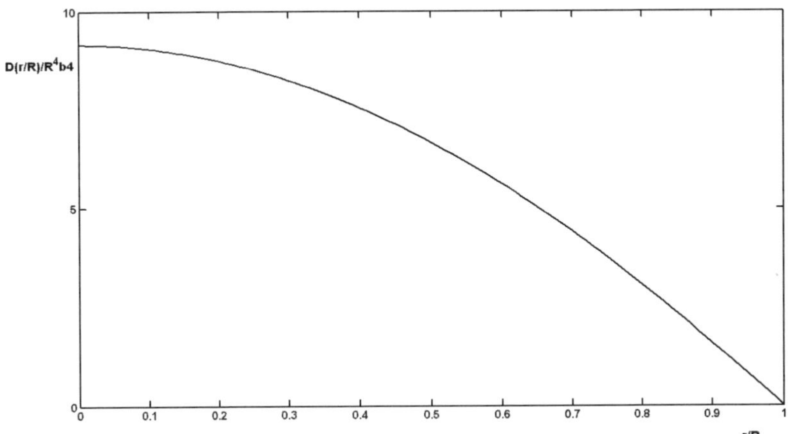

Figure 11.1. Variation of $D(r/R)$ versus non-dimensional radial coordinate r/R for $k = 1$ and Poisson's ratio $v_r = 0.3$ ($\beta = 0$) (hereinafter $D_r(r)$ is denoted as $D(r)$)

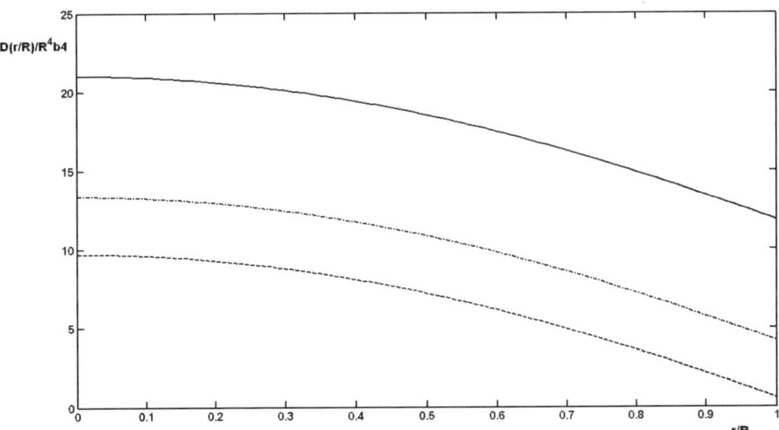

Figure 11.2. Variation of $D(r/R)$ versus non-dimensional radial coordinate r/R for $k = 1$ and Poisson's ratio $v_r = 0.3(- - -\beta = 10; - \cdot -\beta = 100; —\beta = 1000)$

The natural frequency squared ω^2 and the coefficients b_1, b_2, and b_3 remain the same as in Eqs. (11.26)–(11.29). From Eq. (11.14), we get $b_0 = 0$. Substitution of $b_0 = 0$ into Eq. (11.36) and solving for a_0, we get

$$\alpha_0 = \frac{2}{3} \frac{(v_r + 4)R^4}{(1 + v_r)} \tag{11.37}$$

Hence, the mode shape of Eq. (11.15) becomes

$$W(r) = \frac{2}{3}\frac{(v_r - 6)R^4}{(1 + v_r)} - \frac{2R^2(3 + k^2 v_r)}{1 + k^2 v_r}r^2 + r^4 \qquad (11.38)$$

The attendant flexural rigidity reads

$$\frac{D_r(r)}{R^4 b_4} = 4\frac{(v_r + 5)}{1 + v_r} - 3 + 2\frac{(v_r - 3)}{1 + v_r} + \left[-4\frac{(v_r - 3)}{1 + v_r} + 8 \right.$$
$$\left. - 4\frac{R^3(v_r + 5)}{1 + v_r} \right]\frac{r}{R} + \left[2\frac{(v_r - 3)}{1 + v_r} - 6 \right]\left(\frac{r}{R}\right)^2 + \left(\frac{r}{R}\right)^4 \qquad (11.39)$$

Figure 11.3 shows the variation of flexural rigidity $D_r(r)/R^4 b_4$ versus the radial coordinate r/R for $k_w = 0$ and Poisson's ratio $v_r = 0.3$.

Now, when k_w approaches to infinity, a change occurs just in the coefficient b_0:

$$b_0 = \frac{1}{2}\frac{(v_r^2 + 12v_r + 35)}{(1 + v_r)}R^4 b_4 \qquad (11.40)$$

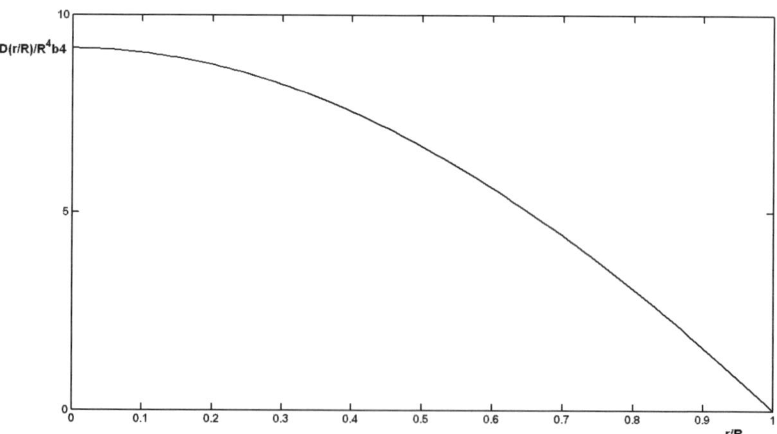

Figure 11.3. Variation of $D(r/R)$ versus non-dimensional radial coordinate r/R for $k = 1$ and Poisson's ratio $v_r = 0.3$ ($k_w = 0$)

Thus, the flexural rigidity becomes

$$\frac{D_r(r)}{R^4 b_4} = \frac{1}{2}\frac{(v_r^2 + 12v_r + 35)}{(1 + v_r)} + 4\frac{(v_r + 5)}{1 + v_r} - 3 + 2\frac{(v_r - 3)}{1 + v_r}$$
$$+ \left[-4\frac{(v_r - 3)}{1 + v_r} + 8 - 4\frac{R^3(v_r + 5)}{1 + v_r} \right]\frac{r}{R}$$
$$+ \left[2\frac{(v_r - 3)}{1 + v_r} - 6 \right]\left(\frac{r}{R}\right)^2 + \left(\frac{r}{R}\right)^4 \tag{11.41}$$

and the mode shape is written as

$$W(r) = -\frac{R}{(1 + v_r)}\left[-\frac{2R^3(3 + v_r)}{1 + k^2 v_r} - \frac{2R^3(3 + v_r)v_r}{1 + v_r} + R^3 + R^3 v_r \right]$$
$$- \frac{2R^2(3 + v_r)}{1 + v_r}r^2 + r^4 \tag{11.42}$$

Figure 11.4 portrays behavior of the flexural rigidity for large values of β. For $\beta = 10,000$, $D(0) = 23.6654$; $\beta = 15,500$, $D(0) = 23.7943$; $\beta = 100,000$, $D(0) = 23.9969$; $\beta = 10^6$, $D(0) = 24.0308$. For $k_w \to \infty$ yields $D(0) = 24.0346$ (Fig. 11.5). As seen, for values of $\beta > 15,500$, the

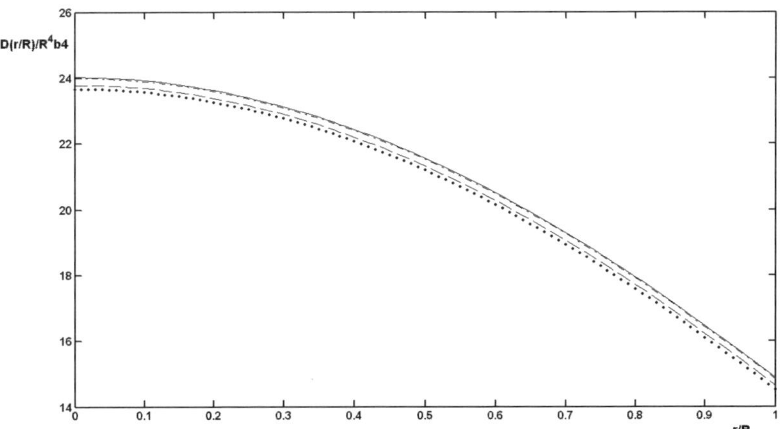

Figure 11.4. Variation of $D_r(r/R)$ versus non-dimensional radial coordinate r/R for $k = 1$ and Poisson's ratio $v_r = 0.3$ ($\cdots\beta = 10,000$; $---\beta = 15,500$; $-\cdot-\beta = 100,000$; $-\beta = 10^6$)

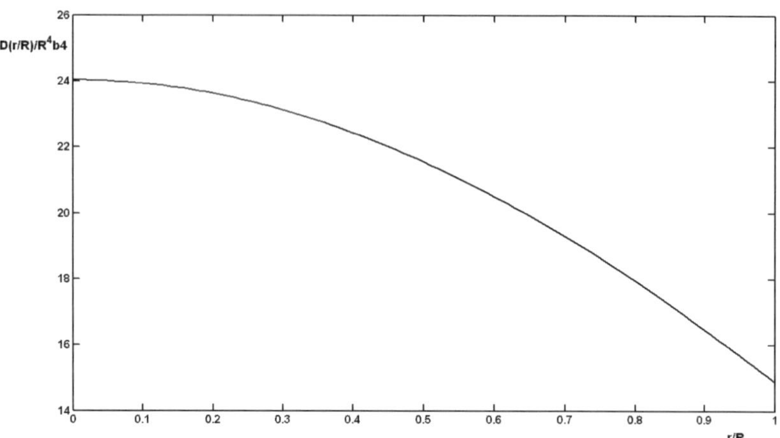

Figure 11.5. Variation of $D(r/R)$ versus non-dimensional radial coordinate r/R for $k = 1$ and Poisson's ratio $v_r = 0.3$ $(k_w \rightarrow \infty)$

value of $D(0)$ differs from 24.0346 by less than 1%. Thus for $\beta > 15,500$ we can consider that k_w equals infinity, implying that the plate is simply supported at the edge.

Semi-inverse method of solution associated with $k = 2$

In this particular case of $k = 2$, the substitution of Eqs. (11.2), (11.3) and (11.13) into the governing differential equation leads to the following fifth-order polynomial equation

$$\Psi_0 + \Psi_1 r + \Psi_2 r^2 + \Psi_3 r^3 + \Psi_4 r^4 + \Psi_5 r^5 = 0 \tag{11.43}$$

where

$$\Psi_0 = -16b_1 R^2 v_r - 96b_4 R^5 - 48b_2 R^3 - 48b_3 R^4 v_r - 192b_3 R^4 v_r^2$$
$$+ 32b_2 R^3 v_r + 24b_1 R^2 + 128b_2 R^3 v_r^2 + 256b_4 R^5 v_r^2 + 72b_3 R^4$$
$$- 64b_1 R^2 v_r^2 + 64b_4 R^5 v_r \tag{11.44}$$

$$\Psi_1 = 40k_w b_0 - 112k_w b_3 R^3 + 160k_w b_0 v_r + 64k_w b_2 R^2 - 800b_4 R^4 v_r$$
$$- 256k_w b_2 R^2 v_r^2 - 160k_w b_1 R v_r + 184k_w b_4 R^4 - 40k_w b_1 R$$

$$+ 320k_w b_3 R^3 v_r + 128\rho h\omega^2 b_0 R v_r + 768k_w b_3 R^3 v_r^2$$
$$- 4\rho h\omega^2 k_w R^4 v_r - 5\rho h\omega^2 R^4 k_w + 56\rho h\omega^2 R b_0$$
$$- 1536k_w b_4 R^4 v_r^2 \tag{11.45}$$

$$\Psi_2 = 288b_3 R^2 - 384b_4 R^3 - 864b_2 R v_r + 432b_1 v_r$$
$$+ 192b_1 v_r^2 + 1536b_4 R^3 v_r^2 + 864b_3 R^2 v_r + 96b_1$$
$$- 384b_2 R v_r^2 - 192b_2 R \tag{11.46}$$

$$\Psi_3 = -1008b_4 R^2 + 8\rho h\omega^2 R^2 v_r + 512b_2 v_r^2 + 2048b_4 R^2 v_r^2$$
$$+ 6\rho h\omega^2 R^2 + 832b_2 v_r + 4160b_4 R^2 v_r - 1536b_3 R v_r^2$$
$$- 2496b_3 R v_r + 176b_2 - 528b_3 R \tag{11.47}$$

$$\Psi_4 = -5440b_4 R v_r - 1120b_4 R + 960b_3 v_r^2 + 1360b_3 v_r$$
$$- 3840b_4 R v_r^2 + 280b_3 \tag{11.48}$$

$$\Psi_5 = 2016b_4 v_r + 1536b_4 v_r^2 + 408b_4 - \rho h\omega^2 - 4\rho h\omega^2 v_r \tag{11.49}$$

Solution of Eq. (11.49) for the natural frequency squared ω^2, give us

$$\omega^2 = 24(16v_r + 17)b_4/\rho h \tag{11.50}$$

From Eq. (11.48), we get the formula for b_3

$$b_3 = 4Rb_4 \tag{11.51}$$

Substitution of Eqs. (11.50) and (11.51) into Eq. (11.47) results in the expression for b_2:

$$b_2 = 4\frac{(16v_r^2 + 4v_r - 21)R^2}{(1 + v_r)(8v_r + 11)}b_4 \tag{11.52}$$

Equation (11.46) gives the expression for b_1:

$$b_1 = -128\frac{(v_r^2 + 3v_r + 2)R^3}{(1 + 4v_r)(8v_r + 11)}b_4 \tag{11.53}$$

The equation resulting from substitution of Eqs. (11.51)–(11.53) into Eq. (11.45) leads to the formula for b_0:

$$b_0 = \frac{256(8v_r^4 + 40v_r^3 + 69v_r^2 + 47v_r + 10)k_w R^4}{(1 + 4v_r)(8v_r + 11)(9216Rb_4v_r + 2856Rb_4 + 5k_w + 20k_w v_r + 6144Rb_4v_r^2)}b_4 \qquad (11.54)$$

The expression for the flexural rigidity becomes:

$$\frac{D_r(r)}{R^4 b_4} = \frac{256(8v_r^4 + 40v_r^3 + 69v_r^2 + 47v_r + 10)\beta}{(1 + 4v_r)(8v_r + 11)(9216v_r + 2856 + 5\beta + 20\beta v_r + 6144v_r^2)}$$

$$+ \left[\frac{-4(16v_r^2 + 4v_r - 21)}{(1 + 4v_r)(8v_r + 11)} + 8 - \frac{128(v_r^2 + 3v_r + 2)}{(1 + 4v_r)(8v_r + 11)} \right] \frac{r}{R}$$

$$+ \left[\frac{-4(16v_r^2 + 4v_r - 21)}{(1 + 4v_r)(8v_r + 11)} - 6 \right] \left(\frac{r}{R} \right)^2 + \left(\frac{r}{R} \right)^4 \qquad (11.55)$$

The mode shape is expressed as

$$W(r) = -\frac{(8v_r^4 + 40v_r^3 + 69v_r^2 + 47v_r + 10)(14336 + 32768v_r)}{(1 + 4v_r)^2(8v_r + 11)(9216v_r + 2856 + 5\beta + 20\beta v_r + 6144v_r^2)}R^4$$

$$+ 2\frac{(3 + 4v_r)}{(1 + 4v_r)}R^4 - R^4 - 2\frac{(3 + 4v_r)}{1 + 4v_r}R^2 r^2 + r^4 \qquad (11.56)$$

Figures 11.6 and 11.7 represent the graph of $D_r(r)/R^4 b_4$ for different values of β and Poisson's ratio $v_r = 0.3$.

Semi-inverse method of solution associated with $k = 3$

For the particular case of $k = 3$, the governing differential equation becomes

$$\Gamma_0 + \Gamma_1 r + \Gamma_2 r^2 + \Gamma_3 r^3 + \Gamma_4 r^4 + \Gamma_5 r^5 = 0 \qquad (11.57)$$

where

$$\Gamma_0 = -972b_3 R^4 v_r^2 - 168b_2 R^3 + 252b_3 R^4 - 336b_4 R^5 - 288b_2 R^3 v_r$$

$$+ 648b_2 R^3 v_r^2 + 1296b_4 R^5 v_r^2 + 144b_1 R^2 v_r - 324b_1 R^2 v_r^2$$

$$+ 432b_3 R^4 v_r + 84b_1 R^2 - 576b_4 R^5 v_r \qquad (11.58)$$

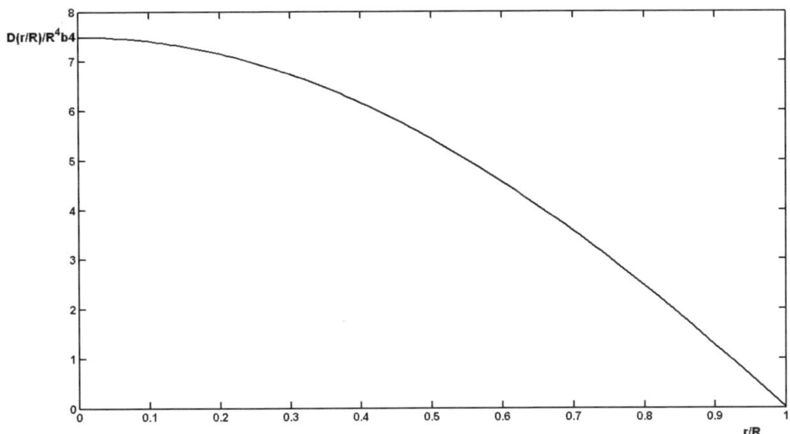

Figure 11.6. Variation of $D(r/R)$ versus non-dimensional radial coordinate r/R for $k = 1$ and Poisson's ratio $v_r = 0.3$ ($\beta = 0$)

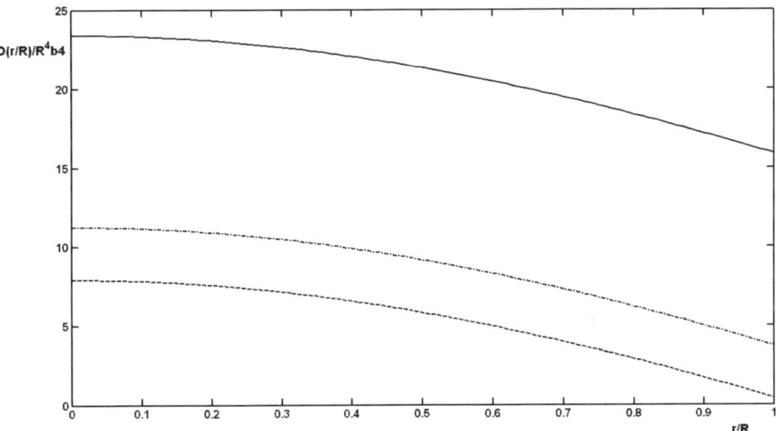

Figure 11.7. Variation of $D(r/R)$ versus non-dimensional radial coordinate r/R for $k = 2$ and Poisson's ratio $v_r = 0.3$ ($---\beta = 10; -\cdot-\beta = 100; —\beta = 1000$)

$$\Gamma_1 = 144k_w b_2 R^2 + 96\rho h\omega^2 b_0 R - 432k_w b_3 R^3 - 9\rho h\omega^2 k_w R^4 v_r$$
$$+ 864k_w b_4 R^4 - 1296k_w b_2 R v_r^2 + 3888k_w b_3 R^3 v_r^2 + 288\rho h\omega^2 b_0 R v_r$$
$$- 7776k_w b_4 R^4 v_r^2 - 5\rho h\omega^2 k_w R^4 \tag{11.59}$$

$$\Gamma_2 = 7776b_4 R^3 v_r^2 + 972b_1 v_r^2 - 1944b_2 R v_r^2 + 36b_1 - 864b_2 R v_r$$
$$+ 432b_1 v_r - 72b_2 R - 864b_4 R^3 + 864b_3 R^2 v_r + 288b_3 R^2 \tag{11.60}$$

$$\Gamma_3 = -288b_3 R + 768b_4 R^2 + 5760b_4 R^2 v_r + 2592b_2 v_r^2 + 10368b_4 R^2 v_r^2$$
$$+ 96b_2 - 7776b_3 R v_r^2 - 3456b_3 R v_r + 1152b_2 v_r + 6\rho h \omega^2 R^2$$
$$+ 18\rho h \omega^2 R^2 v_r \tag{11.61}$$

$$\Gamma_4 = -19440b_4 R v_r^2 - 720b_4 R + 180b_3 + 2160b_3 v_r$$
$$+ 4860b_3 v_r^2 - 8640b_4 R v_r \tag{11.62}$$

$$\Gamma_5 = 288b_4 + 3456b_4 v_r + 7776b_4 v_r^2 - \rho h \omega^2 - 9\rho h \omega^2 v_r \tag{11.63}$$

From Eq. (11.63) the natural frequency squared reads

$$\omega^2 = 288(3v_r + 1)b_4/\rho h \tag{11.64}$$

Solution of Eq. (11.62) yields

$$b_3 = 4Rb_4 \tag{11.65}$$

Substitution of Eqs. (11.64) and (11.65) into Eq. (11.61) result in the formula for b_2:

$$b_2 = 2\frac{(9v_r - 7)R^2}{1 + 9v_r}b_4 \tag{11.66}$$

From Eq. (11.60), we get

$$b_1 = -36\frac{(v_r + 1)R^3}{1 + 9v_r}b_4 \tag{11.67}$$

The equation resulting from substitution of Eqs. (11.64)–(11.66) into Eq. (11.59) yields

$$b_0 = \frac{1}{32}\frac{(9v_r^2 + 14v_r + 5)k_w R^3}{(1 + 9v_r)(3v_r + 1)}b_4 \tag{11.68}$$

In the view of the relationship $\beta = k_w/b_4 R$, the expression for the flexural rigidity reads

$$\frac{D(r)}{R^4 b_4} = \frac{1}{32}\frac{(9v_r^2 + 14v_r + 5)\beta}{(1 + 9v_r)(3v_r + 1)} + 36\frac{(v_r + 1)}{1 + 9v_r} - 3$$
$$+ \left[-4\frac{(9v_r - 7)}{1 + 9v_r} + 8 - 36\frac{(v_r + 1)}{1 + 9v_r} \right]\frac{r}{R}$$
$$+ \left[2\frac{(9v_r - 7)}{1 + 9v_r} - 6 \right]\left(\frac{r}{R}\right)^2 + \left(\frac{r}{R}\right)^4 \tag{11.69}$$

whereas the mode shape is

$$W(r) = -3\frac{(9v_r^2 + 14v_r + 5)}{(1 + 9v_r)(3v_r + 1)}R^4 + 2\frac{(3 + 9v_r)}{(1 + 9v_r)}R^4$$

$$- R^4 - 2\frac{(3 + 9v_r)}{1 + 9v_r}R^2r^2 + r^4 \tag{11.70}$$

The graph of $D_r(r)/R^4 b_4$ is shown in Figs. 11.8 and 11.9 for different values of β and Poisson's ratio $v_r = 0.3$.

Semi-inverse method of solution associated with $k = 4$

Similarly as with the previous cases, the governing differential equation becomes a 5th order polynomial equation

$$\Delta_0 + \Delta_1 r + \Delta_2 r^2 + \Delta_3 r^3 + \Delta_4 r^4 + \Delta_5 r^5 = 0 \tag{11.71}$$

where

$$\Delta_0 = -336b_2 R^3 + 168b_1 R^2 - 3072b_3 R^4 v_r^2 - 2816b_4 R^5 v_r + 4096b_4 R^5 v_r^2$$

$$+ 504b_3 R^4 + 2048b_2 R^3 v_r^2 - 672b_4 R^5 - 1408b_2 R^3 v_r + 704b_1 R^2 v_r$$

$$+ 2112b_3 R^4 v_r - 1024b_1 R^2 v_r^2 \tag{11.72}$$

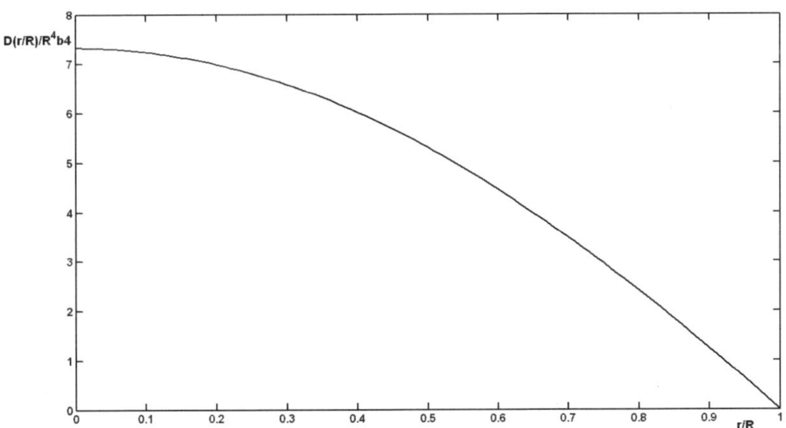

Figure 11.8. Variation of $D(r/R)$ versus non-dimensional radial coordinate r/R for $k = 3$ and Poisson's ratio $v_r = 0.3$ ($\beta = 0$)

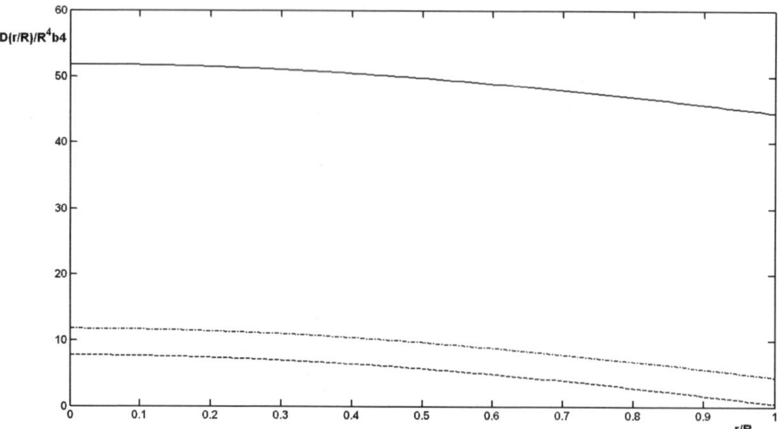

Figure 11.9. Variation of $D(r/R)$ versus non-dimensional radial coordinate r/R for $k = 3$ and Poisson's ratio $\nu_r = 0.3 (- - -\beta = 10; - \cdot -\beta = 100; -\beta = 1000)$

$$\Delta_1 = 896 k_w b_1 R \nu_r + 512 \rho h \omega^2 b_0 R \nu_r - 1792 k_w b_3 R^3 \nu_r + 56 k_w R b_1$$
$$- 56 k_w b_0 + 12288 k_w b_3 R^3 \nu_r^2 - 4096 k_w b_2 R^2 \nu_r^2 + 1816 k_w R^4 b_4$$
$$- 880 k_w b_3 R^3 - 896 k_w b_0 \nu_r + 256 k_w R^2 b_2 - 5 \rho h \omega^2 k_w R^4$$
$$+ 4480 k_w b_4 R^4 \nu_r - 16 \rho h \omega^2 k_w R^4 \nu_r + 152 \rho h \omega^2 R b_0$$
$$- 24576 k_w b_4 R^4 \nu_r^2 \tag{11.73}$$

$$\Delta_2 = 24576 b_4 R^3 \nu_r^2 - 48 b_1 + 3072 b_1 \nu_r^2 - 1536 b_4 R^3 - 576 b_1 \nu_r$$
$$+ 288 b_3 R^2 - 1152 b_3 R^2 \nu_r + 1152 b_2 R \nu_r$$
$$- 6144 b_2 R \nu_r^2 + 96 b_2 R \tag{11.74}$$

$$\Delta_3 = 48 b_3 R + 432 b_4 R^2 + 256 b_2 \nu_r + 32768 b_4 R^2 \nu_r^2 - 768 b_3 R \nu_r$$
$$- 24576 b_3 R \nu_r^2 + 32 \rho h \omega^2 R^2 \nu_r + 1280 b_4 R^2 \nu_r + 8192 b_2 \nu_r^2$$
$$- 16 b_2 + 6 \rho h \omega^2 R^2 \tag{11.75}$$

$$\Delta_4 = -160 b_4 R + 1600 b_3 \nu_r - 6400 b_4 R \nu_r - 61440 b_4 R \nu_r^2$$
$$+ 40 b_3 + 15360 b_3 \nu_r^2 \tag{11.76}$$

$$\Delta_5 = 120 b_4 + 3456 b_4 \nu_r + 24576 b_4 \nu_r^2 - \rho h \omega^2 - 16 \rho h \omega^2 \nu_r \tag{11.77}$$

Solving Eq. (11.77) for the natural frequency squared, we get

$$\omega^2 = 288(3v_r + 1)b_4/\rho h \qquad (11.78)$$

From Eq. (11.76) we get the expression for b_3

$$b_3 = 4Rb_4 \qquad (11.79)$$

Substitution of Eqs. (11.78) and (11.79) into Eq. (11.75) result in the formula for b_2:

$$b_2 = \frac{(2368v_r^2 - 788v_r - 147)R^2}{(1 + 16v_r)(32v_r - 1)}b_4 \qquad (11.80)$$

From Eq. (11.74), we obtained the expression for b_1:

$$b_1 = 2\frac{(v_r^2 - 852v_r + 143)R^3}{(1 + 16v_r)(32v_r - 1)}b_4 \qquad (11.81)$$

The equation resulting from substitution of Eqs. (11.78)–(11.81) into Eq. (11.73) yields

$$b_0 = \frac{(524288v_r^4 + 246784v_r^3 + 261120v_r^2 + 74012v_r + 6313)k_w R^3}{(1 + 16v_r)(32v_r - 1)(-112k_w v_r + 55296b_4 Rv_r^2 + 34848b_4 Rv_r + 5472Rb_4 - 7k_w)}b_4 \qquad (11.82)$$

Hence, the expression for the flexural rigidity reads

$$\begin{aligned}
\frac{D(r)}{R^4 b_4} &= \frac{(524288v_r^4 + 246784v_r^3 + 261120v_r^2 + 74012v_r + 6313)\beta}{(1 + 16v_r)(32v_r - 1)(-112\beta v_r + 55296v_r^2 + 34848v_r + 5472 - 7\beta)} \\
&\quad -2\frac{(320v_r - 852v_r + 1)R^4}{(1 + 16v_r)(32v_r - 1)} - 3 + \frac{(2368v_r^2 - 788v_r - 147)}{(1 + 16v_r)(32v_r - 1)} \\
&\quad + \left[-2\frac{(2368v_r^2 - 788v_r - 147)}{(1 + 16v_r)(32v_r - 1)} + 8 - 2\frac{(320v_r^2 - 852v_r - 143)R}{(1 + 16v_r)(32v_r - 1)}\right]\frac{r}{R} \\
&\quad + \left[\frac{2368v_r^2 - 788v_r - 147}{(1 + 6v_r)(32v_r - 1)} - 6\right]\left(\frac{r}{R}\right)^2 + \left(\frac{r}{R}\right)^4
\end{aligned} \qquad (11.83)$$

The mode shape of the plate for this particular case becomes

$$
W(r) = - \left[152 \frac{\begin{array}{c}(524288\nu_r^4 + 246784\nu_r^3 + 261120\nu_r^2 \\ + 74012\nu_r + 6313)(1 + \nu_r)\end{array}}{\begin{array}{c}(1 + 16\nu_r)^2(32\nu_r - 1)(-112\beta\nu_r + 55296\nu_r^2 \\ + 34848\nu_r + 5472 - 7\beta)\end{array}} \right.
$$

$$
\left. - 2\frac{(3 + 16\nu_r)(1 - 32\nu_r)}{(1 + 16\nu_r)^2} + 1 \right] R^4 - 2\frac{(3 + 16\nu_r)}{1 + 16\nu_r} R^2 r^2 + r^4
$$

$$(11.84)$$

Figure 11.10 depicts the variation of $D_r(r)/R^4 b_4$ versus r/R for different values of β. Note that the minimum value of β, for which the function $D_r(r)/R^4 b_4$ is not negative is, 514.8532 (Fig. 11.11). Therefore, for $k = 3$, $\nu_r = 0.3$, and $\beta > 514.8532$, Eq. (11.84) cannot serve the mode shape of the plate.

11.1.3 Vibration tailoring: Numerical example

Consider a case, where the polar orthotropic circular plate with translational spring has a mode shape, given by Eq. (11.56). Also, the fundamental natural

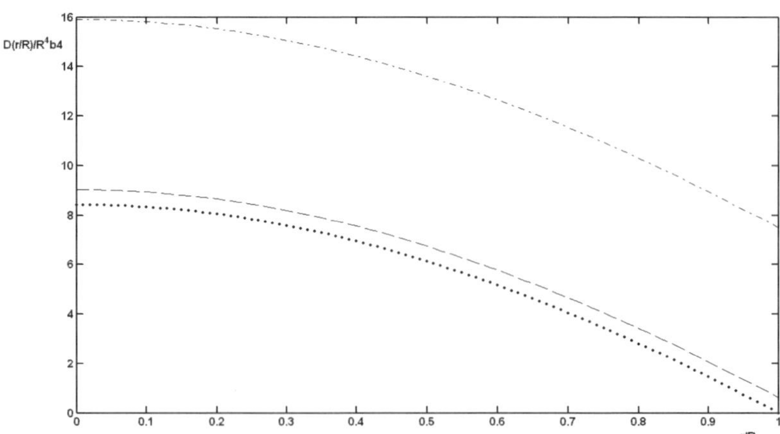

Figure 11.10. Variation of $D(r/R)$ versus non-dimensional radial coordinate r/R for $k = 3$ and Poisson's ratio $\nu_r = 0.3$ ($\cdots \beta = 0; ---\beta = 10; -\cdot-\beta = 100$)

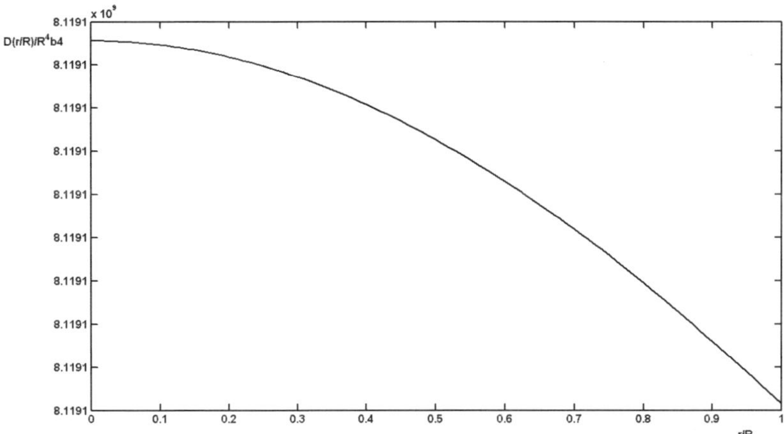

Figure 11.11. Variation of $D(r/R)$ versus non-dimensional radial coordinate r/R for $k = 3$ and Poisson's ratio $v_r = 0.3$ ($\beta = 514.8532$)

frequency is given by Eq. (11.50). We need to examine the behavior of this case by specifying the following properties: $R = 0.1\,\text{m}$, $\rho = 200\,\text{kg/m}^3$, $h = 0.05\,\text{m}$, $\Omega = 120\,\text{Hz}$, and $\beta = 10(k = 2, v_r = 0.3)$. Substitution of these values into Eq. (11.50), yields

$$b_4 = 275.23 \tag{11.85}$$

Thus, the flexural rigidity given by Eq. (11.55), becomes

$$D_r(r) = 0.2181 - 23.37r^2 + 275.23r^4 \tag{11.86}$$

The same procedure can be used for the vibration tailoring of other cases.

Chapter 12

Conclusion

In this book we pursued two objectives: To deal with both traditional, thickness-wise grading, in Part I, along with non-traditional, axial or radial gradings in Part II. These two different approaches open the possibility of multi-directional grading in order to achieve optimal solutions for tailoring the material properties with the desired performance of the structure.

The topic of structures made of functionally graded materials (FGMs) is vast despite the fact that it is relatively new. It appears that encyclopedic covering all aspects of this field is not feasible in a single monograph.

In this book, the vibration tailoring of inhomogeneous elastically restrained vibrating beam has been studied. In this case, it is shown that by postulating the mode shape and by setting the requirement on the desired natural frequency, one obtains a unique solution. Classic formulas for the displacement of simply supported and clamped–clamped homogeneous polar–orthotropic circular plate, that are reported in the monograph by Lekhnitskii (1968), are demanded to serve as an exact mode shape of the vibrating inhomogeneous plates for respective boundary conditions. Twelve closed-form solutions for vibration tailoring are derived for the corresponding natural frequencies. We are unaware of any other closed-form solution for polar–orthotropic plates. This book demonstrates the rich possibilities that the vibrating inhomogeneous polar–orthotropic plates may exhibit, namely, exact coincidence of the mode shape of plate with a specified anisotropy ratio k, with the static displacement of the plate with another anisotropy ratio equal m.

This study furnishes, along the monograph of Elishakoff (2005) and other studies that followed it analytical solutions for vibration tailoring — design of a structure in such a manner so as to possess desired fundamental vibration frequency.

We did not touch many facets, concentrated only on vibration and buckling of columns and plates. Therefore, the book ought to be considered as one of the many possible facets of the general topic of FGM. Some approaches are awaiting their application in structures made of FGMs with promise of success. For example, approaches developed by Vekua (1964, 1965, 1985) as well as by Vashakmadze (2010) for homogeneous structures may prove extremely useful in the context of FGM, and are highly recommended to be explored in the future for structures with grading in thickness direction. In this respect Carrera's Unified Formulation (Carrera *et al.*, 2015) is worth mentioning. It is based on 2 genome to derive any classical and refined theory of structures, including beams, plates, and shells. Shear deformation theories are developed in the classic book by Reddy (2007). Practical problems associated with realization of grading in axial direction or in radial direction ought to be investigated.

Appendix A

A Novel Formulation Leading to Closed-Form Solutions for Buckling of Circular Plates

In this Appendix we study axisymmetric buckling of radially graded circular plates. The flexural rigidity is considered to be a suitably varying function of the radial coordinate. The problem is posed as a semi-inverse one. The buckling mode is selected first, then the variation of the flexural rigidity consistent with the buckling mode is determined. Apparently for the first time in the literature, closed-form solutions are found. Such solutions allow the design of a circular plate whose buckling load is at least the prespecified one. Such a design appears to find much applications in various fields of engineering. This Appendix follows closely the paper by Elishakoff, Ruta and Stavsky (2006).

A.1 Introduction

The buckling problems for plates and shells made of functionally graded materials were tackled by Ng *et al.* (2001), Najafizadeh and Eslami (2002), Javaheri and Eslami (2002) and others. In these papers the elastic modulus varied continuously along the plate's thickness. It is natural to anticipate that the development of functional grading along the radius of a circular plate will follow, resulting in homogeneous plates with variable flexural rigidity.

Note that the buckling of a linearly elastic, homogenous and isotropic circular plate subjected to radial compressive forces in its middle plane, uniformly distributed along its border, was first studied by Bryan (1891) and

215

re-considered by Dinnik (1911). Further contributions are due to Meissner (1933), who discussed the buckling of a circular plate with a hole at the centre. Among the others who studied the subject, credits must be given to Nadai (1915) and Makushin (1960). Vol'mir (1967), among others, reported the approximate determination of the buckling loads by various techniques, and especially via the Rayleigh-Ritz and Boobnov-Galerkin methods.

In all the references above, the buckling is posed as a direct problem: Once the geometry and the constitutive properties of the structural element are given, the critical value of the load multiplier and the buckling mode are determined. In this study the inverse question is posed, in the following form: Is it possible that, once a suitable expression is assumed satisfying the boundary conditions, that an expression will yield an exact buckling mode for an inhomogeneous circular plate, with inhomogeneity in the form of a suitable radial variation of the flexural rigidity? Such a problem appears to be an extremely pertinent one. Indeed, let the buckling load be expressible in the form

$$(N_\rho)_{cr} = D_0 f(R, v),   \text{(A.1)}$$

where D_0 is a parameter that should be chosen, whereas $f(R, v)$ is a certain function of the radius of the plate, and possibly of the material Poisson's ratio. Let the design demand the buckling load $(N_\rho)_{cr}$ to equal at least N_{demanded}:

$$(N_\rho)_{cr} \geq N_{\text{demanded}},   \text{(A.2)}$$

Then, choice of D_0

$$D_0 \geq \frac{N_{\text{demanded}}}{f(R, v)}   \text{(A.3)}$$

produces the plate with the desired buckling behavior. If, in particular, the equality sign holds in Eq. (A.3), so it does also in Eq. (A.2), and thus the structure has been obtained, the buckling load of which attains the demanded value.

An analogous formulation for inhomogeneous columns was given by Elishakoff (2000, 2001). In particular, in the above papers closed-form solutions were derived for the Euler's columns subjected either to concentrated force or to own weight.

In this study, first the direct buckling problem of a linearly elastic, homogenous and isotropic circular plate subjected to radial compression in its middle surface is briefly reviewed, in order to compare it later with an equivalent behavior of the graded plates. Then, assuming that one of the test functions of the Boobnov-Galerkin technique actually is an exact buckling mode, a suitable expression for the radial variation of the flexural rigidity will be looked for. To the best of our knowledge, in this Appendix, for the first time in the literature, several closed-form solutions are derived.

A.2 Buckling of circular plates

Let us consider a linearly elastic, homogeneous and isotropic circular plate of radius R subjected to axisymmetric constraint and external action at its boundary. Then a deflection transforms the middle plane of the plate in a surface of revolution, and any of its diametrical sections will well describe the deflection, because of axial symmetry. Let us fix a cylindrical coordinate system with origin at the center of the middle plane of the plate and ρ, ϕ, ζ the radial, angular and axial coordinates of a point of the plate. The middle plane of the plate will be characterized by $\zeta = 0$. Because of the axial symmetry, all the fields of mechanical interest defined in the middle plane will actually depend on the radial variable ρ only.

The plate will be described by the so-called Krichhoff-Love model, i.e., the material fibers orthogonal to the middle plane in the reference shape remain orthogonal to the deflection surface. Denote by w the deflection of the middle plane of the plate and by θ the angle between the axis of revolution of the deflection surface and any normal to it; by the Kirchhoff-Love hypothesis, θ also coincides with the rotation in the plane $\rho\zeta$ of a material fiber originally orthogonal to the middle plane.

The radial and circumferential curvatures of the middle plane, denoted by χ_ρ, χ_ϕ, are given by Timoshenko (1956)

$$\chi_\rho = \frac{d\theta}{d\rho} = -\frac{d^2 w}{d\rho^2}, \quad \chi_\phi = \frac{\theta}{\rho} = -\frac{1}{\rho}\frac{dw}{d\rho}, \tag{A.4}$$

where, for the sake of simplicity, the dependence of the indicated fields on the radial coordinate ρ has been omitted.

The generalized linear elastic homogeneous and isotropic constitutive relations provide the radial and circumferential bending moments, denoted by M_ρ and M_ϕ, respectively in Timoshenko (1956),

$$M_\rho = -D\left(w'' + \frac{v}{\rho}w'\right) = D\left(\theta' + \frac{v}{\rho}\theta\right), \qquad (A.5.1)$$

$$M_\phi = -D\left(\frac{1}{\rho}w' + vw''\right) = D\left(\frac{\theta}{\rho} + v\theta'\right) \qquad (A.5.2)$$

where to shorten the notation, a prime denotes the derivative of the indicated function with respect to ρ, v is the Poisson's ratio of the material of the plate, and

$$D = \frac{Eh^s}{12(1 - v^2)} \qquad (A.6)$$

is the flexural rigidity of the plate, provided E is the plate's material Young's modulus and h the plate's thickness. Because of axial symmetry, the curvatures provided by Eq. (A.4) are also principal ones, and no torsional moments $M_{\rho\phi}$ arise. Considering a section of the plate in the plane $\rho\zeta$, the variation of the radial moment M_ρ must be balanced by a shearing force, the magnitude of which per unit length will be denoted by Q. The projection on the ρ-axis of the balance of moments in an element of the plate leads to (see Timoshenko, 1956)

$$M_\rho + \rho M'_\rho - M_\phi + \rho Q = 0. \qquad (A.7)$$

By the Kirchhoff-Love hypothesis, the shearing force Q is of reactive nature and it must be determined by statics for any particular case. Let us consider an axisymmetric distribution of compressive forces of magnitude N_ρ per unit length applied at the boundary $\rho = R$ of the plate. In any deflected configuration, simple balance considerations provide

$$Q = N_\rho\theta = -N_\rho w'. \qquad (A.8)$$

If the plate has uniform material characteristics, the substitution of Eqs. (A.5) and (A.8) into the balance Eq. (A.7) leads to the differential equation governing buckling, either in terms of the deflection (Vol'mir, 1967)

$$D\Delta\Delta w + N_\rho\Delta w = 0 \qquad (A.9)$$

or in terms of the rotation angle in Timoshenko (1956, 1961)

$$D\left(\theta'' + \frac{1}{\rho}\theta' - \frac{\theta}{\rho^2}\right) + N_\rho\theta = 0, \tag{A.10}$$

where Δ and $\Delta\Delta$ are the Laplace and bi-harmonic operators, respectively. The formulations provided by Eqs. (A.9) and (A.10) are equivalent, since Eq. (A.9) is nothing but the derivative of Eq. (A.10) with respect to the radial coordinate, and, conversely, by integrating Eq. (A.9) with respect to ρ and imposing the regularity of the primitive at $\rho = 0$, one finds Eq. (A.10).

It is easier to analyze the buckling by considering Eq. (A.10); denoting

$$a^2 = \frac{N_\rho}{D}, \quad u = a\rho, \tag{A.11.1,2}$$

Eq. (A.10) is turned into the well-known Bessel equation of first order

$$u^2\frac{d^2\theta}{du^2} + u\frac{d\theta}{du} + (u^2 - 1)\theta = 0, \tag{A.12}$$

the general solution of which is

$$\theta(u) = A_1 J_1(u) + A_2 Y_1(u), \tag{A.13}$$

where A_1, A_2 are integration constants and J_1, Y_1 are Bessel functions of first order of the first and second kind, respectively. By axial symmetry, it must be $\theta(0) = 0$; and indeed, $J_1(0) = 0$; and indeed, $J_1(0) = 0$ satisfies this condition, while Bessel functions of the second kind have a logarithmic divergence at the origin. It is thus necessary that $A_2 = 0$; the other condition to provide, since Eq. (A.12) is of second order, must be given by considering the constraint at the boundary of the plate.

If the plate has clamped edge the rotation at $\rho = R$ must vanish; hence

$$J_1(aR) = 0. \tag{A.14}$$

The smallest root of the transcendental equation (A.14) is $aR \approx 3.832$, leading to

$$(N_\rho)_{cr} \approx \frac{14.682D}{R^2}. \tag{A.15}$$

If the plate has a simply supported edge the radial bending moment at $\rho = R$ must vanish; hence, from Eq. (A.5.1) one obtains

$$\left| \theta' + \frac{v}{\rho}\theta \right|_{\rho=R} = 0, \tag{A.16}$$

resulting in the transcendental equation

$$(aR)\,J_0(aR) - (1-v)\,J_1(aR) = 0. \tag{A.17}$$

Supposing $v = 1/3$, the minimal root of Eq. (A.17) is $aR \approx 2.069$, and we get

$$(N_\rho)_{cr} \approx \frac{4.282D}{R^2}, \tag{A.18}$$

where use has been made of the derivative formula $dJ_1/du = J_0 - J_1/u$.

$$\tilde{w} = a(R^2 - \rho^2)^2 \tag{A.19}$$

so that

$$\tilde{\theta} = -\tilde{w}' = 4a(R^2 - \rho^2)\rho, \tag{A.20}$$

a being an undetermined amplitude factor. Substituting Eqs. (A.20) and (A.11.2) into Eq. (A.12), multiplying the result by Eq. (A.20) and integrating over the domain $0 \le \rho \le R$, one obtains

$$(\widetilde{N}_\rho)_{cr} = \frac{16D}{R^2}, \tag{A.21}$$

which is some 8.98% above the value provided by Eq. (A.15).

In the case of a simply supported plate, an admissible function is

$$\tilde{w} = b\{R^4(5+v) + \rho^2[(1+v)\rho^2 - 2(3+v)R^2]\} \tag{A.22}$$

with

$$\bar{\theta} = -\bar{w}' = 4b[(3+v)R^2 - (1+v)\rho^2]\rho, \tag{A.23}$$

b being an undetermined amplitude factor. Substituting Eqs. (A.23) and (A.11.2) into Eq. (A.12), multiplying the result by Eq. (A.23) and integrating over $0 \le \rho \le R$, one obtains, supposing

$$(\overline{N}_\rho)_{cr} \approx \frac{4.293D}{R^2} \tag{A.24}$$

which is some 0.24% above the value provided by Eq. (A.18).

So far, direct exact and approximate buckling problems have been considered. In the next Section, semi-inverse buckling problems will be considered.

A.3 Inverse buckling of radially graded circular plates

When we consider radially graded circular plates, with inhomogeneity represented by a variation of the flexural rigidity with the radial abscissa, ρ, Eqs. (A.4)–(A.8) remain unchanged, provided one reads $D = D(\rho)$ in each of them. Then, taking into account the variability of D, the substitution of Eqs. (A.5) and (A.8) into the balance equation (A.7) leads to the differential equation governing buckling, either in terms of the deflection

$$D\rho^3 \Delta\Delta w + D'[2\rho^3 w'' + \rho^2(2+v)w'' - \rho w']$$
$$+ D''(\rho^3 w'' + v\rho^2 w') + N_\rho \rho^3 \Delta w = 0 \qquad (A.25)$$

or in terms of the rotation angle

$$D\left(\theta'' + \frac{1}{\rho}\theta' - \frac{\theta}{\rho^2}\right) + D'\left(\theta' + \frac{v}{\rho}\theta\right) + N_\rho\theta = 0. \qquad (A.26)$$

It is trivial to check that when $D = D_0 = constant$, Eqs. (A.25) and (A.26) return their counterparts for ungraded plates, Eqs. (A.9) and (A.10). Again, the formulations provided by Eqs. (A.25) and (A.26) are equivalent, since Eq. (A.25) is the ρ-derivative of Eq. (A.26) and, conversely, by integrating Eq. (A.25) with respect to ρ and imposing the regularity of the primitive at $\rho = 0$ one finds Eq. (A.26). Remark that both Eqs. (A.25) and (A.26) were derived supposing Poisson's ratio to be constant, hence the flexural rigidity is provided by a radial variation of Young's modulus, supposing the geometry of the plate to be uniform ($h = constant$).

In the following, it will be assumed that the approximate deflections provided by Eqs. (A.19) and (A.22) are the exact buckling modes, in the case of the clamped and simply supported plate, respectively. Since in both cases the approximate deflection is a fourth order polynomial and the approximate rotation (expressed by Eqs. (A.20) and (A.23), respectively) is a third order polynomial, from either Eq. (A.25) or Eq. (A.26) multiplied by ρ^2 it turns out that the last term is a fifth order polynomial. Since all terms in both equations must be homogeneous, thus fifth order polynomials, it is sufficient that the flexural rigidity be a second order polynomial,

$$D = D_0 + D_1\rho + D_2\rho^2. \tag{A.27}$$

Let us begin with the case of the clamped plate. Substituting Eqs. (A.20) and (A.27) into Eq. (A.26), results in the polynomial equation

$$D_1 R^2 (1+v) - \{8D_0 - R^2[N_\rho + 2D_2(1+v)]\}\rho$$
$$- D_1(11+v)\rho^2 - [N_\rho + 2D_2(7+v)]\rho_3 = 0, \tag{A.28}$$

which leads to

$$D = D_0 \left(1 - \frac{2\rho^2}{3R^2}\right) \tag{A.29}$$

with the buckling load given by

$$(N_\rho)_{cr} = \frac{4D_0(7+v)}{3R^2}, \tag{A.30}$$

D_0 being an amplitude coefficient of the same physical dimensions of D. A confrontation between Eqs. (A.27) and (A.29) brings into evidence that $D_1 = 0$, which was to be expected, due to the axial symmetry of the problem. If $v = 1/3$, Eq. (A.30) provides

$$(N_\rho)_{cr} = \frac{9.778D_o}{R^2}, \tag{A.31}$$

which is less than the critical load given by Eq. (A.21) because it is clear from Eq. (A.29) that the flexural rigidity of the plate decreases with respect to that of the ungraded plate.

At this juncture it appears instructive to calculate the equivalent flexural rigidity $D_{\theta q}$ of a uniform circular plate that possesses the buckling load of the original, axially graded plate

$$D_{\theta q} = \frac{1}{\pi R^2} \int_0^{2\pi} \int_0^R D(\rho)\rho d\rho d\theta, \tag{A.32}$$

which turns out to be

$$D_{\theta q} = \frac{2D_0}{3}. \tag{A.33}$$

Now, the homogeneous, ungraded plate has the buckling load given in Eq. (A.15), which in view of Eq. (A.32) yields the buckling load of an ungraded clamped plate with flexural rigidity

$$(N_\rho)_{cr|D=D_{\theta q}} \approx \frac{14.682 \left(\frac{z}{\theta}\right) D_0}{R^2} \approx \frac{9.788 D_0}{R^2},$$
(A.34)

which is about 0.1% above the value $9.778 D_0/R^2$ in Eq. (A.31) for the radially graded plate.

When considering the simply supported plate, substituting Eqs. (A.23) and (A.27) into Eq. (A.26) provides the polynomial equation

$$D_1 R^2 (1+v)(3+v) + \left\{ -8D_0(1+v) + R^2(3+v) \right.$$
$$\times \left. [N_\rho + 2D_2(1+v)] \right\} \rho - D_1(1+v)(11+v)\rho^2 - (1+v)$$
$$\times [N_\rho + 2D_2(7+v)]\rho^3 = 0.$$
(A.35)

which leads to

$$D = D_0 \left(1 - \frac{2(1+v)\rho^2}{3(3+v)R^2} \right),$$
(A.36)

with the buckling load given by

$$(N_\rho)_{cr} = \frac{4D_o(1+v)(7+v)}{3(3+v)R^2}.$$
(A.37)

If $v = 1/3$, Eq. (A.37) provides

$$(N_\rho)_{cr} = \frac{3.911 D_0}{R^2},$$
(A.38)

less than the critical load given by Eq. (A.24) because from Eq. (A.36) the flexural rigidity of the plate decreases with respect to that of the ungraded plate, similarly to what happens for the clamped plate.

The equivalent flexural rigidity, defined by Eq. (A.32), equals

$$D_{\theta q} = \frac{2(4+v)D_0}{3(3+v)}$$
(A.39)

and, for $v = 1/3$,

$$D_{\theta q} = \frac{13}{15} D_0.$$
(A.40)

Equation (A.24) then leads to

$$(N_\rho)_{cr}|_{D=D_{\theta q}} \approx \frac{4.282 D_0}{R^2} \frac{2(4+v)}{3(3+v)}, \tag{A.41}$$

which, for $v = 1/3$, yields

$$(N_\rho)_{cr}|_{D=D_{\theta q}} \approx 3.711 \frac{D_0}{R^2}, \tag{A.42}$$

which is 5.11% below the corresponding value, provided by Eq. (A.38), for the graded plate.

In both cases of the clamped and the simply supported plate, the polynomial equations (A.28) and (A.35) are of third order instead of fifth, as it could be supposed because of the argument provided before Eq. (A.27). This happens because approximate expressions (A.19) and (A.22) miss coefficients of odd powers of ρ, hence a term ρ^2 can be factorized in the polynomial equation resulting from substituting Eqs. (A.19), (A.22) and (A.27) into Eq. (A.12).

Supposing that Poisson's ratio, as well as the flexural rigidity, is a function of the radial coordinate, one obtains, instead of Eq. (A.26),

$$D\left(\theta'' + \frac{1}{\rho}\theta' - (1-v')\frac{\theta}{\rho^2}\right) + D'\left(\theta' + \frac{v}{\rho}\theta\right) + N_\rho\theta = 0. \tag{A.43}$$

By the same considerations made for the variation of the flexural rigidity, it is sufficient that Poisson's ratio be a linear function of the radial coordinate,

$$v(\rho) = v_0 + v_1\rho. \tag{A.44}$$

Operating as above, in both cases of the clamped and simply supported plate it turns out that $v_1 = 0$, i.e., Poisson's ratio must be constant and all the preceding results, especially those provided by Eqs. (A.29) and (A.36), hold true again. The passages are omitted for the sake of brevity, because they add no new results to what already shown. The reason of this result is clear looking at the expression (A.6) of the flexural rigidity: in order to keep $D(\rho)$ a second-order polynomial in the radial coordinate, Poisson's ratio must be a constant.

A.4   Conclusion

In this Appendix novel semi-inverse buckling problems for circular radially graded Kirchhoff-Love plates subjected to uniform radial compression have been considered. We supposed that an admissible polynomial that is usually adopted for the realization of the approximate, Boobnov-Galerkin technique is an exact buckling mode. We then determine the radial variation of the flexural rigidity as a consequence. The clamped and simply supported plate have been considered, and in both cases closed form solutions both for the flexural rigidity expression and for the critical load have been obtained, to the best of our knowledge, for the first time in the literature. These solutions can be utilized first for theoretical purposes as a benchmark, then for the practical purpose of designing plates that possess the specified buckling load.

Appendix B

Inverse Vibration Problem for Inhomogeneous Circular Plate with Translational Spring

The free vibrations of uniform and homogenous circular plates with translational springs have been studied in the literature for some time; although exact solutions have been found, no closed-form solution has been reported yet. In this Appendix, using the semi-inverse method we derive a closed-form solution for the natural frequency via postulating the vibration mode of the plate as a polynomial of the radial coordinate. This Appendix follows closely the paper by Elishakoff and Meyer (2005).

B.1 Introduction

The free vibrations of circular plates with the translational springs was studied by Leissa (1969). He derived the transcendental equations yielding the natural frequency of both axisymmetric and non-symmetric vibrations for the plate with both translational and rotational springs. The uniform plate was considered, with attendant transverse displacement expressed in terms of the Bessel functions. The Boobnov-Galerkin method to this problem was applied by Laura $et\ al.$ (1976). They approximated the transverse displacement as

$$W(r) \approx W_{N_1}(r) = \sum_{j=0}^{N_1} A_j (a_j r^4 + \beta_j r^2 + 1) r^{2j}, \qquad \text{(B.1)}$$

where N_1 denotes the number of terms retained, a_j and β_j are constants chosen so as to satisfy the boundary conditions. Note that the related Appendix

227

on vibrations of circular plates with supports along the circumference is by Laura *et al.* (1975). To the best of our knowledge, there is no closed-form solutions reported to this problem. It may even seem at the first glance that there would be no closed-form solutions. This ambitious goal is addressed in this Appendix.

We consider an inhomogeneous plate, with inhomogeneity in the form of the variable flexural rigidity. We pose and solve an inverse problem: we postulate the mode shape to be fourth-order polynomial, corresponding to the zero value of N_1 in Eq. (B.1). Then we find the flexural rigidity's variation along the radial coordinate, so as the inhomogeneous plate to have a postulated closed-form solution.

B.2 Basic equations

The differential equation governing free small axisymmetric vibrations of circular plates reads

$$D(r)r^3\nabla^2\nabla^2 W + \frac{dD}{dr}\left(2r^3\frac{d^3 W}{dr^3} + r^2(2+v)\frac{d^2 W}{dr^2} - r\frac{dW}{dr}\right)$$
$$+ \frac{d^2 D}{dr^2}\left(r^3\frac{d^2 W}{dr^2} + vr^2\frac{dW}{dr}\right) - \rho h w^2 r^3 W = 0, \tag{B.2}$$

where h is the thickness of the plate, ρ the material density, v the coefficient of Poisson, r the radial coordinate, D the flexural rigidity, W the mode shape and ∇^2 the Laplace operator in polar coordinates,

$$\nabla^2 = \frac{d}{dr^2} + \frac{1}{r}\frac{d}{dr}. \tag{B.3}$$

The transverse displacement W is postulated to be in the form

$$W(r) = a_0 + a_2 r^2 + r^4. \tag{B.4}$$

We set

$$\rho h = \delta(r) \tag{B.5}$$

that we suppose to vary along the radial coordinate r as

$$\delta(r) = \sum_{i=0}^{m} a_i r^i. \tag{B.6}$$

Since W is a fourth-order polynomial expression in terms of r, in view of Eq. (B.6), the last term in the differential equation (B.2) is a polynomial expression of degree $m + 7$. Moreover, the operator $\nabla^2\nabla^2$ in Eq. (B.2) involves the four-fold differentiation with respect to r. In order for the highest degree of the first term's polynomial expression in $Dr^3\nabla^2\nabla^2 W$ to be of order $m + 7$, it is necessary and sufficient for the flexural rigidity to be represented as a polynomial of degree $m + 4$. Thus, the sought flexural rigidity can be put in the form

$$D(r) = \sum_{i=0}^{m+4} b_i (r - R)^i. \tag{B.7}$$

B.3 Boundary conditions

The boundary conditions at the outer boundary $r = R$ consist of the bending moment M_r acting along the circumference sections to vanish, and the shearing force per unit length to be proportional to the deflection of the plate:

$$M_r(R) = 0, \quad Q_r(R) + k_W W(R) = 0, \tag{B.8,9}$$

where

$$M_r(r) = -D(r)\left(\frac{d^2 W}{dr^2} + \frac{v}{r}\frac{dW}{dr}\right), \tag{B.10}$$

and k_W is the stiffness per unit of length of the translational spring. The shearing force per unit of length $Q_r(r)$ is obtained from the equilibrium equations from Timoshenko (1959):

$$Q_r(r) = \frac{1}{r}\left[M_t(r) - M_r(r) - r\frac{dM_r(r)}{dr}\right], \tag{B.11}$$

where M_t denotes the bending moment per unit of length acting along the diametrical section rZ of the plate

$$M_t(r) = -D(r)\left(\frac{1}{r}\frac{dW}{dr} + v\frac{d^2 W}{dr^2}\right). \tag{B.12}$$

The problem is posed as follows: *Determine the variation of the flexural rigidity $D(r)$ so that a plate with such $D(r)$ will possess the vibration mode defined in* Eq. (B.4).

B.4 Method of solution

The application of the boundary conditions given in Eqs. (B.8) and (B.9) permits the determination of the coefficients a_0 and a_2 of the mode shape polynomial expression defined in Eq. (B.4). Indeed, Eqs. (B.8) and (B.9) read

$$(12 + 4v)R^2 + 2a_2(1 + v) = 0, \tag{B.13}$$

$$k_W R^4 + ka_2 R^2 - 32b_0 k_W a_0 R = 0. \tag{B.14}$$

From Eqs. (B.13) and (B.14) we get

$$a_0 = -32\frac{Rb_0}{k_W} + \frac{5 + v}{1 + v}R^4, \tag{B.15}$$

$$a_2 = -2\frac{(3 + v)}{(1 + v)}R^2. \tag{B.16}$$

So that the shape mode is written as

$$W(r) = -32\frac{Rb_0}{k_W} + \frac{5 + v}{1 + v}R^4 - 2\frac{(3 + v)}{(1 + v)}R^2 r^2 + r^4. \tag{B.17}$$

Here it must be noted that the mode shape depends both upon the coefficient b_0 of the flexural rigidity and the stiffness of the translational spring k_W. Yet, it can be argued that it ought be anticipated that the *closed-form solution* would only be attainable for specific values and combinations of the system parameters, and for specific relationships between the mode shape and the system's characteristics.

Further steps involve the substitution of Eqs. (B.6), (B.7) and (B.17) into the governing differential equation (B.2) and demanding the so-obtained polynomial expression to vanish. This implies that all the coefficients in front of power r^i must be zero. This requirement is leading, in turn, to a set of algebraic equations in terms of b_i, and ω^2. We consider various case for the inertial term $\delta(r)$ in Eq. (B.6).

B.5   Constant inertial term ($m = 0$)

As seen from Eq. (B.7) in this particular case, the flexural rigidity is sought as a fourth-order polynomial

$$D(r) = b_0 + b_1(r - R) + b_2(r - R)^2 + b_3(r - R)^3 + b_4(r - R)^4. \quad \text{(B.18)}$$

The differential equation (B.2) becomes

$$\sum_{i=0}^{7} c_i r^i = 0, \quad \text{(B.19)}$$

where

$$c_0 = c_1 = 0,$$

$$c_2 = -4(3 + v)(R^2 b_1 - 2R^3 b_2 + 3R^4 b_3 - 4R^5 b_4),$$

$$c_3 = 64 b_0 - 64 R b_1 + 16(1 - v)R^2 b_2 + 16(5 + 3v)R^3 b_3$$

$$- 32(7 + 3v)R^4 b_4 - \frac{5 + v}{1 + v} a_0 R^4 \omega^2 + 32 \frac{a_0 R}{k_W} b_0 \omega^2,$$

$$c_4 = 12(11 + v)b_1 - 24(11 + v)b_2 + 288 R^2 b_3 - 96(1 - v)R^3 b_4,$$

$$c_5 = 32(7 + v)b_2 - 96(7 + v)b_3 + 128(9 + v)R^2 b_4 + 2\frac{3 + v}{1 + v} a_0 R^2 \omega^2,$$

$$c_6 = 20(17 + 13v)b_3 - 80(17 + 3v)R b_4,$$

$$c_7 = 96(5 + v)b_4 - a_0 \omega^2. \quad \text{(B.20)}$$

Since the left-hand side of the differential equation (B.18) must vanish for any r within $[0; R]$, we demand that all the coefficients c_i to be zero. This leads to a homogenous set of six non-linear algebraic equations for six unknowns. From the requirement $c_7 = 0$, the natural frequency squared is obtained as

$$\omega^2 = \frac{96 b_4 (5 + v)}{a_0} \quad \text{(B.21)}$$

in the desired closed-form solution. Upon substitution of Eq. (B.20) into Eq. (B.19), the remaining equations yield the coefficient in the flexural rigidity

$$b_0 = \frac{b_4 R^4 k_W (5+v)(7+v)}{2(1+v)[48 R b_4 (5+v) + k]},$$

$$b_1 = -4b_4 \frac{5+v}{1+v} R^3, \quad b_2 = -2b_4 \frac{3-v}{1+v} R^2, \quad b_3 = 4b_4 R. \qquad \text{(B.22)}$$

Hence, the flexural rigidity reads

$$D(r) = \frac{5+v}{1+v} \left(\frac{R^4 k_W (5+v)(7+v)}{2(1+v)[48 R b_4 (5+v) + k_W]} - 4\frac{5+v}{1+v} R^3 r \right.$$
$$\left. -2\frac{3-v}{1+v} R^2 r^2 + 4R r^3 + r^4 \right) b_4. \qquad \text{(B.23)}$$

It must be stressed that the determined flexural rigidity of the plate depends on the stiffness k_W of the translational spring in a nonlinear manner. Only when there is a relation between $D(r)$ and k_W is the closed-form polynomial solution for the mode shape possible.

Substituting the expression for b_0 from Eq. (B.21) into the mode shape given by Eq. (B.17), the latter becomes

$$W(r) = \frac{5+v}{1+v} \left(1 - \frac{7+v}{3(5+v) + k_W/b_4 R} \right) R^4 - 2\frac{(3+v)}{(1+v)} R^2 r^2 + r^4.$$
$$\text{(B.24)}$$

We introduce the non-dimensional constant

$$\beta = k_W / b_4 R. \qquad \text{(B.25)}$$

The flexural rigidity and the mode shape are then expressed as

$$D(r) = \left[\begin{array}{c} \dfrac{R^4 \beta (5+v)(7+v)}{2(1+v)[48(5+v) + \beta]} - 4\dfrac{5+v}{1+v} R^3 (r - R) \\[2mm] -2\dfrac{3-v}{1+v} R^2 (r - R)^2 + 4R(r - R)^3 + (r - R)^4 \end{array} \right] b_4, \qquad \text{(B.26)}$$

$$W(r) = \frac{5+v}{1+v} \left(1 - \frac{7+v}{3(5+v) + \beta} \right) R^4 - 2\frac{(3+v)}{(1+v)} R^2 r^2 + r^4. \qquad \text{(B.27)}$$

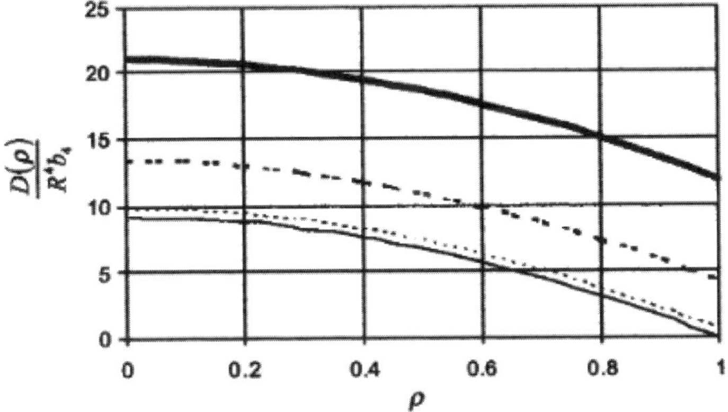

Figure B.1. Variation of the flexural rigidity of an inhomogeneous plate with translational spring on the border, a constant inertial term and a coefficient of Poison fixed to $v = 0.3$, when the non-dimensional coefficient β varies: ——, $\beta = 0$; - - - -, $\beta = 10$; - - -, $\beta = 100$; ——, $\beta = 1000$

Let

$$\rho = \frac{r}{R}. \tag{B.28}$$

We have

$$\frac{D(\rho)}{R^4 b_4} = \frac{(57 + 18v + v^2)\beta + 96(5 + v)(11 + 3v)}{2(1 + v)[48(5 + v) + \beta]}$$

$$- 4\frac{3 + v}{1 + v}\rho^2 + \rho^4, \tag{B.29}$$

$$\frac{W(\rho)}{R^4} = \frac{5 + v}{1 + v}\frac{4(2 + v) + \beta}{3(5 + v) + \beta} - 2\frac{3 + v}{1 + v}\rho^2 + \rho^4. \tag{B.30}$$

As is seen, the coefficient b_4 can be chosen arbitrarily. Thus there are infinite amounts of closed-form solutions. Once one specifies b_{47}, a specific solution is obtained. The Figs. B.1 and B.2 represent the graph of $D(\rho)$ and $W(\rho)$ for different values of β with the coefficient of Poison v fixed to $v = 0.3$ and $b_4 = 1$.

B.6 Linearly varying inertial term ($m = 1$)

In this case, the inertial term is expressed as

$$\delta(r) = a_0 + a_1 r. \tag{B.31}$$

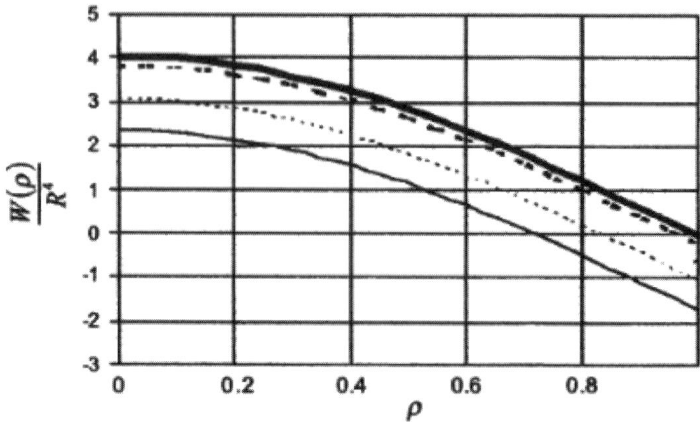

Figure B.2.   Mode shape of an inhomogeneous plate with translational spring on the border, a constant inertial term and a coefficient of Poison fixed to $v = 0.3$ when the non-dimensional coefficient β varies: —, $\beta = 0$; - - - -, $\beta = 10$; - - -, $\beta = 100$; —, $\beta = 1000$

Let us introduce the non-dimensional coefficient γ defined such that

$$\gamma = \frac{a_0}{a_1 R}, \quad a_1 \neq 0. \tag{B.32}$$

Hence, the inertial term can be expressed with the reduced coordinate as

$$\delta(\rho) = a_0 \left(1 + \frac{\rho}{\gamma} \right), \tag{B.33}$$

where ρ is defined in Eq. (B.28).

Instead of the set (B.20), we get here seven algebraic expressions for the coefficients of the flexural rigidity polynomial form defined in Eq. (B.7),

$$c_0 = c_1 = 0,$$

$$c_2 = -4(3 + v)(R^2 b_1 - 2R^3 b_2 + 3R^4 b_3 - 4R^5 b_4 + 5b_5),$$

$$c_3 = 64b_0 - 64Rb_1 + 16(1 - v)R^2 b_2 + 16(5 + 3v)R^3 b_3$$
$$- 32(7 + 3v)R^4 b_4 + 32(13 + 5v)R^5 b_5 - \frac{5 + v}{1 + v} a_0 R^4 \omega^2$$
$$+ 32 \frac{a_0 R}{k_W} b_0 \omega^2,$$

$$c_4 = 12(11 + v)b_1 - 24(11 + v)b_2 + 288R^2 b_3 - 96(1 - v)R^3 b_4$$
$$- 60(7 + 5v)R^4 b_5 + 32 \frac{a_1 R}{k_W} b_0 \omega^2,$$

$$c_5 = 32(7 + v)b_2 - 96(7 + v)b_3 + 128(9 + v)R^2 b_4$$
$$- 1280 R^3 b_5 + 2\frac{3 + v}{1 + v} a_0 R^2 \omega^2,$$

$$c_6 = 20(17 + 3v)b_3 - 80(17 + 3v)Rb_4 + 100(31 + 5v)R^2 b_5$$
$$+ 2\frac{3 + v}{1 + v} a_1 R^2 \omega^2,$$

$$c_7 = 96(5 + v)b_4 - 480(5 + v)Rb_5 - a_0 \omega^2,$$

$$c_8 = 28(23 + 5v)Rb_5 - a_1 \omega^2. \tag{B.34}$$

Since the polynomial expression of the differential equation must vanish for every positive r not greater than R, all the coefficients c_i must be equal to zero. From $c_8 = 0$, the natural frequency squared is obtained as

$$\omega^2 = \frac{287 b_s (23 + 5v)}{a_1}. \tag{B.35}$$

Upon substitution of Eq. (B.35) into $c_5 = 0$, $c_6 = 0$, $c_7 = 0$ of Eqs. (A.34), the remaining equations yield the coefficient b_2, b_3, b_4 of the flexural rigidity:

$$b_2 = \frac{\begin{array}{c}-35(23 + 5v)(3 - v)(17 + 3v)a_0 + 12(5 + v)\\(15v^2 - 296v + 1823)a_1 R\end{array}}{60(5 + v)(17 + 3v)(1 + v)a_1} R^2 b_5,$$

$$b_3 = \frac{\begin{array}{c}35(23 + 5v)(17 + 3v)(1 + v)a_0 + 6(5 + v)\\(105v^2 + 568v - 41)a_1 R\end{array}}{30 a_1 (5 + v)(17 + 3v)(1 + v)} Rb_5,$$

$$b_4 = \frac{120(5 + v)a_1 R + 7(23 + 5v)a_0}{24 a_1 (5 + v)} b_5, \tag{B.36}$$

Let us consider now the set of equations $c_3 = 0$ and $c_4 = 0$ of Eqs. (B.34) substituting the values of ω^2, b_2, b_3, b_4 obtained in Eqs. (B.35) and (B.36). We obtain for b_0 and b_1 the following solution:

$$b_0 = 32 \frac{\begin{array}{c}(7 + v)(11 + v)(17 + 3v)(23 + 5v)a_0\\+(15v^3 + 463v^2 + 3789v + 8885)a_1 R\end{array}}{\begin{array}{c}(1 + v)(17 + 3v)[240(11 + v)a_1 k_W + 3360\\(11 + v)(23 + 5v)a_0 b_5 + 17920(23 + 5v)a_1 b_5)]\end{array}} \times R^4 k_W b_5,$$

$$\begin{aligned}
b_1 = &-\{105(11+v)(23+5v)(17+3v)a_0 a_1 k_W + 6[18375a_0^2 b_5 \\
&+ (60a_1^2 k_W + 475300a_0^2 b_5)v^3 - (1456a_1^2 k_W + 4351690a_0^2 b_s)v^2 \\
&+ (12060a_1^2 kw + 17017700a_0^2 b_5)v + 29816a_1^2 k_W \\
&+ 24236135a_0^2 b_5]Rb_5 + 28(23+5v)(705v^3 + 17633v^2 \\
&+ 130715v + 294723)R^2 a_0 a_1 b_5 + 2688(23+5v)(15v^2 \\
&+ 232v + 721)a_1^2 R^3 b_5 R^3 b_5 /(17+3v)(1+v)[90(11+v)a_1^2 k_W \\
&+ 1260(11+v)(23+5v)a_1 a_0 b_5 + 6720(23+5v)a_1^2 b_5].
\end{aligned}$$ (B.37)

Taking into account the previous results (B.35)–(B.37), equation $c_2 = 0$ from Eq. (B.34) must now be satisfied. Two solutions for b_5 are obtained

$$b_5 = 0 \tag{B.38}$$

or

$$b_5 = \frac{12(15v^3 + 364v^2 + 2619v + 5546)a_1 k_W}{7(23+5v)(165a_0 v^3 + (329a_0 - 1440a_1 R)v^2 \\ -(11441a_0 + 15936a_1 R)v - 37309a_0 - 38688a_1 R)R}. \tag{B.39}$$

The first solution $b_5 = 0$ described in Eq. (B.38) must be dismissed since it leads to the trivial case with flexural rigidity that is identically zero all over the plate.

Eqs. (B.7), (B.7), (B.35)–(B.37) and (B.39) lead to a determinate solution in the case of a linearly varying inertial term. Indeed, the natural frequency squared is obtained by substituting the expression of b_5 into Eq. (B.35)

$$\omega^2 = \frac{48k_W(15v^3 + 364v^2 + 2619v + 5546)}{a_1 R^2[(165v^3 + 329v^2 - 11441v - 37309)\gamma \\ -(1440v^2 + 15936v + 38688)]}. \tag{B.40}$$

The flexural rigidity and mode shape are expressed as follows, respectively

$$D(\rho) = \left[\begin{array}{l} \dfrac{\gamma}{240}\dfrac{(165v^2 + 3314v + 12725)(3+v)^2}{(5+v)(17+v)(1+v)} \\[2ex] -\dfrac{7\gamma}{6}\dfrac{(3+v)(23+5v)}{(1+v)(1+5v)}\rho^2 - \dfrac{9}{5}\dfrac{(33+5v)(3+v)}{(17+3v)(1+v)}\rho^3 \\[2ex] +\dfrac{7\gamma}{24}\dfrac{(23+5v)}{(5+v)}\rho^4 + \rho^5 \end{array} \right] R^5 b_5,$$

$$\tag{B.41}$$

$$\frac{W(\rho)}{R^4} = \frac{81}{35} \frac{(33 + 5v)(3 + v)^2}{(1 + v)(17 + 3v)(23 + 5v)} - 2\frac{3 + v}{1 + v}\rho^2 + \rho^4. \tag{B.42}$$

The physical realizability demands that the expression for the natural frequency ω^2 be positive. Analogously, $D(r)$ must take positive values in the interval $[0, R]$. The investigation of ω^2 shows that a solution exists in either of two cases

$$\gamma < \alpha \text{ for } a_1 > 0, \tag{B.43}$$

$$\gamma > \alpha \text{ for } a_1 < 0, \tag{B.44}$$

where α is defined by the expression

$$\alpha = \frac{1440v^2 + 15936v + 38688}{165v^3 + 329v^2 - 11441v - 37309}. \tag{B.45}$$

Indeed, since the numerator in Eq. (B.40) is positive, so should be the numerator. This implies that a_1 and the expression in the square parentheses should have the same sign. This requirement leads to conditions in Eqs. (B.43) and (B.44).

In the following the value of v is fixed at $v = 0.3$. The flexural rigidity D in Eq. (B.41) is a function of γ and $\rho : D(\gamma, \rho)$. We demand for D not to vanish in the interval $\rho \in [0, 1]$. Moreover, it should not change the sign in that interval. Considering ρ as a parameter, the value of γ that makes D

$$\gamma_1 = \frac{25440\rho^3(-20493 + 2327\rho^2)}{299127609 - 1718133648\rho^2 + 29816100\rho^4}. \tag{B.46}$$

Fig. B.3 represents the variations of γ_1 with ρ. The interpretation of this graph leads us to conclude that $a_1 > 0$ cannot be accepted since for all value of γ such as $\gamma < \alpha$, D vanishes in the interval $[0,1]$. Let us examine the case $a_1 < 0$. The shaded area that represents the admissible values for γ is defined by

$$\beta < \gamma < 0, \tag{B.47}$$

where $\beta \approx -0.3206643660$ is the maximum value of γ on the right of the vertical asymptote when $v = 0.3$. For such values of γ, the first term represented by the square bracket in Eq. (B.41) gets negative values. The

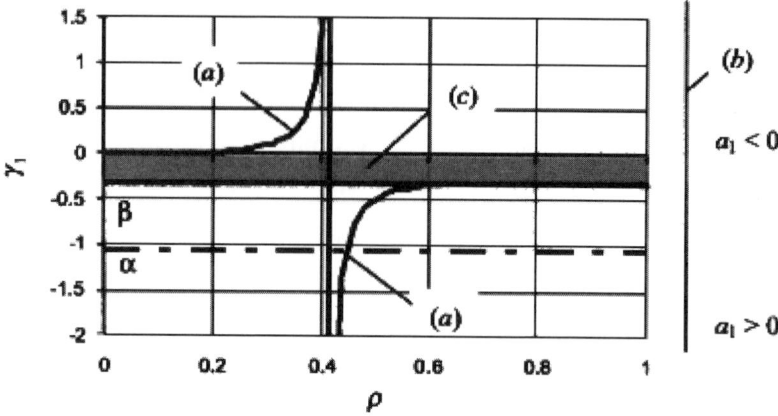

Figure B.3. Values of γ that make the flexural rigidity vanish are given by solid lines (a); (b) represents the relationship between the coefficient a_1 and the coefficient γ that allows a physically acceptable solution for the natural frequency squared ω^2; the shaded area (c) represents the range for γ that allow the flexural rigidity not to vanish

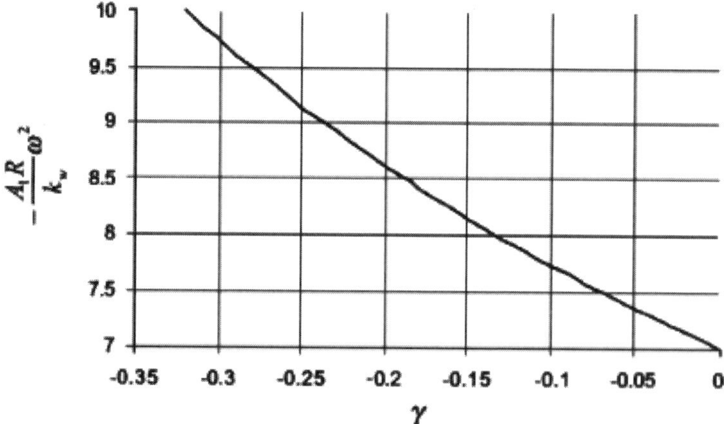

Figure B.4. Variation of the natural frequency squared versus $\gamma \in]\beta, 0[$; $v = 0.3$

condition for D to be positive depends then only upon the sign of b_5 as defined in Eq. (B.39). We can easily observe that

$$b_5 < 0 \quad \text{for } \gamma > a. \tag{B.48}$$

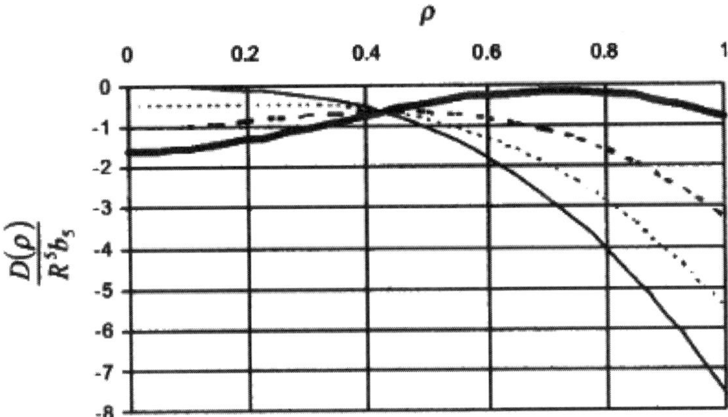

Figure B.5. Variation of the flexural rigidity of an inhomogeneous plate with translational spring on the boundary with a linearly decreased inertial term along the radial coordinate and a coefficient of Poisson fixed to $v = 0.3$, for different value of $\gamma \in]\beta, 0[$: ——, $\gamma = -0.01$; - - - -, $\gamma = -0.1$; - - -, $\gamma = -0.2$; ——, $\gamma = -0.31$

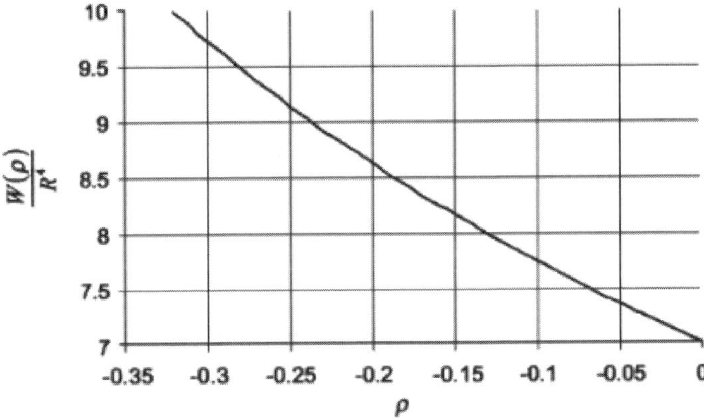

Figure B.6. Mode shape of an inhomogeneous plate with translational spring on the boundary with a linearly decreasing inertial term along the radial coordinate and a coefficient of Poisson fixed to $v = 0.3$

Finally, the solution obtained in Eqs. (B.40)–(B.42) has a physical explication when

$$a_1 < 0 \quad \text{for } \beta < \gamma < 0. \tag{B.49}$$

Figs. B.4–B.6 portray ω^2, $D(\rho)$ and $W(\rho)$ for different values of $\gamma \in [\beta, 0]$.

B.7 Conclusion

Apparently, the first closed-form solution has been derived for the free vibrations of inhomogeneous circular plates supported by a translational spring along plate's boundary.

Appendix C

Apparently First Closed-Form Solutions for Non-Symmetric Vibrations of Inhomogeneous Circular Plates

In this Appendix we report the closed-form solutions reported for non-axisymmetric free vibrations following paper by Storch and Elishakoff (2004). The problem is posed in an inverse setting; i.e. the distribution of the flexural rigidity is determined so as to be compatible with the postulated mode shape in the form of a polynomial function of the polar radius. Closed-form solutions are obtained for two sets of boundary conditions.

C.1 Introduction

The first work dedicated to the free vibrations of circular isotropic plates was apparently due to Chladni (1803) two hundred years ago. The state of the art on this topic was summarized by Leissa (1969). Most papers are devoted to the axisymmetric vibrations. Kovalenko (1976) was apparently the first who provided the exact solutions for non-axisymmetric vibrations for a single nodal diameter of a circular plate with linearly varying thickness. No closed-form solutions have been reported for plates with clamped or simply supported edges until Elishakoff (2000) formulated the axisymmetric vibration frequencies of clamped plates by the semi-inverse method. Here, the apparently first closed-form solutions are reported for the circular plates in the *non-axisymmetric* setting.

C.2   Governing differential equation

In polar coordinates (r, θ), The equation governing the forced vibration of a circular plate with loading $q(r, \theta)$ and flexural rigidity $D(r)$ that varies with the polar radius r is Kovalenko (1976)

$$
D \nabla^4 w + 2 \frac{dD}{dr} \frac{\partial}{\partial r} \nabla^2 w + \nabla^2 D \nabla^2 w - (1 - v)
$$

$$
\times \left[\frac{1}{r^2} \frac{d^2 D}{dr^2} \frac{\partial^2 w}{\partial \theta^2} + \frac{1}{r} \frac{d^2 D}{dr^2} \frac{\partial w}{\partial r} + \frac{1}{r} \frac{dD}{dr} \frac{\partial^2 w}{\partial r^2} \right] + \delta \frac{\partial^2 w}{\partial t^2} = q(r, \theta)
$$

$$\tag{C.1}$$

where $D(r)$ and the Laplacian in polar coordinates, are defined, respectively, as

$$
D = \frac{Eh^3}{12(1 - v^2)}, \quad \nabla^2 = \frac{\partial^2}{\partial r^2} + \frac{1}{r} \frac{\partial}{\partial r} + \frac{1}{r^2} \frac{\partial^2}{\partial \theta^2} \tag{C.2}
$$

Here δ is the mass per unit area of the plate and h is the plate thickness both of which we assume to be a function only of r. The parameters E and v denote the modulus of elasticity and Poisson ratio respectively.

Setting $q = 0$ and

$$
w = \exp[i(\omega t + n\theta)] \, W(r) \quad (n = 0, 1, 2, \ldots) \tag{C.3}
$$

In (C.1), we obtain the following equation on the radial portion of the free vibration mode shape corresponding to the natural frequency ω

$$
D \left[r^4 \frac{d^4 W}{dr^4} + 2r^3 \frac{d^3 W}{dr^3} - (2n^2 + 1) \, r^2 \frac{d^2 W}{dr^2} + (2n^2 + 1)r \frac{dW}{dr} \right.
$$

$$
\left. + n^2(n^2 - 4) W \right] + \frac{dD}{dr} \left[2r^4 \frac{d^3 W}{dr^3} + (v + 2)r^3 \frac{d^2 W}{dr^2} \right.
$$

$$
\left. - (2n^2 + 1)r^2 \frac{dW}{dr} + 3n^2 r W \right] + \frac{d^2 D}{dr^2} \left[r^4 \frac{d^2 W}{dr^2} + vr^3 \frac{dW}{dr} - n^2 v \, r^2 W \right]
$$

$$
- \omega^2 r^4 \delta \, W = 0 \tag{C.4}
$$

Note that when $n = 0$, we recover the equation in Elishakoff (2000) for the axisymmetric case. As in Elishakoff (2000), we pose to find the flexural rigidity $D(r)$ for a given distribution of density $\delta(r)$ and postulated mode shape all of which are polynomials. The results presented below are limited to considerations of uniform density and the first "angular mode" ($n = 1$) for the cases of clamped and simply supported boundary conditions. A closed form expression for the natural frequency is also given.

C.3 Clamped edge

If the plate has radius R, the mode shape W must satisfy the boundary conditions

$$W = dW/dr = 0 \quad \text{at } r = R \tag{C.5}$$

To obtain a candidate mode shape, consider the static deflection of a *uniform* clamped plate under the loading $q = q_0 \cos\theta$ where q_0 is constant. Solving equation (C.1) subject to the above boundary conditions, we find

$$w(r, \theta) = \frac{q_0}{90D} r(r - R)^2 (2r + R) \cos\theta \tag{C.6}$$

Thus we take $n = 1$ and postulate the following mode shape $W(r)$

$$W(r) = r(r - R)^2 (2r + R) \tag{C.7}$$

The problem proceeds as follows: find density and flexural rigidity distributions so that equation (C.4) is identically satisfied. We observe that if $\delta(r)$ is taken as a polynomial of degree m, then $D(r)$ must be of degree $m + 4$. For a uniform density $\delta(r) = a_0$, where a_0, must be positive.

Setting

$$D(r) = \sum_{i=0}^{4} b_i r^i \tag{C.8}$$

we obtain the following set of 5 linear equations on the 6 unknowns $\{b_0, b_1, b_2, b_3, b_4, \omega^2\}$

$$
\begin{aligned}
&R(v + 3)b_1 - 5b_0 = 0 \\
&48R(v + 3)b_2 - 24(v + 9)b_1 + R^3 a_0 \omega^2 = 0 \\
&3R(v + 3)b_3 - (2v + 13)b_2 = 0 \\
&12(3v + 17)b_3 - 48R(v + 3)b_4 + Ra_0 \omega^2 = 0 \\
&21(4v + 21)b_4 - a_0 \omega^2 = 0
\end{aligned}
\tag{C.9}
$$

From the last equation we have

$$\omega^2 = 21(4\nu + 21)b_4/a_0 \tag{C.10}$$

where b_4 is arbitrary but positive. The remaining coefficients are given by

$$
\begin{aligned}
b_0 &= \frac{(\nu + 3)(24\nu^3 + 874\nu^2 + 8495\nu + 21795)R^4 b_4}{40(\nu + 9)(2\nu + 13)(3\nu + 17)} \\[2mm]
b_1 &= \frac{(24\nu^3 + 874\nu^2 + 8495\nu + 21795)R^3 b_4}{8(\nu + 9)(2\nu + 13)(3\nu + 17)} \\[2mm]
b_2 &= -\frac{9(\nu + 3)(4\nu + 33)R^2 b_4}{4(2\nu + 13)(3\nu + 17)} \\[2mm]
b_3 &= -\frac{3(4\nu + 33)R b_4}{4(3\nu + 17)}
\end{aligned}
\tag{C.11}
$$

Figure C.1 depicts the radial variation in the flexural rigidity $(D/R^4 b_4)$ for three values of the Poisson ratio ν.

A second solution can be obtained by selecting as the candidate mode shape the static deflection of a *uniform* clamped plate under the loading $q = q_1(r/R)\cos\theta$ where q_1 is constant (Reddy (1999). Solving equation (C.1) subject to the clamped boundary conditions, we obtain

$$w(r, \theta) = \frac{q_1 R^4}{192D}\left(\frac{r}{R}\right)\left(1 - \frac{r^2}{R^2}\right)^2 \cos\theta \tag{C.12}$$

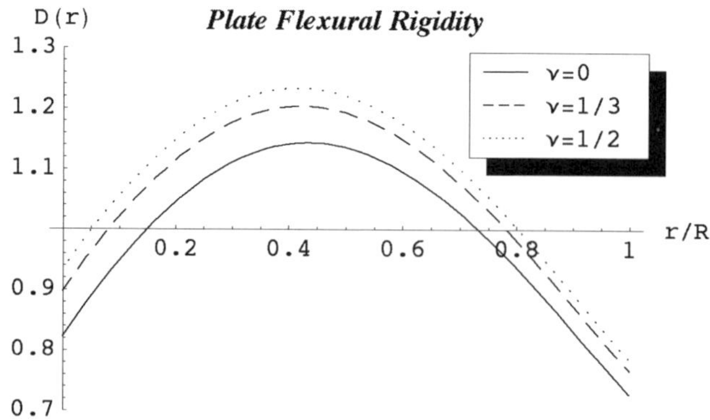

Figure C.1. Variation in Flexural Rigidity for a clamped circular plate corresponding to the mode shape $W = r(r - R)^2(2r + R)\cos\theta$.

We now take the radial portion of the above expression for $W(r)$ in eq. (C.4) with $n = 1$ and again assume a constant density a_0. Assuming the polynomial form $D(r) = \sum_{i=0}^{4} b_i r^i$ for the plate rigidity, we obtain the following system of 6 homogeneous equations in the 6 unknowns $\{b_0, b_1, b_2, b_3, b_4, \omega^2\}$

$$b_1 = 0$$
$$R^4 a_0 \omega^2 + 32[R^2(v+3)b_2 - 6b_0] = 0$$
$$3R^2(v+3)b_3 - (v+17)b_1 = 0$$
$$R^2 a_0 \omega^2 - 24[2R^2(v+3)b_4 - (v+11)b_2] = 0 \tag{C.13}$$
$$b_3 = 0$$
$$128(v+8)b_4 - a_0 \omega^2 = 0$$

Fortunately, the above system has a nontrivial solution. The natural frequency is given by

$$\omega^2 = \frac{128(v+8)}{a_0} b_4 \tag{C.14}$$

where b_4 is arbitrary but positive, while the remaining coefficients are given by

$$b_0 = \frac{1}{9} R^4 (v+33) b_4$$
$$b_1 = 0$$
$$b_2 = -\frac{10}{3} R^2 b_4 \tag{C.15}$$
$$b_3 = 0$$

The corresponding plate flexural rigidity can be expressed in the form

$$D(r) = \frac{1}{9} R^4 [9(r/R)^4 - 30(r/R)^2 + 33 + v] b_4 \tag{C.16}$$

Figure C.2 depicts the radial variation in the flexural rigidity $(D/R^4 b_4)$ for two values of the Poisson ratio v.

C.4 Simply supported edge

The mode shape W must now satisfy the boundary conditions

$$W = M_r = 0 \quad \text{at } r = R \tag{C.17}$$

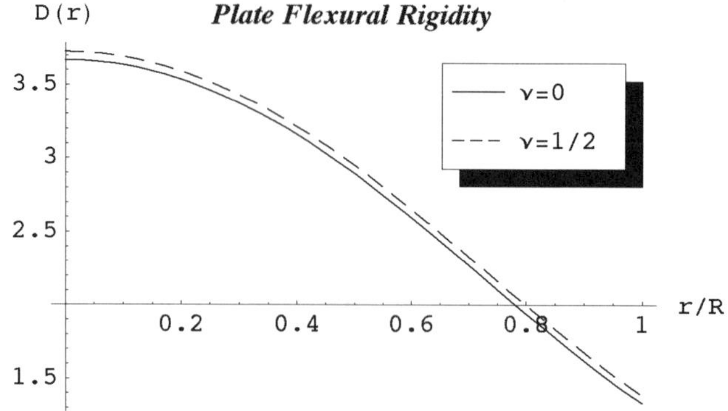

Figure C.2. Variation in Flexural Rigidity for a clamped circular plate corresponding to the mode shape $W = r(1 - r^2/R^2)^2 \cos\theta$.

where the bending moment M_r is given by

$$M_r = -D\left[\frac{\partial^2 W}{\partial r^2} + v\left(\frac{1}{r}\frac{\partial W}{\partial r} + \frac{1}{r^2}\frac{\partial^2 W}{\partial \theta^2}\right)\right] \tag{C.18}$$

To obtain a candidate mode shape, consider the static deflection of a *uniform* simply supported plate under the loading $q = q_0 \cos\theta$ where q_0 is constant. Solving equation (C.1) subject to the above boundary conditions we obtain

$$w(r, \theta) = \frac{q_0 r[2(v + 3)r^3 - 3R(v + 4)r^2 + R^3(v + 6)]}{90D(v + 3)} \cos\theta \tag{C.19}$$

Thus we take $n = 1$ and postulate the following mode shape $W(r)$

$$W(r) = 2(v + 3)r^4 - 3R(v + 4)r^3 + R^3(v + 6)r \tag{C.20}$$

We proceed to find distributions in density and flexural rigidity so that equation (C.4) is identically satisfied. For a uniform density $\delta(r) = a_0$, where a_0 must be positive. Taking $D(r)$ as in Eq. (C.8), we obtain the following set of 5 linear equations on the 6 unknowns $\{b_0, b_1, b_2, b_3, b_4, \omega^2\}$

$$5b_0 - R(v + 4)b_1 = 0$$
$$24(v + 3)\left[(v + 9)b_1 - 2R(v + 4)b_2\right] - R^3(v + 6)a_0\omega^2 = 0$$
$$(2v + 13)b_2 - 3R(v + 4)b_3 = 0 \tag{C.21}$$
$$R(v + 4)a_0\omega^2 + 12(v + 3)[(3v + 17)b_3 - 4R(v + 4)b_4] = 0$$
$$21(4v + 21)b_4 - a_0\omega^2 = 0$$

From the last equation we have

$$\omega^2 = 21(4\nu + 21)b_4/a_0 \tag{C.22}$$

where b_4 is arbitrary but positive. The remaining coefficients are given by

$$b_0 = \frac{R^4(\nu + 4)(24\nu^4 + 1018\nu^3 + 13307\nu^2 + 67761\nu + 118890)b_4}{40(\nu + 3)(\nu + 9)(2\nu + 13)(3\nu + 17)}$$

$$b_1 = \frac{R^3(24\nu^4 + 1018\nu^3 + 13307\nu^2 + 67761\nu + 118890)b_4}{8(\nu + 3)(\nu + 9)(2\nu + 13)(3\nu + 17)}$$

$$b_2 = -\frac{9R^2(\nu + 4)^2(4\nu + 33)b_4}{4(\nu + 3)(2\nu + 13)(3\nu + 17)} \tag{C.23}$$

$$b_3 = -\frac{3R(\nu + 4)(4\nu + 33)b_4}{4(\nu + 3)(3\nu + 17)}$$

Figure C.3 depicts radial variation in the flexural rigidity (D/R^4b_4) for two values of the Poisson ratio ν. Compared to the clamped case, it is rather insensitive to ν.

A second solution can be obtained by selecting as the candidate mode shape the static deflection $w(r, \theta)$ of a *uniform* simply supported plate under the loading $q = q_1 r/R \cos\theta$ where q_1 is constant (Reddy, 1999).

$$w(r, \theta) = \frac{q_1 R^4}{192D(3 + \nu)}\left(\frac{r}{R}\right)\left(1 - \frac{r^2}{R^2}\right)\left[7 + \nu - (3 + \nu)\left(\frac{r}{R}\right)^2\right]\cos\theta \tag{C.24}$$

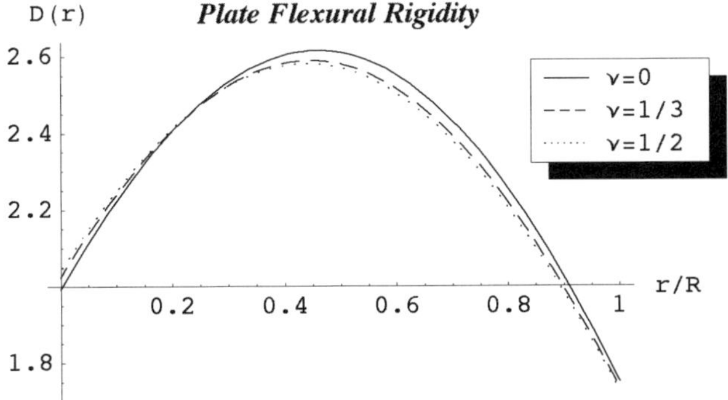

Figure C.3. Variation in Flexural Rigidity for simply supported circular plate with mode shape $W = [2(\nu + 3)r^4 - 3R(\nu + 4)r^3 + R^3(\nu + 6)r]\cos\theta$.

Proceeding as above, we obtain the following system of 6 homogeneous equations in the 6 unknowns $\{b_0, b_1, b_2, b_3, b_4, \omega^2\}$

$$b_1 = 0$$
$$R^4 a_0(v+7)\omega^2 + 32(v+3)[R^2(v+5)b_2 - 6b_0] = 0$$
$$(v+17)b_1 - 3R^2(v+5)b_3 = 0$$
$$R^2 a_0(v+5)\omega^2 + 24(v+3)(v+11)b_2 - 48R^2(v+3)(v+5)b_4 = 0$$
$$b_3 = 0$$
$$128(v+8)b_4 - a_0\omega^2 = 0$$

(C.25)

Fortunately, the above system has a nontrivial solution. The natural frequency is given by

$$\omega^2 = \frac{128(v+8)}{a_0}b_4 \tag{C.26}$$

where b_4 is arbitrary but positive, while the remaining coefficients are given by

$$b_0 = \frac{R^4(v^2 + 40v + 211)}{9(v+3)}b_4$$
$$b_1 = 0$$
$$b_2 = -\frac{10R^2(v+5)}{3(v+3)}b_4$$
$$b_3 = 0$$

(C.27)

The corresponding flexural rigidity of the plate can be expressed in the form

$$D(r) = \frac{R^4[9(v+3)(r/R)^4 - 30(v+5)(r/R)^2 + v^2 + 40v + 211]}{9(v+3)}b_4 \tag{C.28}$$

Figure C.4 depicts the variation in the flexural rigidity $(D/R^4 b_4)$ for three values of the Poisson ratio v.

C.5 Conclusion

It appears remarkable that whereas for the circular plate of *constant* flexural rigidity the mode shapes involve Bessel and trigonometric functions, simple polynomial and trigonometric functions are obtained under a flexural

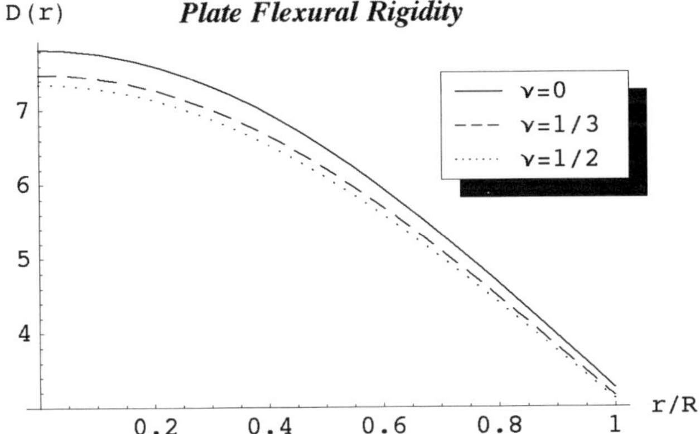

Figure C.4. Variation in Flexural Rigidity for simply supported circular plate with mode shape $W = r(R^2 - r^2)\left[7 + v - (3 + v)\frac{r^2}{R^2}\right]\cos\theta$.

rigidity that varies. The unusual nature of this solution can be explained by our exploitation of the additional parameters arising due to inhomogeneity. The reported solutions can serve, for example, as benchmarks for verifying the accuracy of various numerical techniques.

Appendix D

Closed-Form Solution for Axisymmetric Vibration of Inhomogeneous Simply-Supported Circular Plates

In this Appendix we deal with obtaining the exact, closed-form solution for radially graded inhomogeneous simply supported plates. We closely follow the paper by Elishakoff and Storch (2005).

D.1 Introduction

The free vibration of circular plates with variable thickness or density has received considerable attention in the literature. Conway (1980) found an unusual closed-form solution for a variable-thickness plate on an elastic foundation, in a static setting. Whereas a constant thickness plate involves Kelvin functions, Conway's solution was derived in closed-form. Analogous, closed-form solutions were derived by Harris (1968) for plates of variable thickness, free on their boundary and by Lenox and Conway (1980) who studied plates with arbitrary conditions, and with parabolic thickness variation. In recent papers, Elishakoff (2000, 2000) derived closed-form solutions for inhomogeneous circular plates that are either clamped, or free at the boundary. Here we derive closed-form solutions for the inhomogeneous plates that are simply supported at their boundary.

There appears to be a single monograph (1959) solely devoted to plates of variable thickness. There are several papers dedicated to vibrations of plates with thickness variations. Axisymmetric vibration of circular plates of linearly varying thickness was studied by Prasad *et al.* (1972), whereas plates with double linear thickness were studied by Sing and Sascena (1995).

Various analytical and approximate techniques have been studied (Baraket and Baumann, 1968, Chen and Ren, 1998, Gupta and Ansari, 1998, Liew and Yang, 1999, Singh and Hassan, 1998, Singh and Chakraverty, 1991, 1992, Yang and Xie, 1984, and Yeh, 1994).

In this Appendix, we find closed form solutions to an inverse vibration problem for a simply supported circular plate. Given a candidate mode shape and density distribution, we calculate the plate stiffness so that the governing equation and boundary conditions for the plate mode shape are identically satisfied. We also are able to obtain the expression for the corresponding natural frequency. The solution is characterized here as an unusual one since for its counterpart — the homogeneous plate, transcendental functions are called for whereas here a solution is found in elementary functions, namely, polynomials. The solutions can also serve as benchmarks for validation of numerical techniques.

D.2 Basic equations

The differential equation that governs the free axisymmetric vibration of the circular plate with variable thickness reads in Elishakoff (2000)

$$D(r)r^3 \Delta\Delta W + \frac{dD}{dr}\left(2r^3\frac{d^3W}{dr^3} + (2+v)r^2\frac{d^2W}{dr^2} - r\frac{dW}{dr}\right)$$

$$+\frac{d^2D}{dr^2}\left(r^3\frac{d^2W}{dr^2} + vr^2\frac{dW}{dr}\right) - \rho h\omega^2 r^3 W = 0 \qquad (D.1)$$

where $D(r)$ is the plate flexural rigidity

$$D = \frac{Eh^3}{12(1-v^2)}$$

$h(r)$ is the thickness, v the Poisson ratio, E the Young's modulus, and ρ is the mass density. Here Δ is the Laplacian operator in polar coordinates ($\Delta = d^2/dr^2 + r^{-1}d/dr$) where r denotes the radial coordinate and W the mode shape. Defining the inertial term (mass per unit area) $\delta(r) = \rho h$, we pose to find the stiffness distribution $D(r)$ and natural frequency ω such that Eq.(D.1) is identically satisfied for a specified $\delta(r)$ and $W(r)$.

To obtain a candidate mode shape, consider the static displacement of a *uniform* circular simply supported plate under uniform load q_0 per unit area. From Timoshenko and Woinowsky-Krieger (1959), we have

$$w = \frac{q_0}{64D}(R^2 - r^2)(\theta R^2 - r^2)$$

where R is the radius of the plate and the parameter θ depends solely of the Poisson ratio.

$$\theta = (5 + v)/(1 + v)$$

Thus we seek to determine under what conditions is the function

$$W = (R^2 - r^2)(\theta R^2 - r^2) \tag{D.2}$$

a solution to Eq. (D.1). If we assume that the inertial term is a specified polynomial of degree m,

$$\delta = \sum_{i=0}^{m} a_i r^i \tag{D.3}$$

then it follows from Eq. (D.1) and the observation that W is a 4th degree polynomial, that the stiffness must be of degree $m + 4$

$$D = \sum_{i=0}^{m+4} b_i\, r^i \tag{D.4}$$

In the following sections we find the coefficients b_i and natural frequency ω for the case of a constant, linear and parabolic inertial term.

D.3 Constant Inertial Term ($m = 0$)

We are given $\delta = a_0 > 0$, and the stiffness is sought as a fourth order polynomial

$$D(r) = b_0 + b_1 r + b_2 r^2 + b_3 r^3 + b_4 r^4$$

Inserting this expression along with Eq. (D.2) into the differential equation (D.1), we obtain

$$\sum_{i=0}^{7} c_i r^i = 0$$

where

$$c_0 = 0, \quad c_1 = 0, \quad c_2 = -4(3 + v)R^2 b_1,$$

$$c_3 = 64b_0 - 16(3 + v)R^2 b_2 - a_0\omega^2 R^4 \frac{5 + v}{1 + v}$$

$$c_4 = 12(11 + v)b_1 - 36(3 + v)R^2 b_3 \tag{D.5}$$

$$c_5 = 32(7 + v)b_2 - 64(3 + v)R^2 b_4 + 2a_0\omega^2 R^2 \frac{3 + v}{1 + v}$$

$$c_6 = (340 + 60v)b_3, \quad c_7 = 96(5 + v)b_4 - a_0\omega^2$$

Demanding that the above set of (nontrivial) coefficients vanish, yields a set of 6 homogeneous linear equations on the 6 unknowns $\{b_0, b_1, \ldots, b_4, \omega^2\}$. Fortunately, the determinant of the associated coefficient matrix is zero, hence a non-trivial solution is obtainable. Setting the coefficient c_7 to zero, we obtain the natural frequency

$$\omega^2 = 96(5 + v)/b_4/a_0 \tag{D.6}$$

where b_4 is arbitrary but positive.

Upon substitution of eq. (D.6) into (D.5), the remaining equations yield the coefficients in the stiffness

$$b_0 = \frac{57 + 18v + v^2}{2(1 + v)}R^4 b_4, \quad b_1 = 0, \quad b_2 = -4\frac{3 + v}{1 + v}R^2 b_4, \quad b_3 = 0$$

Hence, the stiffness reads

$$D(r) = \left(\frac{57 + 18v + v^2}{2(1 + v)}R^4 - 4\frac{3 + v}{1 + v}R^2 r^2 + r^4\right) b_4$$

Fig. D.1 depicts the stiffness for various values of the Poisson ratio v.

D.4   Linearly varying inertial term ($m = 1$)

We are given $\delta = a_0 + a_1 r$, and the stiffness is sought as a 5th order polynomial

$$D(r) = b_0 + b_1 r + b_2 r^2 + b_3 r^3 + b_4 r^4 + b_5 r^5$$

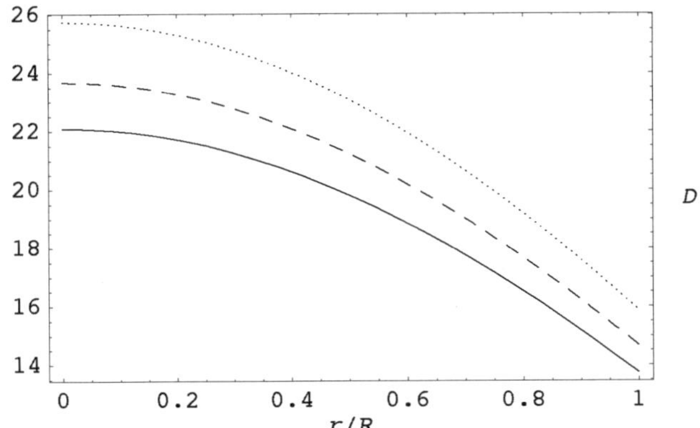

Figure D.1. Stiffnesses for different values of Poisson's ratio, — $\nu = 1/2$; - - - $\nu = 1/3$; - - - $\nu = 1/6$

Instead of the set (D.5), we get here 7 linear algebraic equations in the 7 unknowns $\{b_0, b_1, \ldots, b_5, \omega^2\}$

$$-4(3 + \nu)R^2 b_1 = 0$$

$$64b_0 - 16(3 + \nu)R^2 b_2 - a_0\omega^2 R^4 \frac{5 + \nu}{1 + \nu} = 0$$

$$12(11 + \nu)b_1 - 36(3 + \nu)R^2 b_3 - a_1\omega^2 R^4 \frac{5 + \nu}{1 + \nu} = 0$$

$$32(7 + \nu)b_2 - 64(3 + \nu)R^2 b_4 + 2a_0\omega^2 R^2 \frac{3 + \nu}{1 + \nu} = 0 \qquad \text{(D.7)}$$

$$(340 + 60\nu)b_3 - 100(3 + \nu)R^2 b_5 + 2a_1\omega^2 R^2 \frac{3 + \nu}{1 + \nu} = 0$$

$$96(5 + \nu)b_4 - a_0\omega^2 = 0$$

$$(644 + 140\nu)b_5 - a_1\omega^2 = 0$$

In order to have a nontrivial solution, the determinant of the coefficient matrix must vanish.

$$R^6 \frac{(3 + \nu)(7 + \nu)(5 + \nu)(5546 + 2619\nu + 364\nu^2 + 15\nu^3)}{1 + \nu} a_1 = 0$$

Thus $a_1 = 0$, and we obtain the solution in the previous section.

D.5   Parabolically Varying Inertial Term ($m = 2$)

For $m = 2$, i.e. the plate whose material density varies parabolically

$$\delta(r) = a_0 + a_1 r + a_2 r^2 \tag{D.8}$$

the bending stiffness has to be a sixth order polynomial

$$D(r) = b_0 + b_1 r + b_2 r^2 + b_3 r^3 + b_4 r^4 + b_5 r^5 + b_6 r^6 \tag{D.9}$$

Substitution of Eq. (D.2) in conjunction with Eqs. (D.8) and (D.9) into the governing differential equation (D.1) yields

$$\sum_{i=0}^{9} d_i r^i = 0$$

where

$$d_0 = 0, \quad d_1 = 0, \quad d_2 = -4R^2(3 + v)b_1$$

$$d_3 = 64b_0 - 16R^2(3 + v)b_2 - a_0 \omega^2 R^4 \frac{5 + v}{1 + v}$$

$$d_4 = 12(11 + v)b_1 - 36R^2(3 + v)b_3 - a_1 \omega^2 R^4 \frac{5 + v}{1 + v}$$

$$d_5 = 32(7 + v)b_2 - 64R^2(3 + v)b_4$$
$$+ \omega^2 \left(2a_0 R^2 \frac{3 + v}{1 + v} - a_2 R^4 \frac{5 + v}{1 + v} \right)$$

$$d_6 = 20(17 + 3v)b_3 - 100R^2(3 + v)b_5 + 2a_1 \omega^2 R^2 \frac{3 + v}{1 + v}$$

$$d_7 = 96(5 + v)b_4 - 144R^2(3 + v)b_6 - \omega^2 \left(a_0 - 2a_2 R^2 \frac{3 + v}{1 + v} \right)$$

$$d_8 = (644 + 140v)b_5 - a_1 \omega^2, \quad d_9 = (832 + 192v)b_6 - a_2 \omega^2$$

$$\tag{D.10}$$

As in the case of the constant inertial term, we demand that all $d_i = 0$, thus, we get a set of 8 equations with 8 unknowns (7 coefficients b_i and ω^2). The resulting determinantal equation is

$$(3 + v)(7 + v)a_1 (360490 + 325523v + 113630v^2 + 19024v^3$$
$$+ 1512v^4 + 45v^5)/(1 + v) = 0$$

In order for the homogeneous system to have a non-trivial solution we must demand the coefficient a_1 to vanish. Substituting $a_1 = 0$ into the set (D.10) and using the last equation for the determination of the frequency, we obtain

$$\omega^2 = 64(13 + 3v)b_6/a_2 \tag{D.11}$$

and the stiffness coefficients

$$b_0 = R^4 b_6 [(3285 + 2670v + 744v^2 + 82v^3 + 3v^4)R^2 a_2$$
$$+ (20748 + 14304v + 3496v^2 + 352v^3 + 12v^4)a_0]/12(35 + 47v$$
$$+ 13v^2 + v^3)a_2$$

$$b_1 = 0$$

$$b_2 = \frac{(1095 + 525v + 73v^2 + 3v^3)R^2 a_2 - (2184 + 1544v + 344v^2 + 24v^3)a_0}{3(35 + 47v + 13v^2 + v^3)a_2} R^2 b_6$$

$$b_3 = 0, \quad b_4 = -\frac{(285 + 140v + 15v^2)R^2 a_2 - (52 + 64v + 12v^2)a_0}{6(5 + 6v + v^2)a_2} b_6, \quad b_5 = 0$$

where b_6 is an arbitrary constant. It follows from Eq. (D.11) that the ratio b_6/a_2 must be positive. We have two sub-cases: (D.1) both b_6 and a_2 are positive, or (D.2) both are negative. In the former case (D.1) the necessary condition for positiveness of the stiffness $b_0 \geq 0$ is identically satisfied. In the latter case (D.2) the above inequality reduces to

$$(3285 + 2670v + 744v^2 + 82v^3 + 3v^4)R^2$$
$$+ (20748 + 14304v + 3496v^2 + 352v^3 + 12v^4)a_0/a_2 \leq 0$$

Leading to the following inequality

$$\frac{a_0}{|a_2|R^2} \geq \frac{3285 + 2670v + 744v^2 + 82v^3 + 3v^4}{20748 + 14304v + 3496v^2 + 352v^3 + 12v^4} \tag{D.12}$$

It is interesting to note that in this case, the requirement that $\delta(R)$ be non-negative implies

$$\frac{a_0}{|a_2|R^2} \geq 1$$

which is stronger than the inequality (D.12).

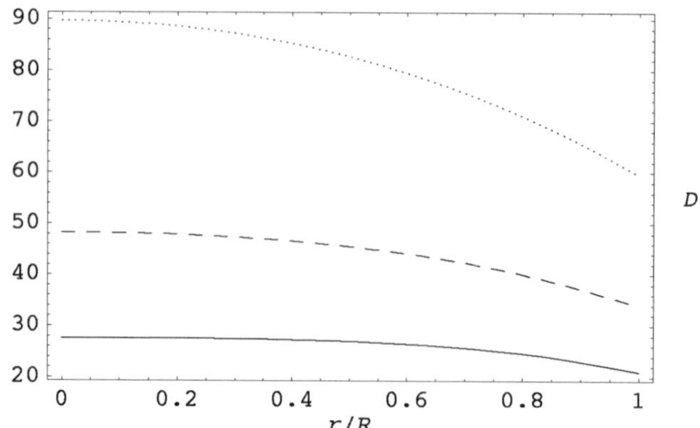

Figure D.2. Variation of flexural rigidity vs. r/R, parabolic inertial term an t different values of t: — $t = 1/2$; - - - $t = 1$; - - - $t = 2$

As an example, for $v = 1/3$, the stiffness reads

$$D(r) = R^6 \left[\left(\frac{3595}{528} + \frac{497a_0}{12R^2a_2} \right) + \left(\frac{719}{88} - \frac{35a_0}{2R^2a_2} \right) \left(\frac{r}{R} \right)^2 \right.$$
$$\left. + \left(-\frac{125}{16} + \frac{7a_0}{4R^2a_2} \right) \left(\frac{r}{R} \right)^4 + \left(\frac{r}{R} \right)^6 \right] b_6$$

D.6   Cubic inertial term ($m = 3$)

Proceeding as before, the following set of 9 linear algebraic equations in the 9 unknowns $\{b_0, b_1, \ldots, b_7, \omega\}$, is obtained

$$-4R^2(3 + v)b_1 = 0$$

$$64b_0 - 16R^2(3 + v)b_2 - a_0\omega^2 R^4 \frac{5 + v}{1 + v} = 0 \qquad\qquad (D.13)$$

$$12(11 + v)b_1 - 36R^2(3 + v)b_3 - a_1\omega^2 R^4 \frac{5 + v}{1 + v} = 0$$

$$32(7 + v)b_2 - 64R^2(3 + v)b_4 + \omega^2 \left(2a_0 R^2 \frac{3 + v}{1 + v} - a_2 R^4 \frac{5 + v}{1 + v} \right) = 0$$

$$20(17 + 3v)b_3 - 100R^2(3 + v)b_5$$
$$+ \omega^2 \left(2a_1 R^2 \frac{3 + v}{1 + v} - a_3 R^4 \frac{5 + v}{1 + v} \right) = 0$$

$$96(5 + v)b_4 - 144R^2(3 + v)b_6 - \omega^2\left(a_0 - 2a_2R^2\frac{3+v}{1+v}\right) = 0$$

$$(644 + 140v)b_5 - 196(3 + v)R^2b_7 - \omega^2\left(a_1 - 2a_3R^2\frac{3+v}{1+v}\right) = 0 \tag{D.14}$$

$$(832 + 192v)b_6 - a_2\omega^2 = 0 \tag{D.15}$$

$$(1044 + 252v)b_7 - a_3\omega^2 = 0$$

The detenninantal equation stemming from it reads

$$(2527200 + 3153105v + 1582173v^2 + 406618v^3 + 56058v^4$$

$$+ 3893v^5 + 105v^6)R^2a_3 + (10454210 + 11963597v + 5573931v^2$$

$$+ 1347106v^3 + 177016v^4 + 11889v^5 + 315v^6)a_1 = 0$$

Thus the coefficients a_1 and a_3 in the density distribution must satisfy the relation

$$a_1 = -\frac{38880 + 31761v + 8865v^2 + 971v^3 + 35v^4}{160834 + 114773v + 28889v^2 + 2983v^3 + 105v^4}R^2a_3$$

Substituting this equation into Eqs. (D.13)–(D.15), we find

$$\omega^2 = 36(29 + 7v)b_7/a_3 \tag{D.16}$$

and obtain the following solution for the coefficients in the stiffness

$$b_0 = 3[(95265 + 100425v + 40266v^2 + 7586v^3 + 661v^4 + 21v^5)R^2a_2$$

$$+ (601692 + 560052v + 201512v^2 + 34680v^3 + 2812v^4 + 84v^5)a_0]$$

$$R^4b_7/64a_3M$$

$$b_1 = 0$$

$$b_2 = 3[(31755 + 22890v + 5792v^2 + 598v^3 + 21v^4)R^2a_2$$

$$- (63336 + 60064v + 20784v^2 + 3104v^3 + 168v^4)a_0]R^2b_7/16a_3M$$

$$b_3 = (64800 + 44295v + 10597v^2 + 1041v^3 + 35v^4)R^4b_7/N$$

$$b_4 = 3b_7[(1508 + 2220v + 796v^2 + 84v^3)a_0 - (8265 + 6055v + 1415v^2$$

$$+ 105v^3)R^2a_2]/32a_3(65 + 93v + 31v^2 + 3v^3)$$

$$b_5 = -2R^2(25527 + 20092v + 5424v^2 + 584v^3 + 21v^4)b_7/N$$

$$b_6 = 9(29 + 7v)a_2b_7/16a_3(13 + 3v)$$

where

$$M = 455 + 716v + 310v^2 + 52v^3 + 3v^4$$

$$N = 5546 + 8165v + 2983v^2 + 379v^3 + 15v^4$$

and b_7 an arbitrary constant with the same sign as a_3 (see eq. (D.16) For the particular case $v = 1/3$, the stiffness equals

$$D(r) = R^6 \left[\frac{168965a_2}{19712a_3} + \frac{3337a_0}{64R^2a_3} + \left(\frac{101379a_2}{9856a_3} - \frac{705a_0}{32R^2a_3} \right) \left(\frac{r}{R} \right)^2 \right.$$

$$+ \frac{21524R}{2295} \left(\frac{r}{R} \right)^3 + \left(-\frac{17625a_2}{1792a_3} + \frac{141a_0}{64R^2a_3} \right) \left(\frac{r}{R} \right)^4$$

$$\left. - \frac{389R}{51} \left(\frac{r}{R} \right)^5 + \frac{141a_2}{112a_3} \left(\frac{r}{R} \right)^6 + R \left(\frac{r}{R} \right)^7 \right] b_7$$

D.7 General Inertial Term ($m \geq 4$)

Consider now the general expression of the inertial term given in eq. (D.3) and the stiffness; in eq. (D.4), for $m \geq 4$. Substitution of eqs. (D.2), (D.3) and (D.4) into the terms of the differential equation (D.1) yields

$$r^3 D(r) \Delta \Delta W = 64r^3 \sum_{i=0}^{m+4} b_i r^i \tag{D.17}$$

$$\frac{dD}{dr} \left(2r^3 \frac{d^3 W}{dr^3} + (2+v)r^2 \frac{d^2 W}{dr^2} - r \frac{dW}{dr} \right)$$

$$= 4[(17 + 3v)r^2 - (3+v)R^2]r^2 \sum_{i=1}^{m+4} ib_i r^{i-1} \tag{D.18}$$

$$\frac{d^2 D}{dr^2} \left(r^3 \frac{d^2 W}{dr^2} + vr^2 \frac{dW}{dr} \right) = 4(3+v)(r^2 - R^2)r^3 \sum_{i=2}^{m+4} i(i-1)b_i r^{i-2}$$

$$\tag{D.19}$$

$$-\rho h \omega^2 r^3 W = -\omega^2 r^3 (R^2 - r^2)(\theta R^2 - r^2) \sum_{i=0}^{m} a_i r^i \qquad (D.20)$$

Demanding the sum of (D.17)–(D.20) to be zero, we obtain the following equation

$$\sum_{i=2}^{m+7} g_i r^i = 0$$

where the coefficients g_i are

$$g_2 = -4(3 + v)b_1$$

$$g_3 = 64b_0 - 16R^2(3 + v)b_2 - \theta a_0 R^4 \omega^2$$

$$g_4 = 12(11 + v)b_1 - 36R^2(3 + v)b_3 - \theta a_1 R^4 \omega^2$$

$$g_5 = 32(7 + v)b_2 - 64R^2(3 + v)b_4 - \omega^2[\theta a_2 R^4 - (1 + \theta)a_0 R^2]$$

$$g_6 = 20(17 + 3v)b_3 - 100R^2(3 + v)b_5 - \omega^2[\theta a_3 R^4 - (1 + \theta)a_1 R^2]$$

$$\text{for } 7 \leq i \leq m + 3$$

$$g_i = 4(i - 1)[(v + 3)(i - 1) + 2(1 - v)]b_{i-3} - 4(i - 1)^2 R^2(3 + v)b_{i-1}$$
$$- \omega^2[\theta R^4 a_{i-3} - (1 + \theta)R^2 a_{i-5} + a_{i-7}]$$

$$g_{m+4} = 4(m + 3)[v + m(v + 3) + 11]b_{m+1} - 4R^2(m + 3)^2(3 + v)b_{m+3}$$
$$- \omega^2[a_{m-3} - (1 + \theta)R^2 a_{m-1}]$$

$$g_{m+5} = 4(m + 4)[m(v + 3) + 2(v + 7)]b_{m+2} - 4R^2(m + 4)^2(3 + v)b_{m+4}$$
$$- \omega^2[a_{m-2} - (1 + \theta)R^2 a_m]$$

$$g_{m+6} = 4(m + 5)[3v + 17 + m(v + 3)]b_{m+3} - \omega^2 a_{m-1}$$

$$g_{m+7} = 4(m + 6)[m(v + 3) + 4(v + 5)]b_{m+4} - \omega^2 a_m$$

We demand all coefficients g_i to be zero, thus, we get a set of $m + 6$ homogeneous linear algebraic equations for the $m + 6$ unknowns $\{b_0, b_1, \ldots, b_{m+4}, \omega^2\}$. In order to find a non-trivial solution the determinant of the associated coefficient matrix must vanish. We expand the determinant along the last column of the matrix of the set, getting a linear algebraic expression with a_i as coefficients. The determinantal equation yields a condition for which the non-trivial solution is obtainable. In this case the general expression of the natural frequency squared is obtained from the equation $g_{m+7} = 0$, resulting in

$$\omega^2 = \frac{4(m+6)[m(v+3)+4(v+5)]}{a_m} b_{m+4} \qquad \text{(D.21)}$$

Note that the formulas pertaining to the cases $m = 0$, $m = 2$ and $m = 3$ are formally obtainable from Eq. (D.21) by appropriate substitution.

D.8 An Alternative Mode Shape

In the previous sections, we were able to obtain the stiffness distribution corresponding to a polynomial density such that the 4th degree polynomial given in Eq. (D.2) served as a mode shape for the simply supported plate. The question naturally arises: can we find a lower order polynomial mode shape? Since there are two boundary conditions that need to be satisfied, we try the second degree polynomial

$$W = (r - R)[vr - R(v + 2)] \qquad \text{(D.22)}$$

which satisfies the simply supported boundary conditions $W = M_r = 0$ on $r = R$ (recall that $M_r = -D(\partial^2 W/\partial r^2 + v/r \partial W/\partial r)$). In connection with this mode shape, we will limit our discussion to the case of a constant density. Thus given $\delta = a_0 > 0$, we want to find $D(r)$ and the natural frequency ω such that (D.22) is a solution of Eq. (D.1). It is sufficient to consider 4th degree polynomials

$$D(r) = b_0 + b_1 r + b_2 r^2 + b_3 r^3 + b_4 r^4$$

Inserting this expression along with (D.22) into the differential equation, we find that the 6 unknowns $\{b_0, b_1, \ldots, b_4, \omega^2\}$ must satisfy the following system of 5 homogeneous linear equations.

$$b_0 = 0$$

$$vb_1 + R(1 - 2v)b_2 = 0$$

$$8v(v + 1)b_2 - 4R(v + 1)(3v - 1)b_3 - R^2 a_0(v + 2)\omega^2 = 0$$

$$9vb_3 + 3R(1 - 4v)b_4 + Ra_0\omega^2 = 0$$

$$32(v + 1)b_4 - a_0\omega^2 = 0$$

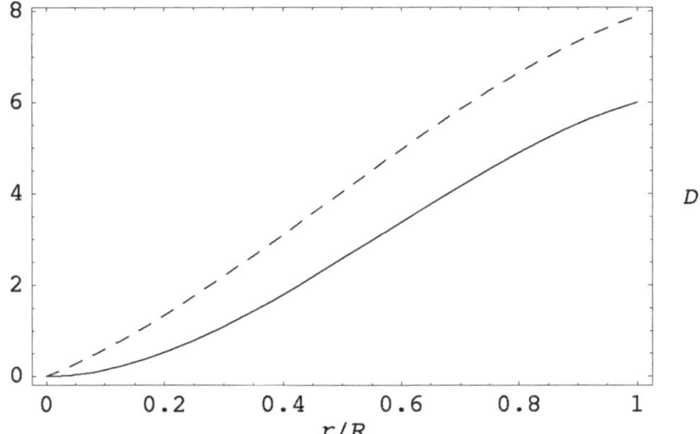

Figure D.3. Variation of flexural rigidity for various values of Poisson's ratio, — $\nu = 1/2$; - - - $\nu = 2/3$

The last equation yields the natural frequency

$$\omega^2 = 32(\nu + 1)b_4/a_0 \tag{D.23}$$

where b_4 is arbitrary but positive. The remaining stiffness coefficients are then given by

$$b_0 = 0, \quad b_1 = R^3(2\nu - 1)(12\nu^2 + 59\nu + 35)b_4/18\nu^3$$

$$b_2 = R^2(12\nu^2 + 59\nu + 35)b_4/18\nu^3, \quad b_3 = -5R(4\nu + 7)b_4/9\nu$$

Since $b_0 = 0$, we must have $b_1 \geq 0$ in order that the stiffness shall be nonnegative. Based upon the above solution, this implies that we must have $\nu \geq 1/2$. The plate stiffness is thus given by

$$D(r) = \left[\xi^4 - \frac{5(4\nu + 7)}{9\nu}\xi^3 + \frac{12\nu^2 + 59\nu + 35}{18\nu^2}\xi^2 \right.$$

$$\left. + \frac{(2\nu - 1)(12\nu^2 + 59\nu + 35)}{18\nu^3}\xi \right] R^4 b_4$$

where $\xi = r/R$.

References

Abid Mian, M. and Spencer, A.J.M., "Exact solutions for functionally graded laminated elastic materials", *Journal of Mechanics of Physics and Solids*, vol. 46, pp. 2283–2295, 1998.

Aboudi, J., Arnold, S.M. and Bednarchuk, B.A., *Micromechanics of Composite Materials: A Generalized Multiscale Analysis Approach*, Elsevier, Oxford, UK, 2013.

Aboudi, J., Arnold, S.M. and Pindera, M.J., "Response of functionally graded composites to thermal gradients", *Composites Engineering*, vol. 4(1), pp. 1–18, 1994.

Aboudi, J., Pindera, M.-J. and Arnold, S.M., "Thermoelastic theory for the response of materials functionally graded in two directions", *International Journal of Solids and Strucutres*, vol. 33(7), pp. 931–966, 1996.

Aboudi, J., Pindera, M.J. and Arnold, S.M., "Higher-order theory for functionally graded materials", *Composites, Part B: Engineering*, vol. 30(8), pp. 777–832, 1999.

Abrate, S., "Vibration of non-uniform rods and beams", *Journal of Sound and Vibration*, vol. 185(4), pp. 703–716, 1995.

Abrate, S., "Free vibration, buckling, and static deflections of functionally graded plates", *Composites Science and Technology*, vol. 66(14), pp. 2383–2394, 2006.

Abrate, S., "Functionally graded plates behave like homogeneous plates", *Composites Part B: Engineering*, vol. 39(1), pp. 151–158, 2008.

Akasaka, T. and Takagishi, T., "Vibration of corrugated diaphragm", *Bulletin of Japanese Society of Mechanical Engineers*, vol. 1, pp. 215–221, 1958.

Akbarzadeh, A.H., Abbasi, M. and Eslami, M.R., "Coupled thermoelasticity of functionally graded plates based on the third-order shear deformation theory", *Thin-Walled Structures*, vol. 53, pp. 141–155, 2012.

Akgöz, B. and Civalek, Ö., "Free vibration analysis of axially functionally graded tapered Bernoulli–Euler microbeams based on the modified couple stress theory", *Composite Structures*, vol. 98, pp. 314–322, 2012.

Akgöz, B. and Civalek, Ö., "Longitudinal vibration analysis of strain gradient bars made of functionally graded materials (FGM)", *Composites Part B: Engineering*, vol. 55, pp. 263–268, 2013.

Akulenko, L.D. and Kostin, G.V., "The perturbation method in problems of the dynamics of inhomogeneous elastic rods", *Journal of Applied Mathematics and Mechanics*, vol. 56(3), pp. 372–382, 1992.

Akulenko, L.D. and Nesterov, S.V., "An effective method of investigating the oscillations of substantially inhomogeneous distributed systems", *Journal of Applied Mathematics and Mechanics*, vol. 61(3), pp. 451–461, 1997a.

Akulenko, L.D. and Nesterov, S.V., "Determination of the frequencies and forms of oscillations of non-uniform distributed systems with boundary conditions of the third kind", *Journal of Applied Mathematics and Mechanics*, vol. 61(4), pp. 531–538, 1997b.

Akulenko, L.D. and Nesterov, S.B., "Free transverse vibrations of an inhomoegenous beam", *Mechanics of Solids*, Issue 3, pp. 179–192, 2003a (in Russian).

Akulenko, L.D. and Nesterov, S.V., "A frequency-parametric analysis of natural oscillations of non-uniform rods", *Journal of Applied Mathematics and Mechanics*, vol. 67(4), pp. 525–537, 2003b.

Akulenko, L.D. and Nesterov, S.V., *High-Precision Methods in Eigenvalue Problems and Their Applications*, Chapman & Hall/CRC Press, Boca Raton, Florida, 2005.

Allahverdizadeh, A., Naei, M.H. and Nikkhah Bahrami, M., "Nonlinear free and forced vibration analysis of thin circular functionally graded plates", *Journal of Sound and Vibration*, vol. 310(4), pp. 966–984, 2008.

Alijani, F., Amabili, M., "Nonlinear dynamic instability of functionality graded plates in thermal environments", *International Journal of a Non-linear Mechanics*, Vol. 50, 109–126, 2013.

Alshorbagy, A.E., Eltaher, M.A. and Mahmoud, F.F., "Free vibration characteristics of a functionally graded beam by finite element method", *Applied Mathematical Modeling*, vol. 35, pp. 412–425, 2011.

Altus, E., "Statistical modeling of heterogeneous microbeams", *International Journal of Solids and Structures*, vol. 38, pp. 5915–5934, 2001.

Altus, E. and Givli, S., "Strength and reliability of statically indeterminate heterogeneous beams", *International Journal of Solids and Structures*, vol. 40, pp. 2069–2083, 2003.

Altus, E., Proskura, A. and Givli, S., "A new functional perturbation method for linear non-homogeneous material", *International Journal of Solids and Structures*, vol. 42, pp. 1577–1595, 2005.

Altus, E. and Totry, E.M., "Buckling of stochastically heterogeneous beams, using a functional perturbation method", *International Journal of Solids and Structures*, vol. 40, pp. 6547–6565, 2003.

Amabili, M., *Nonlinear Vibrations and Stability of Shells and Plates*, Cambridge University Press, Cambridge, UK, 2008.

Ambartsumyan, S.A., *Theory of Anisotropic Plates*, Technomic, Stanford, CT, 1970.

Ameur, M., Tounsi, A., Mechab, I. and Adda Bedia, El A., "A new trigonometric shear deformation theory for bending analysis of functionally graded plates resting on elastic foundation, *KSCE Journal of Civil Engineering*, vol. 15(8), pp. 1405–1414, 2011.

Andrianov, I.V., Awrejrewisz, J. and Diskovsky, A.A., "Optimal design of a functionally graded corrugated rods subjected to longitudinal deformation", *Archives of Applied Mechanics*, vol. 85, pp. 303–314, 2015.

Ansari, R. and Darvizeh, M., "Prediction of dynamic behavior of FGM shells under arbitrary boundary conditions", *Composite Structures*, vol. 85, pp. 284–292, 2008.

Araujo, J.M. and Awruch, A.M., "On stochastic finite elements for structural analysis", *Computers & Structures*, vol. 52, pp. 461–469, 1994.

Ari-Gur, J. and Stavsky, Y., "On rotating polar-orthotropic circular disks", *International Journal of Solids and Structures*, vol. 17, pp. 57–67, 1981.

Arslan, E. and Eraslan, A.N., "Bending of graded curved bars at elastic limits and beyond", *International Journal of Solids and Structures*, vol. 50, pp. 806–814, 2013.

Atay, M.T., "Determination of critical buckling loads for variable stiffness Euler columns using homotopy perturbation method", *International Journal of Nonlinear Science and Numerical Simulation*, vol. 10, pp. 199–206, 2009.

Atmane, H.A., Tounsi, A., Mechab, I. and Adda Bedia, El A., "Free vibration analysis of functionally graded plated resting on Winkler–Pasternak elastic foundations using a new shear deformation theory, *International Journal of Mechanics and Materials in Design*, vol. 6(2), pp. 113–121, 2010a.

Atmane, H.A., Tounsi, A., Meftah, S.A. and Belhadj, H.A., "Free vibration behavior of exponential functionally graded beams with varying cross-section", *Journal of Vibration and Control*, vol. 17(2), pp. 311–318, 2010b.

Ayadoğlu, M., "Semi-inverse method for vibration and buckling of axially functionally graded beams", *Journal of Reinforced Plastics and Composites*, vol. 27(7), pp. 683–689, 2008a.

Ayadoğlu, M., "Conditions for functionally graded plates to remain flat under in-plane loads by classical plate theory", *Composite Structures*, vol. 82, pp. 155–157, 2008b.

Ayadoğlu, M., Maróti, G. and Elishakoff, I., "A note on semi-inverse method for buckling of axially functionally graded beams", *Journal of Reinforced Plastics and Composites*, vol. 32(7), pp. 511–512, 2013.

Ayadoğlu, M. and Taskin, V., "Free vibration analysis of functionally graded beams with simply supported edges, *Materials & Design*, vol. 28(5), pp. 1651–1656, 2007.

Babilio, E., "Dynamics of an axially functionally graded beam under axial load", *The European Physical Journal Special Topics*, vol. 222(7), pp. 1519–1539, 2013.

Baferani, A.H., Saidi, A.R. and Ehteshami, H., "Accurate solution for free vibration analysis of functionally graded thick rectangular plates resting on elastic foundation", *Composite Structure*, vol. 93(7), pp. 1842–1853, 2011.

Bahtui, A. and Eslami, M.R., "Generalized coupled thermoelasticity of functionally graded cylindrical shells", *International Journal of Numerical Methods in Engineering*, vol. 69, pp. 676–697, 2007a.

Bahtui, A. and Eslami, M., "Coupled thermoelasticity of functionally graded cylindrical shells", *Mechanics Research Communications*, vol. 34(1), pp. 1–18, 2007b.

Bakhti, K., Kaci, A., Bousahla, A.A., Houari, M.S.A., Tounsi, A. and Adda Bedia, E.A., "Large deformation analysis for functionally graded carbon nanotube-reinforced composite plates using an efficient and simple refined theory", *Steel and Composite Structures*, vol. 14(4), pp. 335–347, 2013.

Banachour, A., Tehar, H.D., Atmane, H.A., Tounsi, A. and Ahmed, M.S., "A four variable defined plate theory for free vibration of functionally graded plates with arbitrary gradient", *Composites: Part B, Engineering*, vol. 42, pp. 1386–1394, 2011.

Banks-Sills, L., Eliasi, R. and Berlin, Y., "Modeling of functionally graded materials in dynamic analyses", *Composites, Part B*, vol. 33, pp. 7–15, 2002.

Barakat, R. and Baumann, E., "Axisymmetric vibrations of a thin circular plate having parabolic thickness variation, *Journal of the Acoustical Society of America*, vol. 44(2), pp. 641–643, 1968.

Barcilon, V., "Inverse problem of a vibration beam", *Journal of Applied Mechanics and Physics*, vol. 27, pp. 347–358, 1976.

Barcilon, V., "On the multiplicity of solution of the inverse problem for vibrating beam", *SIAM Journal of Applied Mathematics*, vol. 37, pp. 605–613, 1979.

Batra, R.C. and Jin, J., "Natural frequencies of a functionally graded anisotropic rectangular plate", *Journal of Sound and Vibrations*, vol. 282, pp. 509–516, 2005.

Bayat, M., Sahari, B.B., Saleem, M., Aidy, A. and Wong, S.V., "Thermo elastic solution of a functionally graded variable thickness rotating disk with bending based on the first-order shear deformation theory", *Thin-Walled Structures*, vol. 35, pp. 283–309, 2008.

Bayat, M., Sahari, B.B., Saleem, M., Aidy, A. and Wong, S.V., "Bending analysis of functionally graded rotating disk based on the first order shear deformation theory", *Applied Mathematical Modeling*, vol. 33, pp. 4215–4230, 2009a.

Bayat, M., Sahari, B.B., Saleem, M., Hamouda, A.M.S. and Reddy, J.N., "Thermo elastic analysis of functionally graded rotating disks with temperature dependent material properties: Uniforms and variable thickness", *International Journal of Mechanics and Materials in Design*, vol. 5, pp. 263–279, 2009b.

Bayat, M., Saleem, M., Sahari, B., Hamouda, A. and Mahdi, E., "Thermo elastic analysis of a functionally graded rotating disk with small and large deflections", *Thin-Walled Structures*, vol. 45(7–8), pp. 677–691, 2007.

Bayat, M., Saleem, M., Hamouda, A.M.S. and Mahdi, E., "Analysis of functionally graded rotating disks with variable thickness", *Mechanics Research Communications*, vol. 35, pp. 283–309, 2008.

Becquet, R. and Elishakoff, I., "Class of analytical closed-form polynomial solutions for guided-pinned inhomogeneous beams", *Chaos, Solitons Fractals*, vol. 12, pp. 1509–1534, 2001.

Bedjilili, Y., Tounsi, A., Berrabah, H.M., Mechab, I., Adda Bedia, E.A. and Benaissa, S., "Natural frequencies of composite beams with a variable fiber column fraction including rotary inertia and shear deformation", *Applied Mathematics and Mechanics*, vol. 30(6), pp. 717–726, 2009.

Belabed, Z., Houari, M.S.A., Tounsi, A., Mahmoud, S.R. and Anwar Bèg, "An efficient and simple higher order shear and normal deformation theory for functionally graded material (FGM) plates", *Composites Part B: Engineering*, vol. 60, pp. 274–283, 2014.

Benachour, A., Tahar, H.D., Atmane, H.A., Tounsi, A. and Ahmet, M.S., "A 4 variable refined plate theory for free vibrations of functionally graded plated with arbitrary gradient", *Composite Part B: Engineering*, vol. 42(6), pp. 1386–1394, 2011.

Benatta, M.A., Mechab, I., Tounsi, A. and Adda Bedia, El A., "Static analysis of functionally graded short beams including warping and shear deformation effects", *Computational Material Science*, vol. 44(2), pp. 765–773, 2008.

Benatta, M.A., Tounsi, A., Mechab, I. and Bouiadjira, M.B., "Mathematical solution for bending of short hybrid composite beams with variable fibers spacing", *Applied Mathematics and Computation*, vol. 212(2), pp. 337–348, 2009.

Berrabah, H.M., Mechab, I., Tounsi, A., Benyoucef, S., Krour, B., Fekrar, A. and Adda Bedia, E.A., "Electro-elastic stresses in composite active beams with functionally graded layer, *Computational Materials Science*, vol. 48(2), pp. 366–371, 2010.

Bert, C.W., "Nonhomogenous polar-orthotropic circular disks of varying thickness", *Engineering Experiment Station Bulletin No. 190*, The Ohio State University, vol. 31(2), 1962.

Besmann, T.M., Allendorf, M.D., Robinson, McD. and Ulrich, R.K. (eds.), *Proceedings of the Thirteenth International Conference on Chemical Vapor Deposition*, The Electromechanical Society, Inc., 1996.

Bhat, R.B., "Natural frequencies of rectangular plates using characteristic orthogonal polynomials in Rayleigh–Ritz method", *Journal of Sound and Vibration*, vol. 102, pp. 493–499, 1985.

Bi, R.G., Han, X., Jiang, C. and Bai, Y.C., "Uncertain buckling and reliability analysis of the pieroelastic functionally graded cylindrical shells based on the non-probabilistic convex model", *International Journal of Computational Methods*, in press, 2014.

Birman, V., "Stability of functionally graded hybrid composite plates", *Composites Engineering*, vol. 5, pp. 913–921, 1995.

Birman, V.Z., "Stability of functionally graded shape memory alloy sandwich panels", *Smart Materials and Structures*, vol. 6, pp. 282–286, 1997.

Birman, V., *Plate Structures*, Springer, Science + Business Media B.V. Dordrecht, 2011.

Birman, V., "Functionally graded materials", in *Encyclopedia of Thermal Stresses* (R. Hetnarski, ed.), pp. 1858–1865, Springer, Dordrecht, 2014a.

Birman, V., "Modeling and analysis of functionally graded materials and structures", in *Encyclopedia of Thermal Stresses* (R. Hetnarski, ed.), pp. 3104–3112, Springer, Dordrecht, 2014b.

Birman, V. and Byrd, L.W., "Modeling and analysis of functionally graded materials and structurers", *Applied Mechanics Reviews*, vol. 60(5), pp. 195–216, 2007.

Birman, V., Keil, T. and Hosder, S., "Functionally graded materials in engineering," in *Structural Interfaces and Attachments in Biology* (S. Thomopoulos, V. Birman and G.M. Genin, eds.) pp. 19–41, Springer, New York, 2013.

Bokaian, A., "Natural frequencies of beam under compressive axial loads", *Journal of Sound and Vibration*, vol. 126(1), pp. 49–65, 1988.

Boonyachut, N., *Analysis of Thermal Stresses in Functionally Graded Materials*, Department of Metallurgical Engineering, University of Utah, 2002.

Borsuk, K., "Free vibration of rotating of a cylindrically aelotropic circular plates", *Archiwum Mechaniki Stosowanej*, vol. 12, pp. 649–665, 1960.

Bouazza, M., Tounsi, A., Adda Bedia, El A. and Megueni, A., "Buckling analysis of functionally graded plates with simply supported edges", *Leonardo Journal of Sciences*, vol. 8(15), pp. 21–32, 2009.

Bouazza, M., Tounsi, A., Adda Bedia, El A. and Megueni, A., "Thermoelastic stability analysis of functionally graded plates: An analytical approach", *Computational Material Science*, vol. 49(4), pp. 865–870, 2010.

Bouazza, M., Tounsi, A., Adda Bedia, El A. and Megueni, A., "Stability analysis of functionally graded plates subject to thermal loads", *Advanced Structured Materials*, vol. 15, pp. 669–680, 2011a.

Bouazza, M., Tounsi, A., Adda Bedia, El A. and Megueni, A., "Thermal buckling of simply supported FGM square plates", *Applied Mechanics and Materials*, vol. 61, pp. 25–32, 2011b.

Bouchafa, A., Benzair, A., Tounsi, A., Draiche, K., Mechab, I. and Adda Bedia, El A., "Analytical modeling of thermal residual stresses in exponential functionally graded material system", *Materials & Design*, vol. 31(1), pp. 560–563, 2010.

Bouderba, B., Houari, M.S.A. and Tounsi, A., "Thermomechanical bending response of FGM thick plated resting on Winkler–Pasternak elastic foundations", *Steel and Composite Structures*, vol. 14(1), pp. 85–104, 2013.

Bouiadjira, M.B., Houari, M.S.A. and Tounsi, A., "Thermal buckling of functionally graded plates according to a four-variable refined plate theory", *Journal of Thermal Stresses*, vol. 35(8), pp. 677–694, 2012.

Bourada, M., Tounsi, A., Houari, M.S.A. and Adda Bedia, E.A., "A new four-variable refined plate theory for thermal buckling analysis of functionally graded sandwich plates", *Journal of Sandwich Structures and Materials*, vol. 14(1), pp. 5–33, 2012.

Bouremana, M., Tounsi, A., Kaci, A. and Mechab, I., "Controlling thermal deformation by using composite materials having variable fiber volume fraction", *Materials & Design*, vol. 30(7), pp. 2532–2537, 2009.

Brischetto, S. and Carrera, E., "Advanced mixed theories for bending analysis of functionally graded plates", *Computers and Structures*, vol. 88, pp. 1474–1483, 2010.

Bryan, G.H. "On the stability of a plane plate under thrusts on its own plane, with application to the "buckling" of the sides of a ship", *Proc. London Math. Soc.*, vol. 22, pp. 54–67, 1891.

Byrd, L.W. and Birman, V., "Modeling and analysis of functionally graded materials and structures", *Applied Mechanics Reviews*, vol. 60, p. 195, 2007.

Caliò, I. and Elishakoff, I., "Can a harmonic function constitute a closed-form buckling mode of an inhomogeneous column?", *AIAA Journal*, vol. 40, pp. 2532–2537, 2002.

Caliò, I. and Elishakoff, I., "Can a trigonometric function serve both as the vibration and the buckling mode of an axially graded structure", *Mechanics Based Design of Structures and Machines*, vol. 32, pp. 401–421, 2004a.

Caliò, I. and Elishakoff, I., "Closed-form trigonometric solutions for inhomogeneous beam-columns on elastic foundation", *International Journal Structural Stability and Dynamics*, vol. 4, pp. 139–146, 2004b.

Caliò, I. and Elishakoff, I., "Closed-form solutions for axially graded beam-columns, *Journal of Sound and Vibration*, vol. 280, pp. 1083–1094, 2005.

Caliò, I. and Elishakoff, I., "Closed form trigonometric solution of inhomogeneous beam-columns", In *Mechanical Vibration: Where do we Stand?* (Elishakoff, I., ed.), Springer, Vienna, pp. 455–474, 2007.

Caliò, I., Gladwell, G.M.L. and Morassi, A., "Families of beams with a given buckling spectrum", *Inverse Problems*, vol. 27(4), article 045006, 2001.

Çallioğlu, H., "Stress analysis of functionally graded isotropic rotating discs", *Advanced Composites Letters*, vol. 17(5), pp. 147–153, 2008.

Çallioğlu, H., Bektas, N.B. and Sayer, M., "Stress analysis of functionally graded rotating discs: Analytical and numerical solutions", *Acta Mechanica Sinicia*, vol. 27(6), pp. 950–955, 2011.

Candan, S. and Elishakoff, I., "Apparently first closed-form solution for vibrating inhomogeneous beams", *International Journal of Solids and Structures*, vol. 38, pp. 3411–3441, 2001a.

Candan, S. and Elishakoff, I., "Constructing the axial stiffness of longitudinally vibrating rod from fundamental mode shape", *International Journal of Solids and Structures*, vol. 38, pp. 3443–3452, 2001b.

Carlos, E., Rubio, W.M., Paulino, G.M. and Silva, N., "Tailoring vibration mode shapes using topology optimization and functionally graded material concepts", *Smart Materials and Structures*, vol. 20(2), article 025009, 2011.

Carrera, E., Brischetto, S. and Robaldo, A., "Variable Rinematic model for analysis of functionally graded material plates", *AIAA Journal*, vol. 46(1), pp. 194–203, 2008.

Carrera, E., Brischetto, S., Cinefra, M. and Soave, M., "Effects of thickness stretching in functionally graded plates and shells," *Composites Part B: Engineering*, vol. 42(2), 123–133, 2011.

Carrera, E., Pagani, A., Petrolo, M. and Zappino, E., Recent developments on refined theories for beam with applications," *Bulletin of JSME, Mechanical Engineering Reviews*, vol. 2(2), 1–30, 2015.

Caruntu, D.I., "On bending vibrations of some kinds of beams of variable cross-section using orthogonal polynomials", *Revue Roumaine des Sciences Techniques. Série de Mécanique Appliquée*, vol. 41(3–4), pp. 265–272, 1996a.

Caruntu, D.I., "Related studies on factorization of the differential operator in the case of bending vibration of a class of beams with variable cross-section", *Revue Roumaine des Sciences Techniques, Série de Mécanique Appliquée*, vol. 41(5–6), pp. 389–397, 1996.

Caruntu, D.I., Transverse vibrations of an annular plate of parabolical thickness, *Revue Roumaine des Sciences Techniques, Série de Mécanique Appliquée*, vol. 43(1), pp. 127–132, 1998.

Caruntu, D.I., "Factorization method in bending vibrations of rotating nonuniform Euler–Bernoulli beams", in *Proceedings of the Sixth International Congress on Sound and Vibration*, Copenhagen, pp. 2053–2058, 1999.

Caruntu, D.I., "On axisymmetric vibration of annular plate of parabolic thickness variation by factorization of differential equation", *Proceedings of ASME International Mechanical Engineering Congress and Exposition, Design Engineering*, vol. 111, pp. 341–344, New York, 2001.

Caruntu, D.I., "On non-axisymmetrical transverse vibrations of circular plates of convex parabolic thickness variation", *Proceedings of ASME International Mechanical Engineering Congress and Exposition, Applied Mechanics Division*, Anaheim, November, vol. 225, pp. 335–340, 2004.

Caruntu, D.I., "Classical Jacobi polynomials, closed-form solutions for transverse vibrations", *Journal of Sound and Vibration*, vol. 306, pp. 467–494, 2007.

Caruntu, D.I., "Dynamic modal characteristics of transverse vibrations of cantilevers of parabolic thickness", *Mechanics Research Communications*, vo. 36, pp. 391–404, 2009.

Castellazzi, G., Gentilini, C., Krysl, P. and Elishakoff, I., "Static analysis of functionally graded plates using a nodal integrated finite element approach," *Composite Structures*, vol. 103, pp. 197–200, 2013.

Catellani, G. and Elishakoff, I., "Apparently first closed-form solutions of semi-inverse buckling problems involving distributed and concentrated loads", *Thin-Walled Structures*, vol. 42, pp. 1729–1733, 2004.

Celebi, K. and Tutuncu, N., "Free vibration analysis of functionally graded beams using an exact plane elasticity approach", *Proceedings of the Institution of Mechanical Engineers, Part C: Journal of Mechanical Engineering Science*, article 0954406213519974, 2014.

Chakraborty, A., Gopalakrishnan, S. and Reddy, J.N., "A new beam finite element for the analysis of functionally graded materials", *International Journal of Mechanical Sciences*, vol. 45, pp. 519–539, 2003.

Chapman, B.D., *Characterization of Functionally Graded Materials*, Biblioscholar, 2012.

Chen, D.Y. and Ren, B.S., "Finite element analysis of the lateral vibration of this annular and circular plates with variable thickness", *Journal of Vibration and Acoustics*, vol. 120(3), pp. 747–753, 1998.

Chen, W.Q., Bian, Z.G., Lv, C.F. and Ding, H.J., "3D free vibration analysis of a functionally graded piezoelectric hollow cylinder filled with compressible fluid", *International Journal of Solids and Structures*, vol. 41(3), pp. 947–964, 2004.

Chen, W.Q., "Effect of radial inhomogeneity on natural frequencies of an anisotropic hollow sphere", *Journal of Sound and Vibration*, vol. 226(4), pp. 787–794, 1999.

Chen, W.Q. and Ding, H.J., "On the free vibration of functionally graded piezoelectric rectangular plate", *Acta Mechanica*, vol. 153, pp. 207–216, 2002.

Chen, W.Q., Wang, X. and Ding, H.J., "Free vibration of a fluid-filled hollow sphere of a functionally graded material with spherical isotropy", *Journal of the Acoustical Society of America*, vol. 106(5), pp. 2588–2594, 1999.

Chen, W.Q., Wang, L.Z. and Lu, Y., "Free vibrations of functionally graded piezoceramic hollow sphere with radial polarization", *Journal of Sound and Vibration*, vol. 251(1), pp. 103–114, 2002.

Chen, Y., *Co-Extrusion of Cement-Based Materials to Achieve Functionally Graded Microstructure*, ProQuest Information and Learning Company, 2008.

Cheng, Z.-Q. and Batra, R.C., "Three-dimensional thermoelastic deformations of a functionally graded elliptic plate", *Composites Part B*, vol. 31, pp. 97–106, 2000.

Cheng, Z.Q. and Batra, R.C., "Deflection relationships between the homogeneous Kirchhoff plate theory and different functionally graded plate theories", *Archive of Mechanics*, vol. 52, pp. 143–158, 2000.

Cherradi, N., Kawasaki, A. and Gasik, M., "Worldwide trends in functional gradient materials research and development", *Composite Engineering*, vol. 8, pp. 883–894, 1994.

Chi, S. and Chung, Y., "Mechanical behavior of functionally graded material plated under transverse load — Part I: Analysis", *International Journal of Solids and Structures*, vol. 43, pp. 3657–3674, 2006.

Chi, C.-H. and Chung, Y.L., "Mechanical behavior of functionally graded material plates under transverse load", Part II, *International Journal of Solids and Structres*, vol. 43, pp. 3657–3674, 2006.

Chin, E.S.C., Army focused research team on functionally graded armor composites, *Material Science Engineering*, vol. A259, pp. 155–161, 1999.

Chladni, E.F.F., *Entdeckungen über die Theorie des Klanges*, Leipzing, 1787 (in German).

Chladni, E.F.F., *Die Akustik*, Leipzig, 1803 (in German).

Chung, D.D.L., *Functional Materials: Electrical, Dielectric, Electromagnetic, Optical and Magnetic Application* (*with companion solution manual*), vol. 2 of Engineering Materials for Technological Needs, World Scientific, Singapore, 2010.

Chung, Y.L. and Chen, W.T., "Bending behavior of FGM-coated and FGM-undercoated plates with two simply supported opposite edges and two free edges", *Composite Structures*, vol. 81, pp. 157–167, 2007.

Clark, L.G., "Buckling of laminated columns", *Journal of Applied Mechanics*, vol. 22(4), pp. 553–556, 1955.

Conway, H.D., "An unusual closed-form solution for a variable thickness plate on an elastic foundation", *Journal of Applied Mechanics*, vol. 47, p. 204, 1980.

Coskun, S.B., "Analysis of tilt-buckling of Euler columns with varying flexural stiffness using homotopy perturbation method", *Mathematical Modeling and Analysis*, vol. 15, pp. 275–286, 2010.

Coskun, S.B., "Determination of critical buckling loads for Euler columns of variable flexural stiffness with a continuous elastic restraint using homotopy perturbation method", *International Journal of Nonlinear Science and Numerical Simulation*, vol. 10, pp. 191–197, 2009.

Csonka, P., "Buckling of stressed bars of heterogeneous materials", *Proceedings of Hungarian Academy Science*, vol. 9(3–4), pp. 391–404, 1955.

Dai, H.L. and Fu, Y.M., "Magnetothermoelastic interactions in hollow structures of functionally graded material subjected to mechanical loads", *International Journal of Pressure Vessels and Piping*, vol. 84(3), pp. 132–138, 2007.

Dai, H.L., Fu, Y.M. and Dong, Z.M., "Exact solutions for functionally graded pressure vessels in a uniform magnetic field", *International Journal of Solids and Structures*, vol. 41(18–19), pp. 5570–5580, 2006.

Daouadji, T.H., Henni, A.H.H., Tounsi, A. and Adda Bedia, El A., "Elasticity solution of a cantilever functionally graded beam", *Applied Composite Materials*, vol. 20(1), pp. 1–15, 2013.

Daouadji, T.H., Tounsi, A., Hadji, L., Henni, A.H. and Adda Bedia, El A., "A theoretical analysis for static and dynamic behavior of functionally graded plates", *Materials Physics and Mechanics*, vol. 14, pp. 110–128, 2012.

Darbandi, S.M., Firouz-Abadi, R.D. and Haddadpour, H., "Buckling of variable section columns under axial loading" *Journal of Engineering Mechanics*, vol. 136(4), pp. 472–476, 2010.

Darbandi, S.M., "Buckling of variable section columns under axial loading", *Journal of Engineering Mehcanics*, vol. 136, pp. 472–476,

Darbandi, S.M., Firouz-Abadi, R.D. and Haddadpour, H., "Buckling of variable section columns under axial loading", *Journal of Engineering Mechanics*, vol. 136(4), pp. 472–476, 2010.

Dave, E.V., Paulino, G.H. and Buttlar, W.G., "Viscoelastic functionally graded finite-element method using correspondence principle", *Journal of Materials in Civil Engineering*, vol. 23(1), pp. 39–48, 2010.

Dawe, D.J. and Roufaeil, O.L., "Rayleigh–Ritz vibration analysis of Mindlin plates", *Journal of Sound and Vibration*, vol. 69, pp. 345–359, 1980.

Della Croce, L. and Venini, P., "Finite elements for functionally graded Reissner–Mindlin plates", *Computer Methods in Applied Mechanics and Engineering*, vol. 193, pp. 705–725, 2004.

Dhaliwal, R.S. and Singh, B.H., "On the theory of elasticity of inhomogeneous medium", *Journal of Elasticity*, vol. 3, pp. 211–219, 1978.

Di Paola, M., "Probabilistic analysis of truss structures with uncertain parameters (virtual distortion approach)", *Probabilistic Engineering Mechanics*, vol. 19, pp. 321–329, 2004.

Ding, H.J., Wang, H.M. and Chen, W.Q., "Dynamic response of a functionally graded pyroelectric hollow sphere for spherically symmetric problems", *International Journal of Mechanical Science*, vol. 45(6–7), pp. 1029–1051, 2003.

Ding, H.-J., Huang, D.-J. and Chen, W.-Q., "Elastic solution for plane arisotropic functionally graded beams", *International Journal of Solids and Structures*, vol. 44, pp. 176–196, 2007.

Dinnik, A.N., "About stability of a compressed circular plate", *Bulletin Polytech. Inst.*, Kiev, 1911 (in Russian).

Dodds, R.H., Schwalbe, K.-H. and Paulino, G.H., *Fracture of Functionally Graded Materials, Reprinted from Engineering Fracture Mechanics*, vol. 69 (14–16), Elsevier Science Ltd., Oxford, UK, 2002.

Dryden, J., "Bending of inhomogeneous curved bars", *International Journal of Solids and Structures*, vol. 44, pp. 4158–4166, 2007.

Du, H., Lim, M.K. and Liu, K.M., "A power series solution for vibration of a rotating Timoshenko beam", *Journal of Sound and Vibration*, vol. 175(4), pp. 505–523, 1994.

Duan, W.H. and Wang, C.M., "Exact solution for buckling of columns including self-weight", *Journal of Engineering Mechanics*, vol. 134, pp. 116–119, 2008.

Duncan, W.J., "Galerkin's method in mechanics and differential equations", *ARC, Reports and Memoranda*. No. 1738, 1937.

Durodola, J.F. and Attia, O., "Deformation and stresses in functionally graded rotating disks", *Composite of Science and Technology*, vol. 60(7), pp. 987–995, 2000.

Durodola, J.F. and Attia, O., "Deformation and stresses in functionally graded rotating disks", *Composites Science and Technology*, vol. 60, pp. 987–995, 2000.

Dutta, R., Ganguli, R. and Mani, V., "Exploring isospectral cantilever beams using electromagnetism inspired optimization technique", *Swarm and Evolutionary Computation*, 2012.

Ebrahimi, F., Rastgoo, A. and Kargarnovin, M.H., "Analytical investigation on axisymmetric free vibrations of moderately thick circular functionally graded plate integrated with piezoelectric layer", *Journal of Mechanical Science and Technology*, vol. 22, pp. 1058–1072, 2008.

Ebrahimi, F., *Smart Functionally Graded Plates: Vibration Analysis of FGM Plates Coupled with Piezoelectric Layers*, Nova Science Publishers Inc., Hauppauge, NY, 2010.

Ece, M.C., Ayadoğlu, M. and Taskin, V., "Vibration of a variable cross-section beam", *Mechanics Research Communications*, vol. 34(1), pp. 78–84, 2007.

Efraim, E., "Accurate formula for determination of natural frequencies of FGM plates based on frequencies of isotropic plates", *Procedia Engineering*, Issue 10, pp. 242–247, 2011.

Efraim, E. and Eisenberger, M., "Vibrations of variable thickness functionally graded cylindrical shells", *The Third International Conference on Structural Stability and Dynamics* Kissimmee, FL, June 19–22, 2005.

Efraim, E. and Eisenberger, M., "Exact vibration analysis of variable thickness thick annular isotropic and FGM plates", *Journal of Sound and Vibration*, vol. 299, pp. 720–738, 2007.

Efraim, E. and Eisenberger, M., "Dynamic stiffness vibration analysis of thick spherical shell segments with variable thickness", *Journal of Mechanics of Materials and Structures*, vol. 5, pp. 821–835, 2010.

Eisenberger, M., "Exact longitudinal vibration frequencies of a variable cross section rod", *Applied Acoustics*, vol. 34, pp. 123–130, 1991.

Eisenberger, M., "Buckling loads for variable cross-section bars in nonuniform thermal field," *Mechanics Research Communications*, vol. 19(4), pp. 259–266, 1992.

Eisenberger, M., "Dynamic stiffness matrix for variable cross-section Timoshenko beams", *Communications in Numerical Methods in Engineering*, vol. 11, pp. 507–513,1995.

Elishakoff, I., "Adjustable parameter method for vibration of polar orthotropic plates", *Journal of Sound and Vibration*, vol. 116(1), pp. 181–184, 1987.

Elishakoff, I., *Probabilistic Theory of Structures*, Second Edition, Dover Publication, New York, 1999.

Elishakoff, I., "A closed-form solution for the generalized Euler problem", *Proceedings of the Royal Society of London*, vol. 456, pp. 2409–2417, 2000a.

Elishakoff, I., "Axisymmetric vibration of inhomogeneous clamped circular plates an unusual closed-form solution", *Journal of Sound and Vibration*, vol. 233(4), pp. 723–734, 2006.

Elishakoff, I., "Axisymmetric vibration of inhomogeneous free circular plates: an unusual closed-form solution", *Journal of Sound and Vibration*, vol. 234, pp. 167–170, 2000c.

Elishakoff, I., "Apparently first closed-form solution for frequency of beam with rotational spring", *AIAA Journal*, vol. 39, pp. 183–186, 2001a.

Elishakoff, I., "Inverse buckling problem for inhomogeneous columns", *International Journal of Solids Structures*, vol. 38, pp. 457–464, 2001b.

Elishakoff, I., "Some unexpected results in vibration of non-homogeneous beams on elastic foundation", *Chaos, Solitons Fractals*, vol. 12, pp. 2177–2218, 2001c.

Elishakoff, I., "Euler's problem revisited: 222 years later", *Meccanica*, vol. 36, pp. 265–272, 2001d.

Elishakoff, I., *Eigenvalues of Inhomogeneous Structures: Unusual Closed-Form Solutions*, CRC Press, Boca Raton, 2005.

Elishakoff, I., Vibration of Beams and Plates: Review of First Closed Form Solutions in the Past 250 Years, in *Mechanical Vibration: Where Do We Stand?* (I. Elishakoff, ed.), pp. 389–453, Springer, Vienna, 2007.

Elishakoff, I. and Becquet, R., "Closed-form solutions for natural frequencies for inhomogeneous beams with one sliding support and the other clamped", *Journal of Sound and Vibration*, vol. 238, pp. 540–546, 2000.

Elishakoff, I. and Candan, S., "Apparently first closed-form solution for frequencies of deterministically and/or stochastically inhomogeneous simply supported beams", *Journal of Applied Mechanics*, vol. 68, pp. 176–185, 2001a.

Elishakoff, I. and Candan, S., "Apparently first closed-form solution for vibrating inhomogeneous beams", *International Journal of Solids and Structures*, vol. 38, pp. 3411–3441, 2001b.

Elishakoff, I. and Endres, J., "Extension of Euler's problem to axially graded columns: Two hundred and sixty years later", *Journal of Intelligent Material Systems and Structures*, vol. 16(1), pp. 77–83, 2005.

Elishakoff, I. and Gentillini, C., "Three-dimensional flexure of rectangular plates made of functionally grated materials", *Journal of Applied Mechanics*, vol. 72(5), pp. 788–791, 2005a.

Elishakoff, I., Gentilini, C. and Santoro, R., "Some conventional and unconventional educational column stability problems", *International Journal of Structural Stability and Dynamics*, vol. 6(1), pp. 139–151, 2006.

Elishakoff, I., Gentillini, C. and Viola, E., "Forced Vibration of functionally graded plates in the three-dimensional setting", *AIAA Journal*, vol. 43(9), pp. 2000–2007, 2005b.

Elishakoff, I., Gentillini, C. and Viola, E., "Three-dimensional analysis of an all-round clamped plate made of functionally graded materials", *Acta Mechanica*, vol. 180, pp. 21–36, 2005c.

Elishakoff, I. and Guede, Z., "Novel closed-form solutions in buckling of inhomogeneous columns under distributed variable loading", *Chaos Soliton Fractal*, vol. 12, pp. 1075–1089, 2001a.

Elishakoff, I. and Guede, Z., "A remarkable nature of the effect of boundary conditions on closed-form solutions for vibrating inhomogeneous Euler–Bernoulli beams", *Chaos Soliton Fractal*, vol. 12, pp. 659–704, 2001b.

Elishakoff, I. and Guede, Z., "Analytical polynomial solutions for vibrating axially graded beams", *Mechanics of Advanced Material and Structure*, vol. 11, pp. 517–533, 2004.

Elishakoff, I., Hettema, Ch. D. and Wilson, E.L., "Direct superposition of Wilson trial functions by computerized symbolic algebra", *Acta Mechanica*, vol. 74, pp. 69–79, 1988.

Elishakoff, I. and Johnson, V., "Apparently the first closed-form solution of vibrating inhomogeneous beam with a tip mass", *Journal of Sound and Vibration*, vol. 286, pp. 1057–1066, 2005e.

Elishakoff, I. and Meyer, D., "Inverse vibration problem for inhomogeneous circular plate with translational spring", *Journal of Sound and Vibration*, vol. 285, pp. 1192–1202, 2005f.

Elishakoff, I. and Miglis, Y., "Some intriguing results pertaining to functionally graded columns", *Journal of Applied Mechanics*, vol. 1, p. 431, 2012.

Elishakoff, I. and Pentaras, D., "Lekhnitskii's classic formula serving as an exact mode shape of simply supported polar orthotropic inhomogeneous circular plates", *Journal of Sound and Vibration*, vol. 291(3–5), pp. 1239–1254, 2006a.

Elishakoff, I. and Pentaras, D., Apparently the first closed-form solution of inhomogeneous elastically restrained vibrating beams", *Journal of Sound and Vibration*, vol. 298(1–2), pp. 439–445, 2006b.

Elishakoff, I. and Pentaras, D., "Vibration tailoring of a polar orthotropic circular plate with a translational spring", *Journal of Applied Mechanics*, vol. 75(3), pp. 034502–034505, 2008.

Elishakoff, I. and Pentaras, D., "Design of heterogeneous clamped circular plates with specified fundamental natural frequency", *International Journal of Solids and Structures*, vol. 46(10), pp. 1997–2010, 2009.

Elishakoff, I. and Perez, A., "Design of polynomially inhomogenous bar with a tip mass for special mode shape and natural frequency", *Journal of Sound and Vibration*, vol. 287, pp. 1004–1012, 2005.

Elishakoff, I. and Ren, Y.J., "The bird's eye view on finite element method for stochastic structures", *Computer Methods in Applied Mechanics and Engineering*, vol. 168, pp. 51–61, 1999.

Elishakoff, I. and Ren, Y.J., *Finite Element Methods for Structures with Large Stochastic Variations*, Oxford University Press, New York, 2003.

Elishakoff, I., Ren, Y.J. and Shinozuka, M., "Improved finite element method for stochastic problems", *Chaos, Solitons & Fractals*, vol. 5, pp. 833–846, 1995.

Elishakoff, I., Ren, Y.J. and Shinozuka, M., "Some critical observations and attendant new results in the finite element method for stochastic problems", *Chaos, Solitons & Fractals*, vol. 7, pp. 597–609, 1996.

Elishakoff, I. and Rollot, O., "New closed-form solutions for buckling of a variable stiffness column by Mathematica®", *Journal of Sound and Vibration*, vol. 224, pp. 172–182, 1999.

Elishakoff, I., Ruta, G.C. and Stavsky, Y., "A novel formulation leading to closed-form solutions for buckling of circular plates", *Acta Mechanica*, vol. 185(1–2), pp. 81–88, 2006.

Elishakoff, I. and Storch, J., "An unusual exact, closed-form solution for axisymmetric vibration of inhomogeneous simply supported circular plate", *Journal of Sound and Vibration*, vol. 284, pp. 1217–1228, 2005.

Elishakoff, I. and Yost, J., "Vibration tailoring for elastically constrained axially graded bars," *Acta Mechanica Sinica*, vol. 25, pp. 313–316, 2010.

Elishakoff, I., Zaza, N., Curtin, J. and Hashemi, J., "Apparently first closed-form solution for vibration of functionally graded rotating beams", *AIAA Journal*, vol. 52(11), pp. 2587–2593, 2014.

Emuna, N., "Bending theory of functionally graded beams," *Master of Science Thesis*, Faculty of Aerospace Engineering, Technion — Israel Institute of Technology, Haifa, Isarel, March 2014.

Epstein, M. and Elzanowski, M., *Material Inhomogeneities and their Evolution: A Geometric Approach*, Springer, Berlin, 2007.

Erdoğan, F., "Fracture mechanics of functionally graded materials", *Composites Engineering*, vol. 5(7), pp. 753–770, 1995.

Euler, L., *Methodus Inveniedi Lineas Curvas Maximi Minimive Proprietate Gaudentes*, Lausanne: Gent, 1744.

Fahsi, B., Tounsi, A. and Adda Bedia, El A., "A four variable refined plate theory for nonlinear cylindrical bending analysis of functionally graded plates under thermo-mechanical loadings", *Journal of Mechanical Science and Technology*, vol. 26(12), pp. 4073–4079, 2012.

Falsone, G. and Impollonia, N., "A new approach for the stochastic analysis of finite element modelled structures with uncertain parameters", *Computer Methods in Applied Mechanics and Engineering*, vol. 191, pp. 5067–5085, 2002.

Falsone, G. and Impollonia, N., Erratum to "A new approach for the stochastic analysis of finite element modelled structures with uncertain parameters" [*Comput. Methods Appl. Mech. Engrg.* 191 (2002) 5067–5085], *Computer Methods in Applied Mechanics and Engineering*, vol. 192, pp. 2187–2188, 2003.

Falsone, G. and Impollonia, N., "About the accuracy of a novel response surface method for the analysis of finite element modeled uncertain structures", *Probabilistic Engineering Mechanics*, vol. 19, pp. 53–63, 2004.

Fan, J. and Chen, H., *Advances in Heterogeneous Material Mechanics*, DEStech Publications, Inc., 2008.

Fan, J. and Ye, J., "A series solution of the exact solution for thick orthotropic plates", *International Journal of Solids and Structures*, vol. 26, pp. 773–778, 1990.

Fazelzadeh, S.A. and Hosseini, M., "Aerothermoelastic behavior of supersonic rotating thin walled beams made of functionally graded materials", *Journal of Fluids and Structures*, vol. 23(8), pp. 1251–1264, 2007.

Fekrar, A., Meiche, N. El., Bessaim, A., Tounsi, A. and Adda Bedia, E.A., "Buckling analysis of functionally graded hybrid composite plates using a new four variable refined plate theory", *Steel and Composite Structures*, vol. 13(1), pp. 91—107, 2012.

Feldman, E. and Aboudi, J., "Buckling analysis of functionally graded plates subjected to uniaxial loading", *Composite Structures*, vol. 38(1–4), pp. 29–36, 1997.

Ferrante, F.J. and Graham-Brady, L.L., "Stochastic simulation of non-Gaussian/non-stationary proportion on a functionally graded plate", *Computer Methods in Applied Mechanics and Engineering*, vol. 194(12–16), pp. 1675–1692, 2005.

Ferreira, A.J.M., Batra, R.C., Roque, C.M.C., Qian, L.F. and Martins, P.A.L.S., "Static analysis of functionally graded plates using third-order shear deformation theory and meshless method", *Composite Structures*, vol. 69, pp. 449–457, 2005.

Filipich, C.P. and Piovan, M.T., "The dynamics of thick curved beams constructed with functionally graded materials", *Mechanics Research Communications*, vol. 37(6), pp. 565–570, 2010.

Firouzjaei, K.R., "Application of differential transform method in analysis of axially graded beams resting on elastic foundation", In *Proceedings of International Conference on Advances in Engineering and Technology*, India, 2013.

Froes, F.H., "Aerospace materials for the twenty-first century", *Materials and Design*, vol. 10(3), pp. 110–120, 1989.

Fromme, J.A. and Leisssa, A.W., "Free vibration of the rectangular parallelepiped", *Journal of the Acoustical Society of America*, vol. 48, pp. 290–298, 1970.

Fukui, Y., "Fundamental investigation of functionally gradient material manufacturing system using centrifugal force", *JSME International Journal Series* III, vol. 34, pp. 144–148, 1991.

Gang, S.W., Lam, K.Y. and Reddy, J.N., "The elastic response of functionally graded cylindrical shells to low-velocity", *International Journal of Impact Engineering*, vol. 22, pp. 397–417, 1999.

Gao, H. and Herrmann, G., "On estimates of stress intensity factors for cracked beams and pipes", *Engineering Fracture Mechanics*, vol. 41, pp. 695–706, 1992.

Gentilini, C., "Modelling of the static and dynamic behavior of structures with variable parameters", *PhD Thesis*, University of Bologna, Italy, 2005.

Gentilini, C. and Elishakoff, I., "Three-dimensional free vibration analysis of functionally graded rectangular plates", *Technical Report No. 145*, DISTART, University of Bologna, Italy, 2004.

Gentilini, C. and Viola, E., "On the three-dimensional vibration analysis of rectangular plates", in *Problems in Structural Identification and Diagnostics: General Aspects and Applications* (C. Davini and E. Viola, eds.), pp. 149–162, Springer, Vienna, 2003.

Gentillini, C., Viola, E. and Elishakoff, I., "Random vibrations of three-dimensional functionally graded plates", *ICOSSAR, Safety and Reliability of Engineering Systems and Structures*, pp. 2337–2344, 2005.

Gentilini, C., Viola, E. and Ubertini, F., "Probabilistic characterization of linear truss structures with cracked members", *Key Engineering Materials*, vols. 251–252, pp. 141–146, 2003.

Gentilini, C., Viola, E. and Ubertini, F., "Probabilistic analysis of linear elastic cracked structures with uncertain damage", *Probabilistic Engineering Mechanics*, vol. 20(4), pp. 307–323, 2005.

Ghanem, R. and Spanos, P.D., *Stochastic Finite Elements: A Spectral Approach*, Springer, New York, 1991.

Ghavami, K., Rodrigues, C.S. and Paciornik, S., "Bamboo: Functionally graded composite material", *Asian Journal of Civil Engineering (Building And Housing)*, vol. 4, pp. 1–10, 2003.

Ghosh, A., Lannutti, J.J. and Miyamoto, Y., *Functionally Graded Materials: Manufacture, Properties and Applications*, American Ceramic Society, 1997.

Gibson, R.E., "Some results concerning displacements and stresses in a non-homogeneous elastic half-space", *Géotechnique*, vol. 17, pp. 58–69, 1967.

Gilat, R., Calio, I. and Elishakoff, I., "Inhomogeneous beams possessing an exponential mode shape", *Mechanics Research Communications*, vol. 37(4), pp. 417–426, 2010.

Gilhooley, D.F., Batra, R.C., Xiao, J.R., McCarthy, M.A. and Gillespie Jr., J.W., "Analysis of thick functionally graded plated by using higher-order shear and normal deformable plate theory and MLPG method with radial basis functions", *Composite Structures*, vol. 80, pp. 539–552, 2007.

Gong, S.X. and Meguid, S.A., "A general treatment of the elastic field of an elliptical inhomogeneity under antiplane shear", *Journal of Applied Mechanics*, vol. 59, pp. 5131–5135, 1992.

Gong, S.X. and Meguid, S.A., "On the elastic fields of an elliptical inhomogeneity under plane deformation", *Proceedings of the Royal Society (London)*, A, vol. 443, pp. 457–471, 1993.

Goupee, A.J. and Vel, S.S., "Optimization of natural frequencies of bidrectional functionally graded beams", *Structural and Multidisciplinary Optimization*, vol. 32, pp. 473–484, 2006.

Gray, L.J., Kaplan, T., Richardson, J.D. and Paulino, G.H., "Green's functions and boundary integral analysis for exponentially graded materials: heat conduction", *Transactions-American Society of Mechanical Engineers Journal of Applied Mechanics*, vol. 70(4), pp. 543–549, 2003.

Grossi, R.O., Laura, P.A.A. and Narita, Y., "A note on vibrating polar orthotropic circular plates", *Journal of Sound and Vibration*, vol. 102, pp. 181–186, 1986.

Guede, Z. and Elishakoff, I., "Apparently the first closed-form solution for inhomogeneous vibrating beams under axial loading", *Proceedings of Royal Society A*, vol. 457, pp. 623–649, 2001a.

Guede, Z. and Elishakoff, I., "A fifth-order polynomial that serves as both buckling and vibration mode of an inhomogeneous structure", *Chaos, Solitons Fractals*, vol. 12, pp. 1267–1298, 2001b.

Gupta, U.S. and Ansari, A.H., "Free vibration of polar orthotropic circular plates of varibale thickness with elastically restrained edge", *Journal of Sound and Vibration*, vol. 213(3), pp. 429–445, 1998.

Gupta, U.S. and Lal, R., "Effect of elastic foundation on axisymmetric vibrations of polar orthotropic circular plates of variable thickness", *Journal of Sound and Vibration*, vol. 139(3), pp. 503–513, 1990.

Güven, U., Çelik, A. and Baykara, C., "On transverse vibrations of functionally graded polar orthtropic rotating solid disk with variable thickness and constant radial stress", *Journal of Reinforced Plastics and Composites*, vol. 23(12), pp. 1279–1284, 2004.

Haldar, A. and Mahadevan, S., *Reliability Assessment Using Stochastic Finite Element Analysis*, Wiley, Canada, 2000.

Haldar, S., *Crack Propagation in Functionally Graded Materials*, VDM Publishing, 2010.

Hamidi, A., Zidi, M., Houari, M.S. and Tounsi, A., "A new four variable refined plate theory for bending response of functionally graded sandwich plates under thermomechanical loading", *Composites Part B: Engineering*, in press, 2012.

Harris, C.Z., "The normal modes of a circular plate of variable thickness", *Quarterly Journal of Mechanics and Applied Mathematics*, vol. 21, pp. 320–327, 1968.

Han, X., Liu, G.R., Xi, Z.C. and Lam, K.Y., "Transient waves in a functionally graded cylinder", *International Journal of Solids and Structures*, vol. 38(17), pp. 3021–3037, 2001.

Hashin, Z., "On the determination of Ayry polynomial stress functions", Bulleting of the Research Council of Israel, Section C, vol. 8(3), pp. 93–102,

He, M., Zhang, L. and Wang, Q., "Reconstructing cross-sectional physical parameters for two-span beams with overhang using fundamental mode", *Acta Mechanica*, vol. 225, pp. 349–359, 2014.

Heyliger, P.R. and Jilani, A., "The free vibrations of inhomogeneous elastic cylinders and spheres", *International Journal of Solids and Structures*, vol. 29(22), pp. 2689–2708, 1992.

Hibbeler, R.C., "Free vibration of a beam supported by unsymmetrical spring-hinges", *Journal of Applied Mechanics*, vol. 42(2), pp. 501–502, 1975.

Hill, R., "A self-consistent mechanics of composite materials", *Journal of the Mechanics and Physics of Solids*, vol. 13, pp. 213–222, 1965.

Hilton, H.H., "Optimum viscoelastic designer materials for minimizing failure probabilities during composite cure," *Journal of Thermal Stresses.* vol. 26, pp. 547–557, 2003.

Hilton, H.H., "Optimum linear and nonlinear viscoelastic designer functionally graded materials — characterizations and analysis", *Composites Part A: Applied Science and Manufacturing*, vol. 36(10), pp. 1329–1334, 2005a.

Hilton, H.H., "Designer linear viscoelastic materials tailored to maximize aerodynamic noise control, *Proceedings of the Second International Workshop on High Speed*

Transport Noise and Environmental Acoustics (*HSTNEA 2005*) 11–13, Computer Center of the Russian Academy of Sciences, Moscow, Russia, 2005b.

Hilton, H.H., "Designer linear viscoelastic material properties tailored to minimize probabilistic failures or thermal stress induced dynamic column creep buckling", *Journal of Thermal Stresses*, vol. 29, pp. 403–421, 2006a.

Hilton, H.H., "Tailored designer functionally graded materials for minimizing probabilistic creep buckling failures in linear viscoelastic columns with large deformations and follower loads", In *Proceedings 47th AIAA/ASME/ASCE/AHS Structures, Structural Dynamics and Materials Conference*, AIAA Paper AIAA-2006–1629, 2006b.

Hilton, H.H., "A novel approach to structural analysis: Designer/engineered viscoelastic materials vs. 'off the shelf' property selections", *Journal of Vacuum Technology & Coating*, vol. 10, pp. 23–29, 2009a.

Hilton, H.H., "The elusive and fickle viscoelastic Poisson's ratio and its relation to the elastic–viscoelastic correspondence principle", *Journal of Mechanics of Materials and Structures*, vol. 4(7), pp. 1341–1364, 2009b.

Hilton, H.H. and Lee, D.H., "Designer functionally graded viscoelastic materials performance tailored to minimize probabilistic failures in viscoelastic panels subjected to aerodynamic noise", *Proceedings Ninth International Conference on Recent Advances in Structural Dynamics*, 18–39, Southampton, UK, 2006.

Hilton, H.H., Lee, D.H. and El Fouly, A.R.A., "General analysis of viscoelastic designer functionally graded auxetic materials engineered/tailored for specific task performances", *Mechanics of Time-Dependent Materials*, vol. 12, pp. 151–178, 2008.

Hirai, T., "Functionally gradient materials and nano-composites", *Ceramic Transactions, Functionally Gradient Materials*, (J.B. Holt, M. Koizumi, T. Hirai, Z.A. Munir, eds.), The American Ceramic Society, Westerville, Ohio, vol. 34, pp. 11–20, 1993.

Hirai, T. and Chen, L., "Recent and prospective development of functionally graded materials in Japan", *Materials Science Forum*, vols. 308–311, pp. 509–514, 1999.

Horgan, C.O. and Chan, A.M., "The pressurized hollow cylinder or disk problem for functionally graded isotropic linearly elastic materials", *Journal of Elasticity*, vol. 55(1), pp. 43–59, 1999a.

Horgan, C.O. and Chan, A.M., "Vibration of inhomogeneous strings, rods and membranes", *Journal of Sound and Vibration*, vol. 225(3), pp. 502–513, 1999b.

Horgan, C.O. and Chan, A.M., "The stress response of functionally graded isotropic binary elastic rotating disks", *Journal of Elasticity*, vol. 55, pp. 219–230, 1999c.

Hoshiya, M. and Shah, H.C., "Dynamics and eigenvalue analysis of a rectangular plate with stochastic properties", *Proceedings JSCE*, No. 187, pp. 109–118, 1971.

Houari, M.S., Tounsi, A. and Bég, O.A., "Thermoelastic bending analysis of functionally graded sandwich plates using an new higher order shear and normal deformation theory", *International Journal of Mechanical Sciences*, vol. 76, pp. 102–111, 2013.

Hu, J.D., Lim, M.K. and Liew, K.M., "A power series solution for vibration of a rotating Timoshenko beam", *Journal of Sound and Vibration*, vol. 175(4), pp. 505—523, 1994.

Huang, X.L. and Shen, H.S., "Vibration and dynamic response of functionally graded plates with piezoelectric actuators in thermal environments", *Journal of Sound and Vibration*, vol. 289(1), pp. 25–53, 2006.

Huang, Y. and Li, X.F., "Buckling of functionally graded circular columns including shear deformation", *Materials & Design*, vol. 31(7), pp. 3159–3166, 2010a.

Huang, Y. and Li, X.-F., "A new approach for free vibration of axially functionally graded beams with non-uniform cross-section", *Journal of Sound and Vibration*, vol. 329, pp. 2291–2303, 2010b.

Huang, Z.Y., Lu, C.F. and Chen, W.Q., "Benchmark solutions for functionally graded thick plates resting on Winkler–Pasternak elastic foundations", *Composite Structures*, vol. 85(2), pp. 95–104, 2008.

Huang, Y. and Li, X.-F., "A new approach for free vibration of axially functionally graded beams with non-uniform cross-section", *Journal of Sound and Vibration*, vol. 329, pp. 2291–2303, 2010a.

Huang, Y. and Li, X.-F., "Bending and vibration of cylindrical beams with arbitrary radial nonhomogeneity", *International Journal of Mechanical Sciences*, vol. 52, pp. 595–601, 2010b.

Huang, Y. and Li, X.-F., "Buckling analysis of nonuniform and axially graded columns with varying flexural rigidity", *Journal of Engineering Mechanics*, vol. 137(1), pp. 73–81, 2011.

Huang, Y. and Luo, Q.-Z., "A simple method to determine the critical buckling loads for axially inhomogeneous beams with elastic restraint", *Computers and Mathematics with Applications*, vol. 61, pp. 2510–2517, 2011.

Hutchinson, J.R. and Zellmer, S.D., "Vibration of a free rectangular parallelepiped", *Journal of Applied Mechanics*, vol. 50, pp. 123–130, 1983.

Ibrahim, H.H., Yoo, H.H. and Lee, K.S., Supersonic flutter of functionally graded panels subject to acoustic and thermal loads, *Journal of Aircraft*, vol. 46(5), pp. 593–600, 2009.

Ibrahim, R.A., "Structural dynamics with parameter uncertainties", *Applied Mechanics Reviews*, vol. 40, pp. 309–328, 1987.

Impollonia, N. and Sofi, A., "A response surface approach for the static analysis of stochastic structures with geometrical nonlinearities", *Computer Methods in Applied Mechanics and Engineering*, vol. 192, pp. 4109–4129, 2003.

Iyengar, K.T.S., Chandrashekhora, K. and Sebastian, V.K., "On the analysis of thick rectangular plates", *Archives of Applied Mechanics*, vol. 43, pp. 317–330, 1974.

Jabareen, M. and Eisenberger, M., "Free vibration of non-homogeneous circular and annular membranes", *7th International Congress on Sound and Vibration*, Garmisch-Partenkirchen, Germany, CD Proceedings, July 2000a.

Jabareen, M. and Eisenberger, M., "Vibration frequencies of variable thickness circular plates", *Proceedings of 1st International Conference of Structural Stability and Dynamics*, ICSSD 2000, Taipe, Taiwan, Invited paper, pp. 199–204, December 2000b.

Jabbari, M., Bahtui, A. and Eslami, M.R., "Axisymmetric mechanical and thermal stresses in thick short length FGM cylinders", *International Journal of Pressure Vessels and Piping*, vol. 86, pp. 296–306, 2009.

Jabbari, M., Sohrabpour, S. and Eslami, M.R., "Mechanical and thermal stresses in a functionally graded hollow cylinder due to radially symmetric loads", *International Journal of Pressure Vessels and Piping*, vol. 79(7), pp. 493–497, 2002.

Jabbari, M., Sohrabpour, S. and Eslami, M. R., "General solution for mechanical and thermal stresses in a functionally graded hollow cylinder due to nonaxisymmetric steady-state loads", *ASME Journal of Applied Mechanics*, vol. 70(1), pp. 111–118, 2003.

Jabbari, M., Sohrabpour, S. and Eslami, M.R., "Axisymmetric mechanical and thermal stresses in thick long FGM cylinders", *Journal of Thermal Stresses*, vol. 29(7), pp. 643–662, 2006.

Jaiani, G., "Arbitrarily loaded elastic half-plane, modulus of elasticity of which obeys the power low", *Materials of the First All-Union School on Theory and Numerical Methods in Analysis of Shells and Plates*, pp. 351–361, Tbilesi University Press, 1975 (in Russian).

Jaiani, G., "Cylindrical bending of a rectangular plate with power law changing stiffness", *Proceedings of Conference of Young Scientists in Mathematics*, pp. 49–52, Tbilisi University Press, 1976 (in Russian).

Jaiani, G. "Bending of a plate with power law changing stiffness", *Annuaire Des Ecoles Superieures Mechanique Technique*, vol. XII (2), pp. 15–19, 1977 (in Russian).

Jaiani, G., "On a mathematical model of a bar with variable rectangular cross-section, Preprint 98/21, Universitaet Potsdam, Institut fuer Mathematik, Potsdam, 1998a.

Jaiani, G., "On a model of a bar with variable thickness", *Bull. of TICMI*, vol. 2, 36–40, 1998b (for electronic version see: http://www.viam.science.tsu.ge/others/ticmi).

Janghorban, M. and Zare, A., "Free vibration analysis of functionally graded carbon nanotubes with variable thickness by differential quadrature method", *Physica E: Low-Dimensional Systems and Nanostructures*, vol. 43(9), pp. 1602–1604, 2011.

Jankowski, J. and Kowal-Michalska, K., "Dynamic response of FGM thin — walled plate structure subjected to a thermal pulse loading", in *Shell Structures: Theory and Applications*, (W. Pietraskiewicz and J. Górski, eds.), pp. 297–300, CRC Press, Leiden, 2014.

Jaroszewicz, J. and Zoryj, L., "Flexural vibrations and stability of beams with variable coefficients", *International Applied Mechanics*, vol. 30(9), pp. 713–719, 1994.

Javaheri, R. and Eslami, M.R., "Buckling of functionally graded plates under in-plane compressive loading", *ZAMM*, vol. 82, pp. 277–283, 2002a.

Javaheri, R. and Eslami, M.R., "Thermal buckling of functionally graded plates", *AIAA Journal*, vol. 40(1), pp. 162–169, 2002b.

Jha, B.B., Galgali, R.K. and Misra, V.N., *Futuristic Materials*, Allied Publishers Pvt. Ltd., New Delhi, 2004.

Jha, D.K., Kant, T. and Singh, R.K., "A critical review of recent research on functionally graded plates", *Composite Structures*, vol. 96, pp. 883–849, 2013.

Jha, D.K., Kant, T. and Singh, R.K., "Free vibration response of functionally graded thick plates with shear and normal deformation effects", *Computers and Structures*, vol. 96, pp. 799–823, 2013.

Jin, Z.H., Paulino, G.H. and Dodds, R.H., "Cohesive fracture modeling of elastic–plastic crack growth in functionally graded materials", *Engineering Fracture Mechanics*, vol. 70(14), pp. 1885–1912, 2003.

Jobareen, M. and Eisenberger, M., "Free vibrations of non-homogeneous circular and annular membranes", *Journal of Sound and Vibration*, vol. 240, pp. 409–429, 2001.

Johns, D.J., Comments on "An approximate expression for the fundamental frequency of vibration of elastic plates", *Journal of Sound and Vibration*, vol. 41(3), pp. 385–387, 1975.

Jones, R., "An approximate expression for the fundamental frequency of vibration of elastic plates", *Journal of Sound and Vibration*, vol. 38(4), pp. 503–504, 1975.

Kadoli, R. and Ganesan, N., "Buckling and free vibration analysis of functionally graded cylindrical shells subjected to a temperature-specified boundary condition", *Journal of Sound and Vibration*, vol. 289(3), pp. 450–480, 2006.

Kadoli, R., Akhtar, K. and Ganesan, N., "Static analysis of functionally graded beams using higher order shear deformation theory", *Applied Mathematical Modelling*, vol. 32(12), pp. 2509–2525, 2008.

Kambampati, S., Ganguli, R. and Mani, V., "Determination of isospectral nonuniform rotating beams", *Journal of Applied Mechanics*, vol. 79(6), p. 061016, 2012.

Kambampati, S., Ganguli, R. and Mani, V., "Non-rotating beams isospectral to a given rotating uniform beam", *International Journal of Mechanical Sciences*, vol. 66, pp. 12–21, 2013a.

Kambampati, S., Ganguli, R. and Mani, V., "Rotating beams isospectral to axially loaded nonrotating uniform beams", *AIAA Journal*, vol. 51(5), pp. 1189–1202, 2013b.

Kant, T., Jha, D.K. and Singh, R.K., "A higher-order shear and normal deformation functionally graded plate model: some recent results", *Acta Mechanica*, vol. 225, pp. 2865–2876, 2014.

Kapuria, S., Bhattacharyya, M. and Kumar, A.N., "Bending and free vibration response of layered functionally graded beams: A theoretical model and its experimental validation", *Composite Structures*, vol. 82(3), pp. 390–402, 2008.

Kardomateas, G.A., "Bending of a cylindrically orthotropic curved beam with linearly distributed elastic constants", *The Quarterly Journal of Mechanics and Applied Mathematics*, vol. 43(1), pp. 43–55, 1990.

Karnovsky, I.A. and Lebed, O.I, *Formulas for Structural Dynamics: Tables, Graphs, and Solution*, Chapter 6: Bernoulli–Euler uniform one span beams with elastic supports, pp. 159–194. New York: Mc Graw-Hill, 2000.

Karşun, A., Topçu, M. and Yücel, U., "Stress analysis of a rotating FGM circular disk with exponentially varying properties", *Paper No. IMECE 2012-8559*, pp. 659–663, ASME International Mechanical Engineering Congress and Exposition, vol. 8, ASME Press, New York, 2012.

Kashtalyan, M., "Three-dimensional elasticity solution for bending of functionally graded rectangular plates", *European Journal of Mechanics-A/Solids*, vol. 23(5), pp. 853–864, 2004.

Kawasaki, A., Niino, M. and Kumakawa, A., *Multiscale, Multifunctional and Functionally Graded Materials: Selected, Peer Reviewed Papers from the 10th International Symposium on Mm & Fgms, 22nd–25th, September 2008, Sendai, Japan*, Trans Tech Publications, 2010.

Ke, L.L., Yang, J. and Kitipornchai, S., "Flexural vibration and elastic buckling of a cracked Timoshenko beam mode of functionally graded materials", *Mechanics of Advanced Materials and Strucutres*, vol. 16(6), pp. 488–502, 2009.

Ke, L.L., Yang, J., Kitipornchai, S. and Xiang, Y., "Flexural vibration and elastic buckling of a cracked Timoshenko beam made of functionally graded materials", *Mechanics of Advanced Materials and Structures*, vol. 16, pp. 209–225, 2009.

Keles, I. and Tutuncu, N., "Exact analysis of axisymmetric dynamic response of functionally graded

Khalfi, Y., Houari, M.S.A. and Tounsi, A., "A refined and simple shear deformation theory for thermal buckling of solar functionally graded plates on elastic foundation", *International Journal of Computational Methods*, DOI: 10.1142/S0219876213500771, 2013.

Khan, K.A. and Hilton, H.H., "On inconstant Poisson's ratios in non-homogeneous elastic media", *Journal of Thermal Stresses*, vol. 33(1), pp. 29–36, 2009.

Khoma, I.Yu, "On general solution of the system of equilibrium equations of constant thickness plates", *Doklady AN SSSR*, vol. 213(1), pp. 59–62, 1973 (in Russian).

Khulief, Y.A., "Vibration frequencies of a rotating tapered beam with end mass", *Journal of Sound and Vibration*, vol. 134, pp. 87–97, 1989.

Khvoles, A.R., "Basic boundary problems for equilibrium equations of the moment theory for circular plate of the constant thickness", *Proceedings of the Applied Mathematics Institute*, Tbilisi State University, vol. 2, pp. 113–120, 1969 (in Russian).

Kim, J. and Paulino, G.H., "Finite elements evaluation of mixed mode stress intensity factors in functionally graded materials", *International Journal for Numerical Methods in Engineering*, vol. 53, pp. 1903–1935, 2002a.

Kim, J.H. and Paulino, G.H., "Mixed-mode fracture of orthotropic functionally graded materials using finite elements and the modified crack closure method", *Engineering Fracture Mechanics*, vol. 69(14), pp. 1557–1586, 2002b.

Kim, J.H. and Paulino, G.H., "Consistent formulations of the interaction integral method for fracture of functionally graded materials", *Urbana*, vol. 51, p. 61801, 2003a.

Kim, J.H. and Paulino, G.H., "The interaction integral for fracture of orthotropic functionally graded materials: Evaluation of stress intensity factors", *International Journal of Solids and Structures*, vol. 40(15), pp. 3967–4001, 2003b.

Kim, J.H. and Paulino, G.H., "An accurate scheme for mixed-mode fracture analysis of functionally graded materials using the interaction integral and micromechanics models", *International Journal for Numerical Methods in Engineering*, vol. 58(10), pp. 1457–1497, 2003c.

Kim, J.H. and Paulino, G.H., "T-stress, mixed-mode stress intensity factors, and crack initiation angles in functionally graded materials: a unified approach using the interaction integral method", *Computer Methods in Applied Mechanics and Engineering*, vol. 192(11), pp. 1463–1494, 2003d.

Kim, Y.W., "Temperature dependent vibration analysis of functionally graded rectangular plates", *Journal of Sound and Vibration*, vol. 284(3), pp. 531–549, 2005.

Kitipornchai, S., Ke, L.L., Yang, J. and Xiang, Y., "Nonlinear vibration of edge cracked functionally graded Timoshenko beams", *Journal of Sound and Vibration*, vol. 324(3), pp. 962–982, 2009.

Kitipornchai, S., Yang, J. and Liew, K.M., "Random vibration of the functionally graded laminates in thermal environments", *Computer Methods in Applied Mechanics and Engineering*, vol. 195(9), pp. 1075–1095, 2006.

Kleiber, M. and Hien, T.D., *Stochastic Finite Element Method: Basic Perturbation Technique and Computer Implementation*, Wiley, Chichester, 1992.

Knoppers, R., Gunnik, J.W., van den Hout, J. and van Vliet, W., "The reality of functionally graded material products", *TNO Science and Industry*, The Netherlands, pp. 38–43, 2005.

Kocatürk, T., Şimşek, M. and Akbaş, Ş.D., "Large displacement static analysis of a cantilever Timoshenko beam composed of functionally graded material", *Science and Engineering of Composite Materials*, vol. 18(1–2), pp. 21–34, 2011.

Koizumi, M., "The concept of FGM", in*Ceramic Transitions, Functionally gradient Materials*, The American Ceramic Society (Holt, J.B., Koizumi, M., Mirai, F. and Munir, Z.A., eds.), vol. 34, pp. 3–10, 1993.

Koizumi, M., "FGM activities in Japan", *Composites Part B*, vol. 28, pp. 1–4, 1997.

Kordkheili Hosseini, S.A. and Naghdabadi, R., "Thermoelastic analysis of functionally graded cylinders under axial loading", *Journal of Thermal Stresses*, vol. 31, pp. 1–17, 2008.

Kovalenko, A.D., *Selected Works*, Naukova Dumka Publishers, Kiev, 1976 (in Russian).

Krenk, S., "Theories for elastic plates via orthogonal polynomials", *Journal of Applied Mechanics*, vol. 48, 900–904, 1981.

Krylov, V. and Sorokin, S.V., "Dynamics of elastic beams with controlled distributed stiffness parameters", *Smart Materials and Structures*, vol. 6, pp. 573–582, 1997.

Kumar, A. and Ganguli, R., "Iso-spectral rotating and non-rotating beams", in *IUTAM Symposium on Multi-Functional Material Structures and Systems*, pp. 261–268, Springer Netherlands, 2010.

Kuo, S.-Y. and Shiau, L.-C., "Buckling and vibration of composite laminated plates with variable fiber spacing", *Composite Structures*, vol. 90(2), pp. 196–200, 2009.

Lanche, W., "Thermal buckling of a simply-supported moderately thick rectangular FGM plate", *Composite Structures*, vol. 64, pp. 211–218, 2004.

Lannutti, J.J., "Functionally graded materials: Properties, potential and design guidelines", *Composites Engineering*, vol. 4(1), pp. 81–94, 1994.

Larbi, L.O., Kaci, A., Sid, M., Houari, A. and Tounsi, A., "An efficient shear deformation beam theory based on neutral surface position for bending and free vibration of functionally graded beams", *Mechanics Based Design of Structures and Machines: An International Journal*, vol. 41(4), pp. 421–433, 2013.

Larson, R.A., *A Novel Method for Characterizing the Impact Response of Functionally Graded Plates*, ProQuest LLC, 2009.

Laura, P.A.A. and Duran, R., "A note on forced vibrations of a clamped rectangular plate", *Journal of Sound and Vibration*, vol. 42, pp. 129–135, 1975.

Laura, P.A.A. and Guttierez, R.H., "Vibration of an elastically restrained cantilever beam of varying cross-section with tip mass of finite length", *Journal of Sound and Vibration*, vol. 108, pp. 123–131, 1986.

Laura, P.A.A., Luizoni, L.E. and Lopez, J.J., "A note on free and forced vibrations of circular plates: the effect of support flexibility", *Journal of Sound and Vibration*, vol. 47(2), pp. 287–291, 1976.

Laura, P.A.A, Paloto, J.C. and Santos, R.D., "A note on the vibration and stability of circular plate elastically restrained against rotation", *Journal of Sound and Vibration*, vol. 41(2), pp. 177-180, 1975.

Lee, S.Y. and Kuo, Y.H., "Exact solution for the analysis of generally elastically restrained non-uniform beams", *Journal of Sound and Vibration*, vol. 59, pp. 205–212, 1992.

Leissa, A.W., *Vibration of Plates*, NASA SP-160, U.S. Government Printing Office, Washington, D.C., 1969.

Leissa, A.W., "The free vibration of rectangular plates", *Journal of Sound and Vibration*, vol. 31, pp. 257–293, 1973.

Leissa, A.W., "Recent research in plate vibration: classical theory", *Shock and Vibration Digest*, vol. 9(10), pp. 13–24, 1977.

Leissa, A.W., "The analysis of forced vibrations of plates having damping", *Proceedings of the Ninth Southeastern Conference on Theoretical and Applied Mechanics*, vol. 9, pp. 183–193, Plenum Press, New York, 1978a.

Leissa, A.W., "A direct method for analyzing the forced vibrations of continuous systems having damping", *Journal of Sound and Vibration*, vol. 56, pp. 313–324, 1978b.

Leissa, A.W., "Plate vibration research, 1976–1989: Complicating effects", *The Shock and Vibration Digest*, vol. 13(10), p. 25, 1981.

Leissa, A.W. and Zhang, Z., "On the three-dimensional vibrations of the cantilevered rectangular parallelepiped", *Journal of Acoustical Society of America*, vol. 73, pp. 2013–2021, 1983.

Leissa, A.W. and Young, T.H., "Extensions of the Ritz–Galerkin method for the forced, damped vibrations of structural elements", *Proceedings of the Vibration Damping Workshop*, Wright-Patterson Air Force Base, Ohio, pp. EE1–EE22, 1984.

Lekhnitskii, S.G., *Anisotropic Plates. Chapter 10: Bending of plates by normal load*, pp. 318–393. New York: Gordon and Breach Science Publishers, 1968.

Lenox, T.A. and Conway, H.D., "An exact closed-form solution for the flexural vibration of a thin annular plate having a parabolic thickness variation", *Journal of Sound and Vibration*, vol. 68, pp. 231–239, 1980.

Lepik, Ü. and Hein, H., "Free vibrations on non-uniform and axially functionally graded Euler–Bernoulli beams", in *Haar Wavelets* (pp. 167–176), Springer International Publishing, 2014.

Li, Q., Lu, V.P. and Kou, K.P., "Three-dimensional vibration analysis of functionally graded material sandwich plates", *Journal of Sound and Vibration*, vol. 311(1), pp. 498–515, 2008.

Li, Q.S., "Exact solutions for buckling of non-uniform columns under axial concentrated and distributed loading", *European Journal of Mechanics: A/Solids*, vol. 20, pp. 485–500, 2001.

Li, Q.S., "Exact solution of the generalized Euler's problem", *Journal of Applied Mechanics*, vol. 76, article, 041015, 2009.

Li, S.R. and Batra, R.C., "Buckling of axially compressed thin cylindrical shells with functionally graded middle layer", *Thin Walled Structures*, vol. 44(10), pp. 1039–1047, 2006.

Li, S.-R. and Batra, R.C., "Relations between buckling loads of functionally graded Timoshenko and homogeneous Euler–Bernoulli beams, *Composite Structures*, vol. 95, pp. 5–9, 2013.

Li, X.D., "Nanoscale structural and mechanical characterization of natural nanocomposites: seashells", *Journal of Materials*, vol. 59(3), pp. 71–74, 2007.

Li, X.F., "A unified approach for analyzing static and dynamic behaviors of functionally graded Timoshenko and Euler–Bernoulli beams", *Journal of Sound and Vibration*, vol. 318, pp. 1210–1229, 2008.

Li, X.-F. and Peng, X.L., "A pressurized functionally graded hollow cylinder with arbitrarily varying material properties", *Journal of Elasticity*, vol. 96, pp. 81–95, 2009.

Li, X.F., Xi, L.Y. and Huang, Y., "Stability analysis of composite columns and parameter optimization against buckling", *Composites Part B: Engineering*, vol. 42(6), pp. 1337–1345, 2011.

Li, X.Y., Dingt, H.J. and Chen, W.Q., Axisymmetric elasticity solutions for a uniformly loaded annular plate of transversely isotropic functionally graded materials, *Acta Mechanica*, vol. 196(3–4), pp. 139–159, 2008.

Librescu, L. and Maalavi, K.Y., "Material grading for improved aeroelastic stability in composite wings", *Journal of Mechanics of Materials and Structures*, vol. 2(7), pp. 1381–1394, 2007.

Librescu, L., Oh, S.-Y. and Song, O., "Thin-walled beams made of functionally graded materials and operating in high temperature environment", *Journal of Thermal Stresses*, vol. 28(6–7), pp. 649–712, 2005.

Librescu, L., Oh, S.Y. and Song, O., "Spinning thin-cralled beams made of functionally graded materials: Modeling, vibration and instability", *European Journal of Mechanics-A/Solids*, vol. 23(3), pp. 499–515, 2004.

Liew, K.M. and Teo, T.M., "Three-dimensional vibration analysis of rectangular plates based on differential quadrature method", *Journal of Sound and Vibration*, vol. 220, pp. 577–599, 1999.

Liew, K.M., Hung, K.C. and Lim, M.K., "A continuum three-dimensional vibration analysis of thick rectangular plates", *International Journal of Solids and Structures*, vol. 30, pp. 3357–3379, 1993.

Liew, K.M., Hung, K.C. and Lim, M.K., "Three-dimensional vibration of rectangular plates: variance of simple support and influence of in-plane inertia", *International Journal of Solids and Structures*, vol. 31, pp. 3233–3247, 1994.

Liew, K.M., Hung, K.C. and Lim, M.K., "Three-dimensional vibration of rectangular plates: Effects of thickness and edge constraints", *Journal of Sound and Vibration*, vol. 182, pp. 709–727, 1995a.

Liew, K.M., Hung, K.C. and Lim, M.K., "Vibration of Mindlin plates using boundary characteristic orthogonal polynomials", *Journal of Sound and Vibration*, vol. 182, pp. 77–90, 1995b.

Liew, K.M., Kitipornchai, S., Zhang, X.Z. and Lim, C.W., "Analysis of the thermal stress behaviour of functionally graded hollow circular cylinders", *International Journals of Solids and Structures*, vol. 40, pp. 2355–2380, 2003.

Liew, K.M., Teo, T.M. and Han, J.-B., "Three-dimensional static solutions of rectangular plates by variant differential quadrature method", *International Journal of Mechanical Sciences*, vol. 43, pp. 1611–1628, 2001.

Liew, K.M. and Yang, B., "Three-dimensional solutions for free vibrations of circular plates by polynomials — Ritz analysis", *Computer Methods in Applied Mechanics and Engineering*, vol. 175(1–2), pp. 189–201, 1999.

Liew, K.M., Yang, J. and Wu, Y.F., "Nonlinear vibration of a coating-FGM-substrate cylindrical panel subjected to a temperature gradient", *Computer Methods in Applied Mechanics and Engineering*, vol. 195(9–12), pp. 1007–1026, 2006.

Liew, K.M., Hung, K.C. and Lim, M.K., "Vibration of Minddin plates using boundary characteristic orthogonal polynomials", *Journal of Sound and Vibration*, vol. 182(1), pp. 77–90, 1995.

Liew, K.M., Xiang, Y. and Kitipornchai, S., "Transverse vibration of thick rectangular plates" — II. Inclusion of oblique internal line supports, *Computers and Structures*, vol. 49, pp. 31–58, 1993a.

Liew, K.M., Xiang, Y. and Kitipornchai, S., "Transverse vibration of thick rectangular plates — III. Effects of multiple eccentric internal ring supports", *Computers and Strucutres*, vol. 49, pp. 59–68, 1993b.

Lim, C.W., "Three-dimensional vibration analysis of a cantilevered parallelepiped: Exact and approximate solutions, *Journal of the Acoustical Society of America*, vol. 106(6), pp. 3375–3383, 1999.

Lim, C.W., Kitipornchai, S. and Liew, K.M., "Numerical aspects for free vibration of thick plates. Part II: Numerical efficiency and vibration frequencies", *Computer Methods in Applied Mechanics and Engineering*, vol. 156(1–4), pp. 31–44, 1998.

Lim, C.W., Liew, K.M. and Kitipornchai, S., "Numerical aspects for free vibration of thick plates. Part I: Formulation and verification", *Computer Methods in Applied Mechanics and Engineering*, vol. 156(1–4), pp. 15–29, 1998.

Lipton, R., "Design of FG composite structures in the presence of stress constraints", *International Journal of Solids and Structures*, vol. 39(9), pp. 2475–2586, 2002.

Liu, W.H. and Chen, K.S., "Effects of lateral support on the fundamental natural frequencies and buckling coefficients", *Journal of Sound and Vibration*, vol. 129(1), pp. 155–160, 1989.

Liu, D.Y., Wang, C.Y. and Chen, W.Q., "Free vibration of FGM plates with in-plane material inhomogeneity", *Composite Structures*, vol. 92, pp. 1047–1051, 2010.

Lizarev, A.D., "Free vibration of beams with elastically clamped ends", *Izvestiya VUZOV, Stoitelstvo i Arkhitektura*, No. 10, 1959 (in Russian).

Lowe, B., "On the construction of an Euler–Bernoulli beam from spectral data", *Journal of Sound and Vibration*, vol. 163, pp. 165–171, 1993.

Loy, C.T., Lam, K.Y. and Reddy, J.N., "Vibration of functionally graded cylindrical shells", *International Journal of Mechanical Science*, vol. 41(3), pp. 309–324, 1999a.

Lurie, A.I., *Three-Dimensional Problems of the Theory of Elasticity*, Interscience Publishers, New York, 1964.

Maalawi, K.Y., "Optimization of elastic columns using axial grading concept", *Engineering Structures*, vol. 31(12), pp. 2922–2929, 2009.

Maalawi, K., "Stability, dynamic and aeroelastic optimization of functionally graded composite structures", in *Advances in Computational Stability Analysis* (S.F. Coşkun, ed.), pp. 17–43, Intechopen.com, 2012.

Maeda, Y., Nishiwaki, S., Izui, K., Yoshimura, M., Matsui, K. and Terada, K., "Structural topology optimization of vibrating structures with specified eigenfrequencies and eigenmode shapes", *International Journal for Numerical Methods in Engineering*, vol. 67, pp. 597–628, 2006.

Mahamood, R.M. and Akinlabi, E., "Functioanlly graded material: An overview", *Proceedings of the World Congress on Engineering*, vol. 3, WCE 2012, London, UK, 2012.

Mahi, A., Adda Bedia, El A., Tounsi, A. and Mechab, I., "An analytical method for temperature-dependent free vibration analysis of functionally graded beams with general boundary conditions", *Composite Structures*, vol. 92(8), pp. 1877–1887, 2010.

Makushin, V.M., "Critical values of intensity of radial compressive forces in thin plates", Strength Analysis, vol. 4, pp. 270–298, (1959); vol. 5, pp. 236–248, (1960); vol. 6, pp. 171–181, (1960) (in Russian).

Malekzadeh, P., "Three-dimensional free vibration analysis of thick functionally graded plates on elastic foundations", *Composite Structures*, vol. 89(3), pp. 367–373, 2008.

Malik, M. and Bert, C.W., "Three-dimensional elasticity solutions for free vibrations of rectangular plates by the differential quadrature method", *International Journal of Solids and Structures*, vol. 35, pp. 299–318, 1998.

Mania, R.Y., "Dynamic response of FGM thin plate subjected to combined loads", in *Shell Structures: Theory and Applications* (W. Pietraskiewicz and J. Górski, eds.), pp. 317–320, CRC Press, Leiden, 2014.

Manneth, V., *Numerical Studies on Stress Concentration in Functionally Graded Materials*, University of Rhode Island, 2009.

Manohar, C.S. and Ibrahim, R.A., "Progress in structural dynamics with stochastic parameter variations", *Applied Mechanics Reviews*, vol. 52, pp. 177–197, 1999.

Mareishi, S., Mohammadi, M. and Rafiee, M., "An Analytical Study on Thermally Induced Vibration Analysis of FG Beams Using Different HSDTs", *Applied Mechanics and Materials*, vol. 249, pp. 784–791, 2013.

Markov, K. and Preziosi, L., *Heteregoeneous Media: Mictromechanics Modeling Methods and Simulations*, Birkhauser, Boston, 2000.

Markworth, A.J. and Saunders, J.H., "A model of structure optimization for a functionally graded material", *Materials Letters*, vol. 22, pp. 103–107, 1995.

Markworth, A.J., Ramesh, K.S. and Parks, W.P. Jr., "Review: Modeling studies applied to functionally graded: materials", *Journal of Materials Science Review*, vol. 30, pp. 2183–2193, 1995.

Maróti, G., "Finding closed form solutions of beam vibrations", *Pollack Periodica*, vol. 6(1), pp. 141–154, 2011.

Maróti, G., "Closed form solution for bending oscillations of beams", *Pollack Periodica*, vol. 8(3), pp. 111–118, 2013.

Maróti, G., "Short remark on lateral vibration of functionally graded beams", available at http://seth.asc.tuwien.ac.at/proc12/full_ paper/Contribution487.pdf, 2013.

Maróti, G. and Elishakoff, I., "On buckling of axially functionally graded beams", *Pollack Periodica*, vol. 7(1), pp. 3–13, 2012.

Marur, P., *Fracture Behaviour of Functionally Graded Materials*, Lambert Academic Publishing, Saarbruecken, FRG, 2010.

Matsunaga, H., "Free vibration and stability of functionally graded circular cylindrical shells according to a 2D higher-order deformation theory", *Journal of Composite Structures*, vol. 88, pp. 519–531, 2009a.

Matsunaga, H., "Stress analysis of functionally graded plated subjected to thermal and mechanical loadings", *Composite Structures*, vol. 87, pp. 344–357, 2009b.

Matthies, H.G., "Stochastic finite elements: computational approaches to stochastic differential equations, *Zeitschrift fuer angewandte Mathematik und Mekhanik*, vol. 88, pp. 849–873, 2008.

Maugin, G.A., *Material Inhomogeneities in Elasticity*, Chapman & Hall, London, 1993.

Maurizi, M.J., Rossi, R.E. and Reyes, J.A., "Comments on "A note of generally restrained beams" ", *Journal of Sound and Vibration*, vol. 147(1), pp. 167–171, 1991.

Mazzei Jr., A.J. and Scott, R.A., "On the effects of non-homogeneous materials on the vibrations and static stability of tapered shafts", *Journal of Vibration and Control*, vol. 19(5), pp. 771–786, 2013.

Mazzei, A.J. Jr., "On the effect of functionally graded materials on resonances of rotating beams", *Shock and Vibration*, vol. 19, pp. 1315–1326, 2012.

Mechab, I., Atmane, H.A., Tounsi, A., Belhadj, H.A. and Adda Bedia, El A., "A two variable refined plate theory for the bending analysis of functionally graded plates", *Acta Mechanica Sinica*, vol. 26(6), pp. 941–949, 2010.

Meguid, S.A. and Zhu, Z.H., "A novel finite element for treating inhomogeneous solids", *International Journal for Numerical Methods in Engineering*, vol. 38, pp. 1579–1592, 1995.

Meiche, El, N., Tounsi, A., Ziane, N., Mechab, I. and Adda Bedia, El A., "A new hyperbolic shear deformation theory for buckling and vibration of functionally graded sandwich plate, *International Journal of Mechanical Sciences*, vol. 53(4), pp. 237–247, 2011.

Meissner, E., "Ueber das Knicken Kreisformiger Platten", Schweiz. Bauztg., vol. 101, pp. 87–100, 1933 (in German).

Menaa, R., Tounsi, A., Mouaici, F., Mechab, I., Zidi, M. and Adda Bedia, El A., "Analytical solutions for static shear correction factor of functionally graded rectangular beams", *Mechanics of Advanced Materials and Structures*, vol. 19(8), pp. 641–652, 2012.

Merrett, C.G. and Hilton, H.H., "Generalized linear servo-aero-viscoelasticity: Theory and applications", *49th AIAA/ASME/ASCE/ AHS/ASC Structures, Structural Dynamics, and Materials Conference 16th AIAA/ASME/AHS Adaptive Structures Conference*, Schaumburg, IL, 2008.

Mian, M.A. and Spencer, A.J.M., "Exact solutions for functionally graded and laminated elastic materials", *Journal of the Mechanics and Physics of Solids*, vol. 46, pp. 2283–2295, 1998.

Mindlin, R.D., "Influence of rotatory inertia and shear on flexural motions of isotropic, elastic plates", *Journal of Applied Mechanics*, vol. 18, pp. 31–38, 1951.

Minkarah, I.A. and Hoppman II, W.H., "Flexural vibrations of cylindrically aelotropic circular plates", *Journal of Acoustical Society of America*, vol. 36, pp. 470–475, 1964.

Miyamoto, Y., *Functionally Graded Materials: Design, Processing and Applications*, Kluwer Academic Publishers, Boston, 1999.

Mohanty, S.C., Dash, R.R. and Rout, T., "Static and dynamic stability analysis of a functionally graded Timoshenko beam", *International Journal of Structural Stability and Dynamics*, vol. (1294), 2012, paper 1250025.

Mori, T. and Tanaka, K., Average stress in matrix and average elastic energy of materials with misfitting inclusions, *Acta Metallurgica*, vol. 21, pp. 571–574, 1973.

Muller, W.H., Herrmann, G. and Gao, H., "Elementary strength theory of cracked beams", *Theoretical and Applied Fracture Mechanics*, vol. 18, pp. 163–177, 1993.

Müller, E., Drešar, Č., Schliz, J. and Keysser, N.A., "Functionally graded materials for sensor and energy applications", *Materials Science and Engineering, Series A*, vol. 35(6), pp. 806–813, 1971.

Murakami, Y., *Handbook of Stress Intensity Factors*, Pergamon Press, Oxford, 1987.

Murthy, K.N. and Christiano, P., "Optimal design for prescribed buckling loads", *Journal of Structural Division, Proceedings ASCE*, vol. 100, pp. 2175–2190, 1974.

Na, K.S. and Kim, J.H., "Three-dimensional thermo-mechanical buckling analysis for functionally.

Nadai, A., "Ueber das Ausbeuben von Kreisformigen Platten", Zeitschrift VDI, vol. 59, pp. 169–177, and pp. 221–224, 1915 (in German).

Najafizadeh, M.M. and Eslami, M.R., "Buckling analysis of circular plates of FGM under uniform radial compression", *International Journal of Mechanical Sciences*, vol. 44(12), pp. 2479–2493, 2002a.

Najafizadeh, M.M. and Eslami, M.R., "First-order-theory–based thermoplastic stability of functionally graded material circular plates", *AIAA Journal*, vol. 40(7), pp. 1444–1450, 2002b.

Najafizadeh, M.M. and Heydari, H.R., "Higher-order theory for buckling of functionally graded circular plates", *AIAA Journal*, vol. 45(6), pp. 1153–1160, 2007.

Najafizadeh, M.M. and Heydari, H.R., "An exact solution for buckling of functionally graded circular plates based on higher order shear deformation plate theory under uniform radial compression", *International Journal of Mechanical Sciences*, vol. 50(3), pp. 603–612, 2008.

Najafizadeh, M.M. and Isvandzibaei, M.R., "Vibration of functionally graded cylindrical shells based on higher order shear deformation plate theory with ring support", *Acta Mechanica*, vol. 191(1–2), pp. 75–91, 2007.

Nakagiri, S. and Hisada, T., *Introduction to Stochastic Finiute Element Methods*, Baifu-kan, Tokyo, 1985 (in Japanese).

Nasoskin, V.D., "On distribution of stresses in statistically inhomogeneous half-plane", *Stroitelnaya Mekhanika i Raschet Sosruzhenii*,   Issue 1, pp. 20–23, 1968 (in Russian).

Neubrand, A. and Rödel, J., "Gradient materials: An overview of a novel concept", *Zeitschrift Fur Metallkunde*, vol. 88(5), pp. 358–371, 1997.

Neuringer, J. and Elishakoff, I., "Inhomogeneous beams that may posses a prescribed polynomial second mode", *Chaos Soliton Fractal*, vol. 12, pp. 881–896, 2001.

Ng, T.Y., Lam, K.Y., Liew, K.M. and Reddy, J.N., "Dynamic stability analysis of functionally graded cylindrical shells under periodic axial loading", *International Journal of Solids and Structures*, vol. 38(8), pp. 1295–1309, 2001.

Nguyen, T.K., Sab, K. and Bonnet, G., "First-order shear deformation plate models for functionally graded materials", *Composite Structures*, vol. 83(1), pp. 25–36, 2008.

Nie, G.J. and Zhong, Z., "Semi-analytical solution for three-dimensional vibration of functionally graded circular plates", *Computer Methods in Applied Mechanics and Engineering*, vol. 196(49), pp. 4901–4910, 2007.

Nie, G.J. and Zhong, Z., "Vibration analysis of functionally graded annular sectorial plates with simply supported radial edges", *Composite Structures*, vol. 84(2), 167–176, 2008.

Niino, M., Kisara, K. and Mori, M., "Feasibility study of FGM technology in space solar power systems (SPSS)", *Material Science Forum*, vol. 492, pp. 163–168, 2005.

Nogata, F. and Takahashi, H., "Intelligent functionally graded material: Bamboo", *Composites Engineering*, vol. 5(7), pp. 743–751, 1995.

Noor, A.K., Vinneri, S.L., Paul, D.B. and Hopkins, M.A., II "Structures technology for future aerospace systems", *Computers and Structures*, vol. 74(5), pp. 507–519, 2000.

Okamura, H., Watanabe, K. and Takano, T., "Applications of the compliance concept in fracture mechanics, progress in flaw growth and fracture toughness testing", ASTM STP 536, *American Society for Testing and Materials*, pp. 423–438, 1973.

Okamura, H., Watanabe, K. and Takano, T., "Deformation and strength of cracked member under bending moment and axial force", *Engineering Fracture Mechanics*, vol. 7, pp. 531–539, 1975.

Ootao, Y. and Tanigawa, Y., "Transient thermoelastic problem of functionally graded thick strip due to nonuniform heat supply", *Journal of Composite Structures*, vol. 63, pp. 139–146, 2004.

Oral, A. and Anlas, G., "Effects of radially varying moduli on stress distribution of nonhomogeneous anisotropic cylindrical bodies", *International Journal of Solids and Structures*, vol. 42, pp. 5568–5588, 2005.

Ostoja-Starzewski, M., *Microstructural Randomness and Scaling in Mechanics of Materials*, Chapman & Hall/CRC, Boca Raton, FL, 2008.

Pan, E., Alkasawneh, W. and Chen, E., *An Exploratory Study on Functionally Graded Materials with Application to Multilayered Pavement Design*, University of Akron, Department of Civil Engineering, 2007.

Pan, E., "Exact solution for functionally graded anisotropic composite laminates", *Journal of Composite Materials*, vol. 37, pp. 1903–1920, 2003.

Pan, W., *Funtionally Graded Materials 2002*, Trans Tech Publ., 2003

Pandalai, K.A.V. and Patel, A.S., "Natural frequencies of orthotropic circular plates", *AIAA Journal*, vol. 3, pp. 780–781, 1965.

Papargyri-Beskou, S., Tsepoura, K.G., Polzos, D. and Beskos, D.E., "Bending and stability of gradient elastic beams", *International Journal of Solids and Structures*, vol. 40, pp. 385–400, 2003.

Parameswaran, V. and Shukla, A., "Processing and characterization of a model functionally gradient material", *Journal of Materials Science*, vol. 35, pp. 21–29, 2000.

Park, J.S. and Kim, J.H., "Thermal postbuckling and vibration analyses of functionally graded plates", *Journal of Sound and Vibration*, vol. 289(1), pp. 77–93, 2006.

Parker, D.F. and England, A.H., *Arisotropy, Inhomogeneity and Nonlinearity in Solid Mechanics*, Kluwer Academic Publishers, Dordrecht, 1995.

Patel, B.P., Gupta, S.S., Loknath, M.S.B. and Kadu, C.P., "Free vibration analysis of functionally graded elliptical cylindrical shells using higher-order theory", *Journal of Composite Structures*, vol. 69, pp. 259–270, 2005.

Paulino, G.H. (ed.), *Fracture of Functionally Graded Materials*, Elsevier Science, Kidlington, 2002a.

Paulino, G.H., "Fracture of functionally graded materials", *Engineering Fracture Mechanics*, vol. 69(14), pp. 1519–1520, 2002b.

Paulino, G.H., "Isoparametric graded finite elements for nonhomogeneous isotropic and orthotropic materials", *Urbana*, vol. 51, 61801, 2002c.

Paulino, G.H., "The elastic–viscoelastic correspondence principle for functionally graded materials revisited", *Journal of Applied Mechancics*, vol. 70, pp. 359–363, 2013.

Paulino, G.H. and Dodds, Jr. R.H., "Finite element investigation of quasi-static crack growth in functionally graded materials using a novel cohesive zone fracture model" *Urbana*, vol. 51, article 61801, 2002b.

Paulino, G.H., Jin, Z.H. and Dodds, Jr., R.H., "Failure of functionally graded materials", *Comprehensive Structural Integrity*, vol. 2(13), pp. 607–644, 2003.

Paulino, G.H. and Jin, Z.H., "Correspondence principle in viscoelastic functionally graded materials", *Transactions-American Society of Mechanical Engineers Journal of Applied Mechanics*, vol. 68(1), pp. 129–131, 2001.

Paulino, G.H. and Kim, J.H., "A new approach to compute T-stress in functionally graded materials by means of the interaction integral method", *Engineering Fracture Mechanics*, vol. 71(13), pp. 1907–1950, 2004.

Paulino, G.H., Silva, E.C.N. and Le, C.H., "Optimum design of periodic functionally graded composites with prescribed properties", *Structural and Multidisciplinary Optimization*, vol. 38(5), pp. 469–489, 2009.

Pei, Y.T. and De Hosson, J.Th.M., "Functionally graded materials produced by laser ladding", *Acta Materialia*, vol. 48, pp. 2617–2624, 2000.

Pelletier, Jacob L. and Vel, S.S., "An exact solution for the steady state thermo–elastic response of functionally graded orthotropic cylindrical shells", *International Journals of Solids and Structures*, vol. 43, pp. 1131–1158, 2006.

Peng, X.-L. and Li, X.-F., "Elastic analysis of rotating functionally graded polar orthotropic disks", *International Journal of Mechanical Sciences*, vol. 60, pp. 84–91, 2012.

Pentaras, D., Vibration Tailoring of Inhomogeneous Beams and Circular Plates, *PhD dissertation*, Florida Atlantic University, 2006.

Pentaras, D. and Elishakoff, I., "Polar orthotropic inhomogeneous circular plates: vibration tailoring", *Journal of Applied Mechanics*, vol. 77(3), pp. 031019–031027, 2010.

Petrova, I.S. and Rikards, R.B., "Optimization of a column with variable modulus of elasticity", *Mechanics of Polymers*, Issue 2, pp. 277–284, 1974 (in Russian).

Pierce, A.D., "Physical interpretation of the WKB or eikonal approximation for waves and vibrations in inhomogeneous beams and plates", Journal of the Acoustical Society of America, vol. 48, pp. 275–282, 1970.

Pindera, M.J. and Dunn, P., "Evaluation of the higher-order theory for functionally graded materials via the finite-element method", *Composites Part B: Engineering*, vol. 28(1), pp. 109–119, 1997.

Pindera, M.J., Khatam, H., Drago, A.S. and Bansal, Y., Micromechanics of spatially uniform heterogeneous media: A critical review, and emerging approaches, *Composites: Part B*, vol. 40, pp. 349–378, 2009.

Pindera, M.-J., Aboudi, J. and Arnold, S.M., Thermo-inelastic analysis of functionally graded materials: inapplicability of the classical micromechanics approach, in *Inelasticity and Micromechanics of Metal Matrix Composites* (G.Z. Voyadjis and J.W. Zu, eds.), pp. 273–305, Elsevier, Amsterdam, 1994.

Pinto, Z.H.M., *Fabrication and Characterization of Functionally Graded Aluminum/aluminum Diboride Matrix Composites for High Wear Aerospace Applications Using Centrifugal Casting*, University of Puerto Rico, Mayaguez (Puerto Rico), 2006.

Pitakthapanaphong, S. and Busso, E.P., "Self-consistent elasto-plastic stress solutions for functionally graded material systems subjected to thermal transients", *Journal of Mechanics and Physics of Solids*, vol. 50(4), pp. 695–716, 2002.

Plevako, V.P., "On the theory of elasticity of inhomogeneous media", *Journal of Applied Mathematics and Mechanics*, vol. 35(6), pp. 806–813, 1971.

Poisson, S.D., "L' equilibre et le mouvement des corps elastiques", *Mem. Acad. Roy. Des Su De L'Inst. France*, vol. 8(2), p. 357, 1829 (in French).

Pompe, W., Worch, H., Epple, M., Friess, W., Gelinsky, M., Greit, P., Hempel, U., Scharnweber, D. and Schlte, K., "Functionally graded materials for biomedical applications", *Materials Science and Engineering, Series A*, vol. 362(1–2), pp. 40–60, 2005.

Pradhan, S.C., Loy, C.T., Lam, K.Y. and Reddy, J.N., "Vibration characteristics of functionally graded cylindrical shells under various boundary conditions", *Applied Acoustics*, vol. 61(1), pp. 111–129, 2000.

Pradhan, S.C. and Sarkar, A., "Analyses of tapered FGM beams with nonlocal theory", *Structural Engineering and Mechanics*, vol. 32, pp. 811–833, 2009.

Pradyumna, S. and Bandyopadhyay, J.N., "Free vibration analysis of functionally graded curved panels using a higher-order finite element formulation" *Journal of Sound and Vibration*, vol. 318(1), pp. 176–192, 2008.

Prakash, T., Singha, M.K. and Ganapathi, M., "Influence of neutral surface position on the nonlinear stability behavior of functionally graded plated", *Composite Structures*, vol. 43, pp. 341–350, 2009b.

Prasad, C., Jain, R.K., and Soni, S.R., "Axisymmetric vibrations of circular plates of linearly varying thickness", *Zeitschrift fuer angewandte Mathematik und Physik*, vol. 23(6), pp. 941–948, 1972.

Prathap, G. and Varadan, T.K., "Axisymmetric vibration of polar orthotropic circular plates", *AIAA Journal*, vol. 14, pp. 1639–1640, 1976.

Praveen, G.N. and Reddy, J.N., "Nonlinear transient thermoelastic analysis of functionally graded ceramic-metal plates", *International Journal of Solids and Structures*, vol. 35, pp. 4457–4476, 1998.

Proulx, T., *Dynamic Behavior of Materials Volume 1: Proceedings of the 2010 Annual Conference on Experimental and Applied Mechanics*, Society for Experimental Mechanics, Springer, 2011.

Qian, L.F. and Batra, R.C., "Design of bidirectional functionally graded plate for optimal natural frequencies", *Journal of Sound and Vibration*, vol. 280, pp. 415–424, 2005.

Qian, L.F., Batra, R.C. and Chen, L.M., "Static and dynamic deformations of thick functionally graded elastic plated by using higher-order shear and normal deformable plate theory and meshless local Petrov–Galerkin method", *Composites, Part B: Engineering*, vol. 35(6), 685–697, 2004.

Rafiee, M. and Kalhori, H., "Large amplitude free and forced oscillations of functionally graded beams", *Mechanics of Advanced Materials and Structures*, vol. 21, 255–262, 2014.

Raj, A. and Sujith, R.I., "Closed-form solutions for the free longitudinal vibration of inhomogeneous rods", *Journal of Sound and Vibration*, vol. 283, pp. 1015–1030, 2005.

Rajasekaran, S. and Tochaei, E.N., "Free vibration analysis of axially functionally graded tapered Timoshenko beams using differential transformation element method and differential quadrature element method of lowest-order", *Meccanica*, pp. 1–15, 2013.

Ram, Y. and Elishakoff, I., "Can one reconstruct the cross-section of an axially vibrating rod from one of its mode shapes?", *Proceedings of the Royal Society of London*, vol. 460, pp. 1583–1596, 2004.

Raman, V.M., "On analytical solutions of vibration frequencies of a variable cross-section rod", *Applied Mathematical Modeling*, vol. 7, pp. 356–361, 1983.

Ramirez, F., Heyliger, P.R. and Pan, E., "Discrete layer solution to free vibrations of functionally graded magneto-electro-elastic plates", *Mechanics of Advanced Materials and Structures*, vol. 13, pp. 249–266, 2006.

Ran, M., "Note on the transverse vibration of an isotropic circular plate with density varying parabolically", *Indian Journal of Theoretical Physics*, vol. 24(4), pp. 179–182, 1976.

Rao, S.S., *Mechanical Vibration*, Pearson Education, Upper Saddle River, NY, 2004.

Ravikiran, K., Kashif, A. and Ganesan, G., "Static analysis of functionally graded beams using higher order shear deformation theory", *Applied Mathematical Modeling*, vol. 32, pp. 2509–2525, 2008.

Ray, A.K., Mondal, S., Das, S.K. and Ramachandrarao, P., "Bamboo — a functionally graded composite-correlation between microstructure and mechanical strength, *Journal of Materials Science*, vol. 43, pp. 5249–5253, 2005.

Reddy, J.N., "A simple higher-order theory for laminated composite plates", *Journal of Applied Mechanics*, vol. 51, pp. 745–752, 1984a.

Reddy, J.N., *Energy and Variational Methods in Applied Mechanics*, Wiley, New York, 1984b.

Reddy, J.N., *Theory and Analysis of Elastic Plates and Shells*, CRC Press, New York, 2007.

Reddy, J.N., "Analysis of functionally graded plates", *International Journal for Numerical Methods in Engineering*, vol. 47, pp. 663–684, 2000.

Reddy, J.N. and Cheng, Z.-Q., "Three-dimensional solutions of smart functionally graded plates", *ASME Journal of Applied Mechanics*, vol. 68, pp. 234–241, 2001a.

Reddy, J.N. and Cheng, Z.-Q., "Three-dimensional thermo-mechanical deformations of functionally graded rectangular plates", *European Journal of Mechanics-A/Solids*, vol. 20, pp. 841–855, 2001b.

Reddy, J.N. and Cheng, Z.-Q., "Frequency of functionally graded plates with three-dimensional asymptotic approach", *Journal of Engineering Mechanics*, vol. 129, pp. 896–900, 2003.

Reddy, J.N. and Chin, C.D., "Thermomechanical analysis of functionally graded cylinders and plates", *Journal of Thermal Stresses*, vol. 21, pp. 593–602, 1998.

Reddy, J.N. and Khdeir, A.A., "Buckling and vibration of laminated composite plates using various plate theories", *AIAA Journal*, vol. 27, pp. 1808–1817, 1989.

Reddy, J.N., Wang, C.M. and Kitipornchai, S., Axisymmetric bending of functionally graded circular and annular plates, *European Journal of Mechanics, A: Solids*, vol. 18(2), pp. 185–199, 1999.

Reimanis, I.E., Functionally graded materials, in *Handbook of Advanced Materials: Enabling New Designs* (J.K. Wessel, ed.), John Wiley & Sons, Inc., pp. 465–486, 2004.

Reissner, E. and Stavsky, Y., "Bending and stretching of certain types of heterogeneous aelotropic elastic plates", *Journal of Applied Mechanics*, vol. 28, pp. 402–408, 1961.

Reynolds, N.J., *Functionally Graded Materials*, Nova Science Publishers Inc., Hauppauge, NY, 2011.

Rhee, R.S., *Multi-Scale Modeling of Functionally Graded Materials (FGMs) Using FiniteElement Methods*, ProQuest Information and Learning Company, 2008.

Rogers, T.G., Watson, P. and Spencer, A.J.M., "Exact three-dimensional elasticity solutions for bending of moderately thick inhomogeneous and laminated strips under normal pressure", *International Journal of Solids and Structures*, vol. 32, pp. 1659–1673, 1995.

Rooney, F.J., "Torsion flexure of functionally graded materials", *Composites Engineering*, vol. 5, pp. 753–770, 1995.

Rooney, E.T. and Ferrari, M., "Torsian and flexure of inhomogeneous elements", *Composites Engineering*, vol. 5, pp. 901–911, 1995.

Rowe, D.M., *Thermoelectrics Handbook: Macro to Nano*, Taylor & Francis Group, 2006.

Rubio, W.M., Paulino, G.H. and Silva, E.C.N., "Tailoring vibration mode shapes using topology optimization and functionally graded material concepts", *Smart Materials and Strucutres*, vol. 20, article 025009, 2011.

Ruocco, E. and Minutolo, V., "Two-dimensional stress analysis of multiregion functionally graded materials using a field boundary element model", *Composites Part B: Engineering*, vol. 43(2), pp. 663–672, 2012.

Sahraee, R., "Bending analysis of functionally graded sectorial plates using Levinson plate theory", *Composite Structures*, vol. 88, pp. 548–557, 2009.

Saidi, H., Houari, M.S.A., Tounsi, A. and Adda Bedia, E.A., "Thermo-mechanical bending response with stretching effect of functionally graded sandwich plates using a novel shear deformation theory," *Steel and Composite Structures*, vol. 15, pp. 221–245, 2013.

Salazar, J.W.Al, *Analysis of Smart Functionally Graded Materials Using an Improved Third Order Shear Deformation Theory*, Texas A&M University, 2006.

Sallai, B.-Q., Tounsi, A., Mechab, I., Bachir, B.M., Meradjah, M. and Adda Bedia, El A., "A Theoretical analysis of flexional bending of Al/Al$_2$O$_3$ S-FGM thick beams", *Computational Material Science*, vol. 44(4), pp. 1344–1350, 2009.

Sankar, B.V., "An elasticity solution for functionally graded beams", *Composites Science and Technology*, vol. 61(2), pp. 689–696, 2001.

Sankar, B.V. and Tzeng, J.T., "Thermal stresses in functionally graded beams", *AIAA Journal*, vol. 40(6), pp. 1228–1232, 2002.

Santare, M.H. and Lambros, J., "Use of graded finite elements to model the behavior of nonhomogeneous materials," *Journal of Applied Mechanics*, vol. 67, pp. 819–822, 2000.

Santoro, R. and Elishakoff, I., "Design of axially graded columns under a central force,"*Journal of Applied Mechanics*, vol. 81(2), article 021001, 2014.

Sarkar, B.V., "An elasticity solution for functionally graded beams", *Composite Science and Technology*, vol. 61, pp. 689–696, 2001.

Sarkar, K., Closed-form solutions for rotating and non-rotating beams: An inverse problem approach, *Master of Science (Engineering) Thesis*, Department of Aerospace Engineering, Indian Institute of Science, Bangalore, 2012.

Sarkar, K. and Ganguli, R., "Closed-form solution for non-uniform Euler–Bernoulli free–free beams", *Journal of Sound and Vibration*, vol. 332, pp. 6078–6092, 2013a.

Sarkar, K. and Ganguli, R., "Modal Tailoring and Closed-Form Solutions for Rotating Beams", Presented in the 69th American Helicopter Society: International Annual forum, Phoenix, AZ, May 21–23, 2013b, pp. 262–274 (paper is available at the site: https://vtol.org/store/product/modal-tailoring-and-closedform-solutions-for-rotating-beams-8815.cfm).

Sarkar, K. and Ganguli, R., "Closed-form solutions for axially functionally graded Timoshenko beams having uniform cross-section and fixed–fixed boundary condition", *Composites Part B: Engineering*, vol. 58, pp. 361–370, 2014a.

Sarkar, K. and Ganguli, R., "Closed-form solutions for axially graded Timoshenko beams having uniform cross-section and fixed-fixed bounding condition", *Composites, Part B: Engineering*, vol. 581, pp. 361–370, 2014b.

Sarkar, K. and Ganguli, R., "Modal tailoring and closed-form solutions for rotating non-uniform Euler–Bernoulli beams", *International Journal of Mechanical Sciences*, vol. 88, pp. 208–220, 2014c.

Sarkar, K. and Ganguli, R., "Analytical test functions for free vibration analysis of rotating non-homogeneous Timoshenko beams", *Meccanica*, vol. 49, pp. 1469–1477, 2014d.

Schwarz, B., "On the extrema of the frequencies of nonhomogeneous strings with equimeasurable density", *Journal of Mathematics and Mechanics*, vol. 10(3), pp. 401–422, 1961.

Sepahi, O., Forouzan, M.R. and Malekzadeh, P., "Thermal buckling and postbuckling analysis of functionally graded annular plates with temperature-dependent material properties", *Materials & Design*, vol. 32(7), pp. 4030–4041, 2011.

Shaffer, B.W., "Orthotropic annular disks in plane stress", *Journal of Applied Mechanics*, vol. 34, pp. 1027–1029, 1969.

Shahba, A. and Rajasekaran, S., "Free vibration and stability of tapered Euler–Bernoulli beams made of axially functionally graded materials, *Applied Mathematical Modelling*, vol. 36, pp. 3094–3111, 2012.

Shahba, A., Attarnejad, R. and Hajilar, S., "A mechanical-based solution for axially functionally graded tapered Euler–Bernoulli beams", *Mechanics of Advanced Materials and Structures*, vol. 20(8), pp. 696–707, 2013.

Shahba, A., Attarnejad, R. and Zarrinzadeh, H., "Free vibration analysis of centrifugally stiffened tapered functionally graded beams", *Mechanics of Advanced Materials and Structures*, vol. 20(5), pp. 331–338, 2013.

Shahba, A., Attarnejad, R., Marvi, M.T. and Hajilar, S., "Free vibration and stability analysis of axially functionally graded tapered Timoshenko beams with classical and non-classical boundary conditions, *Composites Part B: Engineering*, vol. 42(4), pp. 801–808, 2011.

Shahrjerdi, A., Bayat, M., Mustapha, F., Sapuan, S.M. and Zahari, R., "Second-order shear deformation theory to analyze stress distribution for solar functionally graded plates", *Mechanics Based Design of Structures and Machines*, vol. 38, pp. 348–361, 2010.

Shakeri, M., Akhlaghi, M. and Hoseini, S.M., "Vibration and radial wave propagation velocity in functionally graded thick hollow cylinder", *Composite Structures*, vol. 76(1–2), pp. 174–181, 2006.

Shankar, B.V. and Tzeng, J.T., "Thernal stresses in functionally graded beams", *AIAA Journal*, vol. 40(6), pp. 1228–1232, 2002.

Shao, Z.S., "Mechanical and thermal stresses of a functionally graded circular hollow cylinder with finite length", *International Journal of Pressure Vessels and Piping*, vol. 82(3), pp. 155–163, 2005.

Shao, Z.S. and Ma, G.W., "Thermo-mechanical stresses in functionally graded circular hollow cylinder with linearly increasing boundary temperature", *Composite Structures*, vol. 82, pp. 259–265, 2008.

Shen, H.-S. and Noda, N., "Post-buckling of FGM cylindrical shell under combined axial and radial mechanical loads in thermal environments", *International Journal of Solids Structures*, vol. 42(16), pp. 4641–4662, 2005.

Shen, H.-S., "A comparison of buckling and postbuckling behavior of FGM plates with piezoelectric fiber reinforced composite actuators", *Journal of Composite Structures*, vol. 91, pp. 375–384, 2009a.

Shen, H.-S., *Functionally Graded Materials: Nonlinear Analysis of Plates and Shells*, CRC Press, Boca Raton, 2009b.

Shen, L.H., Ostojoa-Starzewski, M. and Porcu, E., "Bernoulli–Euler beams with random field properties under random field loads: Fractal and Hurst effects", *Archives of Applied Mechanics*, vol. 84, pp. 1595–1626, 2014.

Shiota, I. and Miyamoto, Y. (eds.), *Functionally Graded Materials*, Elsevier Science B.V., Amsterdam, 1996.

Siginer, A., "Buckling of columns of variable flexural rigidity", *Journal of Engineering Mechanics*, vol. 118(3), pp. 543–560, 1992.

Sih, G.C., *Handbook of Stress-Intensity Factors*, Institute of Fracture and Solid Mechanics, Bethlehem, Lehigh University, 1973.

Silva, E.C.N., Walters, M.C. and Paulino, G.H., "Modeling bamboo as a functionally graded material: lessons for the analysis of affordable materials", *Journal of Materials Science*, vol. 41, pp. 6991–7004, 2006.

Silva, E.C.N., Walters, M.C. and Paulino, G.H., "Modeling bamboo as a functionally graded material", *AIP Conference Proceedings*, pp. 754–759, 2008.

Şimşek, M., "Static analysis of a functionally graded beam under a uniformly distributed load by Ritz Method", *International Journal of Engineering and Applied Sciences*, vol. 1(3), pp. 1–11, 2009.

Şimşek, M., "Fundamental frequency analysis of functionally graded beams by using different higher-order beam theories", *Nuclear Engineering and Design*, vol. 240, pp. 697–705, 2010a.

Şimşek, M., "Vibration analysis of a functionally graded beam under a moving mass by using different beam theories", *Composite Structures*, vol. 92, pp. 904–917, 2010b.

Şimşek, M., "Non-linear vibration analysis of a functionally graded Timoshenko beam under action of a moving harmonic load", *Composite Structures*, vol. 92(10), pp. 2532–2546, 2010c.

Şimşek, M., "Static bending of a functionally graded microscale Timoshenko beam based on the modified couple stress theory, *Composite Structures*, vol. 95, pp. 740–747, 2013.

Şimşek, M. and Cansiz, S., "Dynamics of elastically connected double-functionally graded beam systems with different boundary conditions under action of a moving harmonic load", *Composite Structures*, vol. 94(9), pp. 2861–2878, 2012.

Şimşek, M. and Kocatürk, T., "Free and forced vibration of a functionally graded beam subjected to a concentrated moving harmonic load", *Composite Structures*, vol. 90, pp. 465–473, 2009.

Şimşek, M., Kocatürk, T. and Akbaş, Ş.D., "Dynamic behavior of an axially functionally graded beam under action of a moving harmonic load", *Composite Structures*, vol. 94(8), pp. 2358–2364, 2012.

Sina, S.A., Navazi, H.M. and Haddadpour, H. "An analytical method for free vibration analysis of functionally graded beams", *Materials and Design*, vol. 30, pp. 741–747, 2009a.

Sing, R. and Sascena, V., "Axisymmetric vibration of circular plate with doubly linear variable thickness, *Journal of Sound and Vibration*, vol. 179(1), pp. 879–897, 1995.

Singh, B. and Hassan, S.M., "Transverse vibration of a circular plate with arbitrary thickness variation", *International Journal of Mechanical Sciences*, vol. 40(11), pp. 1089–1104, 1998.

Singh, K.V. and Li, G., "Buckling of functionally graded and elastically restrained non-uniform columns", *Composites: Part B*, vol. 40, pp. 393–403, 2009b.

Singth, S.R. and Chakraverty, S., "Transverse vibration of circular and elliptical plates with quadratically varying thickness", *Applied Mathematics Modelling*, vol. 16, pp. 269–274, 1992.

Singth, S.R. and Chakraverty, S., "Transverse vibration of circular and elliptical plates with variable thickness", *Indian Journal of Pure and Applied Mathematics*, vol. 22(9), pp. 787–803, 1991.

Siu, C.C. and Bert, C.W., "Sinusoidal response of composite-material plates with material damping", *Journal of Engineering for Industry*, vol. 96, pp. 603–610, 1974.

Snyder, M.A., *Chebyshev Methods in Numerical Approximation*, Prentice-Hall, Englewood Cliffs, NJ, 1966.

Sobczak, J. and Drenchev, L.B., *Metal Based Functionally Graded Materials*, Bentham Science Publisher, 2009.

Sobczyk, K., "Random vibrations of statistically inhomogenous elastic systems, *Proceedings of Vibration Problems*, vol. 4(11), pp. 369–381, 1970.

Sobhani, Arag, B. and Yas, M.H., "Static and free vibration analyses of continuously graded fiber-reinforced cylindrical shells using generalized power-law distributions", *Acta Mechanica*, vol. 215(1–4), pp. 155–173, 2010.

Sobolev, D.N., "Towards design of structures resting on statistically inhomogeneous foundation, pages *Stroitelnaya Mekhanika i Raschet Sosruzhenii*, Issue 1, 1965 (in Russian).

Sofiyev, A.H. and Avcar, M., "The stability of cylindrical shells containing an FGM layer subjected to axial load on the Pasternak foundation", *Engineering*, vol. 2(4), pp. 228–236, 2010.

Sofiyev, A.H., "Dynamic buckling of functionally graded cylindrical shells under non-periodic impulsive loading", *Acta Mechanica*, vol. 165(3–4), pp. 153–162, 2003.

Sofiyev, A.H., "Vibration and stability of composite cylindrical shells containing at FG layer subjected to various loads", *Structural Engineering and Mechanics an International Journal*, vol. 27(3), pp. 265–391, 2007.

Sofiyev, A.H., Deniz, A., Akcay, I.H. and Yusufoglu, E., "The vibrations and stability of a three-layered conical shell containing a FGM layer subjected to axial compressive load", *Acta Mechanica*, vol. 183(3–4), pp. 129–144, 2006.

Sohn, K.J. and Kim, J.H., "Nonlinear thermal flutter of functionally graded panels under a supersonic flow", *Composite Structures*, vol. 88, pp. 380–387, 2009.

Soldatos, K., "Review of three dimensional dynamic analyses of circular cylinders and cylindrical shells", *Applied Mechanics Review*, vol. 47(10), pp. 501–516, 1994.

Soldatos, K.P. and Hadjigeorgiu, V.P., "Three-dimensional solution for the free vibration problem of homogeneous inotropic shells and panels", *Journal of Sound and Vibration*, vol. 137(3), pp. 369–384, 1990.

Srinivas, S. and Rao, A.K., "Flexure of thick rectangular plates", *ASME Journal of Applied Mechanics*, vol. 39, pp. 298–299, 1973.

Srinivas, S., Joga Rao, C.V. and Rao, A.K., "An exact analysis for vibration of simple-supported homogeneous and laminated thick rectangular plates", *Journal of Sound and Vibration*, vol. 12(2), pp. 187–199, 1970.

Srivatsan, T.S., *Composites and functionally graded materials: Presented at the 1997 ASME International Mechanical Engineering Congress and Exposition, November 16–21, 1997, Dallas, Texas*, American Society of Mechanical Engineers, 1997.

Stavsky, Y., "On the theory of heterogeneous aristropic plates", *Ph.D. Thesis*, MIT, Cambridge, MA, 1959.

Stavsky, Y., "On the theory of symmetrically heterogeneous plates having the same thickness variation of the elastic moduli," in *Topics in Applied Mechanics* (D. Abir, F. Ollendorff and M. Reiner, eds.), pp. 105–116, Elsevier, Newyork, 1965.

Stefanou, G., "The stochastic finite element method; past, present and future", *Computer Methods in Applied Mechanics and Engineering*, vol. 198, pp. 1031–1051, 2009.

Steinberg, M.A., "Materials for aerospace", *Scientific American*, vol. 256, pp. 66–72, 1986.

Stokes, V.K., "Design with nonhomogeneous material — Part I: Pure bending of prismatic bars", *Transaction of the ASME*, vol. 109, pp. 82–86, 1987.

Storch, J.A. and Elishakoff, I., "Apparently first closed-form solutions of inhomogeneous circular plates in 200 years after Chladni", *Journal of Sound and Vibration*, vol. 276, pp. 1108–1114, 2004.

Storti, D. and Aboelnaga, Y., "Bending vibrations of a class of rotating beams with hypergeometric solutions", *Journal of Applied Mechanics*, vol. 54, pp. 311–314, 1987.

Sudret, B. and Der Kiureghian, A., "Comparison of finite element reliability methods", *Probabilistic Engineering Mechanics*, vol. 17, pp. 337–348, 2002.

Sundararajan, N., Prakash, T. and Ganapathi, M., "Nonlinear free flexural vibrations of functionally graded rectangular and skew plates under thermal environments", *Finite Elements in Analysis and Design*, vol. 42(2), pp. 152–168, 2005.

Suresh, S. and Mortensen, A., *Fundamentals of Functionally Graded Materials: Processing and Thermomechanical Behaviour of Graded Metals and Metal-Ceramic Composites*, IOM Communications Ltd., London, 1998.

Sutradhar, A. and Paulino, G.H., "The simple boundary element method for transient heat conduction in functionally graded materials", *Computer Methods in Applied Mechanics and Engineering*, vol. 193(42), pp. 4511–4539, 2004.

Tada, H., Paris, P.C. and Irwin, G.R., *The stress analysis of cracks handbook*, Paris Production Inc, St. Louis, 1985.

Taeprasartsit, S., "A buckling analysis of perfect and imperfect functionally graded columns", *Proceedings of the Institution of Mechanical Engineers, Part L: Journal of Materials Design and Applications*, vol. 226(1), pp. 16–33, 2012.

Tan, T., Rahbar, N., Allameh, S.M., Kwofie, S., Dissmore, D., Ghavami, K. and Soboyejo, W.O., "Mechanical properties of functionally graded hierarchial bamboo structures", *Acta Biomaterialia*, vol. 7(10), pp. 3796–3803, 2011.

Tanaka, K., "Mechanics of functionally gradient materials: material tailoring on the micro- and macro-levels", *Proceedings of the XIXth International Congress of Theoretical and Applied Mechanics*, Kyoto, Japan, August 25–31, 1996, (Tatsumi, T., Watanabe, F. and Kambe, T. (eds)), New York Elsevier, pp. 253–268, 1997.

Tanigawa, Y., "Some basic thermoelastic problems for nonhomogeneous structural materials, *Applied Mechanics Reviews*, vol. 48(6), pp. 287–300, 1995.

Tarn, J.Q., "Exact solutions for functionally grated anisotropic cylinders subjected to thermal and mechanical loads", *International Journal of Solids and Structures*, vol. 38(46–47), pp. 8189–8206, 2001.

Taylor, J.E., "Minimum mass bar for axial vibration at specified natural frequency", *AIAA Journal*, vol. 5(10), pp. 1911–1913, 1967.

Thai, H.-T. and Choi, D.-H., "A refined plate theory for functionally graded plates resting on elastic foundation", *Composite Science and Technology*, vol. 71(16), pp. 1850–1858, 2011.

Thai, H.-T. and Choi, D.-H., "A refined shear deformation theory for free vibration of functionally graded plated on elastic foundation", *Composites Part B: Engineering*, vol. 43(5), pp. 2335–2347, 2012.

Thai, H.-T. and Kim, S.-E., "A simpler higher-order shear deformation theory for bending and free vibration analysis of functionally graded plates", *Composite Structures*, vol. 96, pp. 165–173, 2013.

Thai, H.-T. and Vo, T.P., "A new sinusoidal shear deformation theory for bending, buckling, and vibration of functionally graded plates", *Applied Mathematical Modeling*, vol. 37(5), pp. 3269–3281, 2013.

Tharp, T.M., "A finite element for edge-cracked beam columns", *International Journal for Numerical Methods in Engineering*, vol. 24, pp. 1941–1950, 1987.

Thomson, W.T., "Parameter uncertainty in dynamic systems", *The Shock and Vibration Digest*, vol. 7(8), pp. 3–9, 1975.

Thornton, E.A. (ed.), *Aerospace Thermal Structures and Materials for a New Era*, vol. 168, American Institute of Aeronautics and Astronautics, Washington DC, 1995.

Timoshenko, S.P., *Strength of Materials*, 3rd ed., Princeton, D. van Norstrand, 1956.

Timoshenko, S.P. and Gere, J.M., *Theory of Elastic Stability*, New York, McGraw-Hill, 1961.

Timoshenko, S. and Woinowsky-Krieger, S., *Theory of Plates and Shells*, Chapter 3: Symmetrical bending of circular plates, 51–78, 2nd edition, New York, McGraw-Hill, 1959.

Tornabene, F. and Reddy, J.N., "FGM and laminated doubly-curved and degenerate shells resting on nonlinear elastic foundation: A GDQ solution for static analysis with *a posteriori* stress and strain recovery", *Journal of Indian Institute of Science*, vol. 93(4), pp. 635–688, 2013.

Tornabene, F., Viola, E. and Inman, D.J., "2-D differential quadrature solution for vibration analysis of functionally graded conical, cylindrical shell and annular plate structures", *Journal of Sound and Vibration*, vol. 328(3), pp. 259–290, 2009.

Torquato, S., *Random Heretogeneous Materials: Miscrostructure and Macroscopic Properties*, Springer, New York, 2001.

Touloukian, Y.S., *Properties of High Temperature Solid Materials*, Macmillan, New York, 1967.

Tounsi, A., Adda Bedia, El. A., Mahmoud, S.R. and Amziane, S., "Mathematical modeling and optimization of functionally graded structures", *Mathematical Problems in Engineering*, vol. 2013, Article ID 536867, 2013.

Tounsi, A., Bouazza, M. and Adda-Bedia, E., "Computation of transient hygroscopic stresses in unidirectional laminated composite plates with cyclic and asymmetrical environmental conditions", *International Journal of Mechanics and Materials in Design*, vol. 1(3), pp. 271–286, 2004.

Trapezon, A.G., "Group approaches to the vibration and static deformation analysis for bars and disks", *Strength of Materials*, vol. 33(4), pp. 380–391, 2001.

Trumble, K. (ed.)., *Functionally Graded Materials 2000: Proceedings of the 6th International Symposium on Functionally Graded Materials*, Ceramics Transactions, Wiley, New York, 2001.

Tutuncu, N. and Ozturk, M., "The exact solution for stresses in functionally graded pressure vessels", *Composites Part B*, vol. 32, pp. 683–691, 2001.

Tvergaard, V., "Effect of thickness inhomogeneities in internally pressurized elastic-plastic spherical shells", *Journal of the Mechanics and Physics of Solids*, vol. 24(5), 1976.

Uymaz, B. and Aydoğlu, M., "Three-dimensional vibration analyses of functionally graded plates under various boundary conditions", *Journal of Reinforcement Plastics and Composites Technology*, vol. 26(18), pp. 1847–1863, 2007.

Vaicaitis, R., "Free vibrations of beams with random characteristics", *Journal of Sound and Vibration*, vol. 35(1), pp. 13–21, 1974.

van der Biest, O., Gasik, M. and Vleugels, J., *Functionally Graded Materials VII: Proceeding of the 8th International Symposium on Multifunctional and Functionally Graded Materials (FGM 2004)*, Leuven, Belgium, 11–14 July 2004, Trans Tech Publications, 2005.

Vanmarke, E. and Grigoriu, M., "Stochastic finite element analysis of simple beams", *Journal of Engineering Mechanics*, vol. 115, pp. 1035–1053, 1989.

Vanmarke, E., Shinozuka, M., Nakagiri, S., Schueller, G. and Grigoriu, M., "Random fields and stochastic finite elements, *Structural Safety*, vol. 3, pp. 143–166, 2012.

Vashakmadze, T.S., *The Theory of Anisotropic Elastic Plates*, Springer Science + Business Media, B.V., 2010.

Vekua, I.N., *On a Version of Theory of Thin Shallow Shells*, Novosibirsk University Press, Novosibirsk, 1964 (in Russian).

Vekua, I.N., "Theory of thin–shallow shells of variable thickness", *Proceedings of the Razmadze Mathematics Institute*, Tbilisi, vol. 30, pp. 3–103, 1965 (in Russian).

Vekua, I.N., *Shell Theory: General Methods of Construction*, Addison–Wesley Longman, New York, 1985.

Vel, S.S. and Batra, R.C., "Exact thermoelasticity solution for functionally graded thick rectangular plates", *American Institute of Aeronautics and Astronautics (AIAA) Journal*, vol. 40, pp. 1421–1433, 2002a.

Vel, S.S. and Batra, R.C., "Exact solution for the cylindrical bending vibration of functionally graded plates", *Proceedings of the American Society of Composites*, Seventh Technical Conference, October 21–23, West Lafayette, Indiana, Purdue University, 2002b.

Vel, S.S. and Batra, R.C., "Exact solution for thermoelastic deformations of functionally graded thick rectangular plates", *AIAA Journal*, vol. 40(7), pp. 1421–1433, 2002c.

Vel, S.S. and Batra, R.C., "Three-dimensional analysis of transient thermal stresses in functionally graded plates", *International Journal of Solids and Structures*, vol. 40, pp. 7181–7196, 2003.

Vel, S.S. and Batra, R.C., "Three-dimensional exact solution for the vibration of functionally graded rectangular plates", *Journal of Sound and Vibration*, vol. 272, pp. 703–730, 2004.

Venkataraman, S. and Sankar, B., "Elasticity solution for stresses in a sandwich beam with functionally graded core", *AIAA Journal*, vol. 41(12), pp. 2501–2505, 2003.

Viola, E. and Tornabene, F., "Free vibrations of three parameter functionally graded parabolic panels of revolution", *Mechanics Research Communications*, vol. 36(5), pp. 587–594, 2009.

Vlasov, B.F., On one case of bending of rectangular thick plate, *Vestnik Moskovskogo Universiteta, Seria Matematika, Mekhanika i Astronomia*, Issue 2, pp. 25–34, 1957 (in Russian).

Vo, T.P., Thai, H.-T., Nguyen, T.-K. and Inam, F., "Static and vibration analysis of functionally graded beams using refined shear deformation theory", *Meccanica*, vol. 49, pp. 155–168, 2014.

Wahlborg, J.R., On the development and implementation of a functionally graded materials representation for implicit solid modeling, University of Washington, 2003.

Walters, M.C., Paulino, G.H. and Dodds, Jr. R.H., "Stress-intensity factors for surface cracks in functionally graded materials under mode-I thermomechanical loading", *International Journal of Solids and Structures*, vol. 41(3), pp. 1081–1118, 2004.

Wang, C.M., Wang, C.Y. and Reddy, J.N., *Exact Solutions for Buckling of Structural Members*, CRC Press, Boca Raton, 2005.

Wang, C.Y. and Wang, C.M., *Structural Vibrations: Exact Solutions for Strings, Membranes, Beams, and Plates*, CRC Press, Boca Raton, Florida, 2013.

Wang, H.M. and Ding, H.J., "Spherically symmetric transient responses of functionally graded magneto-electro-elastic hollow sphere", *Structural Engineering and Mechanics*, vol. 23(5), pp. 525–542, 2006.

Wang, J., "Generalized power series solutions of the vibration of classical circular plates with variable thickness", *Journal of Sound and Vibration*, vol. 202(4), pp. 593–599, 1997.

Wang, M. and Liu, Y., "Elasticity solutions for orthotropic functionally graded curved beams", *European Journal of Mechanics-A/Solids*, vol. 37, pp. 8–16, 2013.

Wang, Q., "On buckling of column structures with a pair of piezoelectric layers", *Engineering Structures*, vol. 24, pp. 199–205, 2002.

Wang, Q.S., Liu, M.H., Zhang, L.H. and He, M., "Inverse problems of single mode about some indeterminate beams", *Applied Mechanics and Materials*, vol. 432, pp. 139–143, 2013.

Wang, X., Synthesis of Functionally Graded Materials via Electrophoretic Deposition and Sintering, University of California, San Diego and San Diego State University, 2006.

Watanabe, R., Nishida, T. and Hirai, T., "Present states of research on design and processing of functionally graded materials", *Metals and Materials International*, vol. 9(6), pp. 513–519, 2003.

Weaver, P.M. and Ashby, M.F., "The optimal selection of material and section shape," *Journal of Engineering Design*, vol. 7, pp. 129–150,1996.

Wen, P.H., Sladek, J. and Sladek, V., "Three-dimensional analysis of functionally graded plates", *Intentional Journal for Numerical Methods in Engineering*, vol. 87, pp. 923–942, 2011.

Wikipedia, "Functionally graded material", available at http://en.wikipedia.org/wiki/Functionally_graded_material (downloaded on 25 May, 2014).

Williams, F.W., "Designing to achieve target values for eigenfrequencies and critical loads", *Preprint, Euromech 112, Colloquium on Bracketing of Eigenfrequencies of Continuous Structures*, Mátrafüred, Hungary, 1979.

Wittrick, W.H., "Analytical three-dimensional elasticity solutions to some plate problems, and some observations on Mindlin's plate theory", *International Journal of Solids and Structures*, vol. 23, pp. 441–464, 1987.

Wolff, E.G., *Introduction to the Dimensional Stability of Composite Materials*, DEStech Publications, Inc., 2004.

Woo, J. and Meguid, S.A., "Nonlinear analysis of functionally graded plates and shallow shells", *International Journal of Solids and structures*, vol. 38(42), pp. 7409–7421, 2001.

Woo, J., Meguid, S.A. and Ong, L.S., "Nonlinear free vibration behavior of functionally graded plates", *Journal of Sound and Vibration*, vol. 289(3), pp. 595–611, 2006.

Wu, L., Wang, Q. and Elishakoff, I., "Semi-inverse method for axially functionally graded beams with an anti-symmetric vibration mode", *Journal of Sound and Vibration*, vol. 284(3), pp. 1190–1202, 2005.

Wu, L., Zhang, L., Wang, Q. and Elishakoff, I., "Reconstructing cantilever beams via vibration mode with a given node location", *Acta Mechanica*, vol. 217, pp. 135–148, 2011.

Xie, L., Ma, X., Jordan, E.H., Padture, N.P., Xiao, D.T. and Gell, M., "Identification of coating deposition mechanisms in the solution-precursor plasma-spray process using model spray experiments", *Materials Science and Engineering: A*, vol. 362(1), pp. 204–212, 2003.

Yaghoobi, H. and Fereidoon, A., "Influence of neutral surface position on deflection of functionally graded beam under uniformly distributed load", *World Applied Science Journal*, vol. 10(3), pp. 337–341, 2010.

Yamanouchi, M., Koizumi, M., Hirai, T. and Shiota, I. (eds.), *Proceedings of the First International Symposium on Functionally Graded Materials*, Sendai, Japan, 1990.

Yamazaki, F., Shinozuka, M. and Dasgupta, G., "Neumann expansion for stochastic finite element analysis", *Journal of Engineering Mechanics*, vol. 114, pp. 1335–1354, 1988.

Yang, C.Y. and Wang, C.M., *Structural Vibration: Exact Solutions for Strings, Membranes, Beams, and Plates*, CRC Press, Boca Raton, 2014.

Yang, J. and Chen, Y., "Free vibration and buckling analyses of functionally graded beams with edge cracks", *Composite Structures*, vol. 83(1), pp. 48–60, 2008.

Yang, J. and Shen, H.-S., "Dynamic response of initially stressed functionally graded rectangular thin plates", *Composite Structures*, vol. 54, pp. 497–508, 2001a.

Yang, J. and Shen, H.-S., "Free vibration and parametric resonance of shear deforchmable functionally graded cylindrical panels", *Journal of Sound and Vibration*, vol. 261, pp. 871–893, 2001b.

Yang, J. and Shen, H.-S., "Vibration characteristics and transient response of shear-deformable functionally graded plates in thermal environments", *Journal of Sound and Vibration*, vol. 255, pp. 579–602, 2002.

Yang, J. and Shen, H.-S., "Nonlinear bending analysis of shear deformable functionally graded plates subjected to thermo-mechanical loads under various boundary conditions", *Composites Part B*, vol. 34, pp. 103–115, 2003a.

Yang, J. and Shen, H.S., "Free vibration and parametric resonance of shear deformable functionally graded cylindrical panels", *Journal of Sound and Vibration*, vol. 261, pp. 871–892, 2003b.

Yang, J., Chen, Y., Xiang, Y. and Jia, X.L., "Free and forced vibration of cracked inhomogeneous beams under an axial force and a moving load", *Journal of Sound and Vibration*, vol. 312(1–2), pp. 166–181, 2008.

Yang, J., Liew, K.M. and Kitipornchai, S., "Dynamic stability of laminated FGM plates based on higher-order shear deformation theory", *Computational Mechanics*, vol. 33, pp. 305–315, 2004.

Yang, J.S. and Xie, Z., "Pertubitation method in the problem of large deflections of circular plates with nonuniform thickness", *Applied Mathematics and Mechanics*, vol. 5, pp. 1237–1242, 1984.

Yeh, K.Y., "Analysis of high-speed rotating disks with variable thickness and inhomogeneity", *Journal of Applied Mechanics*, vol. 61(1), pp. 186–192, 1994.

Yilmaz, Y., Girgin, Z. and Evran, S., "Buckling analyses of axially functionally graded nonuniform columns with elastic restraint using a localized differential quadrature method", *Mathematical Problems in Engineering*, vol. 2013, article 793062, 2013.

Yin, H.M., Sun, L.Z. and Paulino, G.H., "Micromechanics-based elastic model for functionally graded materials with particle interactions", *Acta Materialia*, vol. 52(12), pp. 3535–3543, 2004.

Ying, J. and Wang, H.M., "Magnetoelectroelastic fields in rotating multiferroic composite cylindrical structures", *Journal of Zhejiang University Science A*, vol. 10(3), pp. 319–326, 2009.

Zafarmand, H. and Hassani, B., "Analysis of two-dimensional functionally graded rotating thick disks with variable thickness," *Acta Mechanica*, vol. 225, pp. 453–464, 2014.

Zarrinzadeh, H., Attarnejad, R. and Shahba, A., "Free vibration of rotating axially functionally graded tapered beams", *Proceedings of the Institution of Mechanical Engineers, Part G: Journal of Aerospace Engineering*, vol. 226(4), pp. 363–379, 2012.

Zaslavsky, A., "Nonhomogeneous beams with vertically variable modulus of elasticity", *Israel Journal of Technology*, vol. 8(4), pp. 385–387, 1970.

Zenkour, A.M., "A comprehensive analysis of functionally graded sandwich plates: Part 2 — Buckling and free vibration", *International Journal of Solids and Structures*, vol. 42(18), pp. 5243–5258, 2005a.

Zenkour, A.M., "On vibration of functionally graded plates according to a refined trigonometric plate theory", *International Journal of Structural Stability and Dynamics*, vol. 5(2), pp. 279–297, 2005b.

Zenkour, A.M., "Generalized shear deformation theory for bending analysis of functionally graded plates", *Applied Mathematical Modeling*, vol. 30, pp. 67–84, 2006.

Zenkour, A.M., "Elastic deformation of the rotating functionally graded annular disk with rigid casing", *Journal of Materials Science*, vol. 42(23), pp. 9717–9724, 2007.

Zhan, X.-H., Han, J.-C., He, X.-D.K. and Vanin, V.L., "Combustion synthesis and thermal stress analysis of TiC–Ni functionally graded materials", *Journal of Materials Synthesis and Processing*, vol. 8, pp. 29–34, 2000.

Zhang, X.D., Liu, D.Q. and Ge, C., "Thermal stress analysis of axial symmetry functionally gradient materials under steady temperature field", *Journal of Functionally Graded Materials*, vol. 25, pp. 452–455, 1994.

Zhang, Z.J. and Paulino, G.H., "Cohesive zone modeling of dynamic failure in homogeneous and functionally graded materials", *International Journal of Plasticity*, vol. 21(6), pp. 1195–1254, 2005.

Zhang, C. and Zhong, Z., "Three-dimensional analysis of a simply supported functionally graded plates based on Haar wavelet method", *Acta Mechanica Solida Sinica*, vol. 28, pp. 217–223, 2007.

Zhao, X., Lee, Y.Y. and Liew, K.M., "Thermoelastic and vibration analysis of functionally graded cylindrical shells", *International Journal of Mechanical Sciences*, vol. 51, pp. 694–707, 2009a.

Zhao, X., Lee, Y.Y. and Liew, K.M., "Free vibration analysis of functionally graded plates using the element-free *kp*-Ritz method", *Journal of Sound and Vibration*, vol. 319(3), pp. 918–939, 2009b.

Zhao, L., Chen, W.Q. and Lu, C.F., "New Assessment on the Saint–Venant solutions for functionally graded beams", *Mechanics Research Communications*, vol. 43, pp. 1–6, 2012.

Zhavoronok, S.I., "Variational formulations of Vekua — type shell theories and some of their applications", in *Shell Structures: Theory and Applications*, (W. Pietraskiewicz and J. Górski, eds.), pp. 341–346, CRC Press, Leiden, 2014.

Zhgenti, V.S. and Khvoles, A.P., "General solution of the plate equations by I.N. Vekua", *Mechanics, Third Congress*, Varna, Bulgaria, pp. 415–418, 1997 (in Russian).

Zhgenti, V.S. and Khvoles, A.R., "General solution of the system of equilibrium equations for the V.N. Vekua plate of the constant thickness", *Seminar of I.N. Vekua Institute of Applied Mathematics*, vol. 15, pp. 33–39, 1981 (in Russia).

Zhong, Z. and Shang, E., "Closed-form solutions of three-dimensional functionally graded plates", *Mechanics of Advanced Materials and Structures*, vol. 15(5), pp. 355–363, 2008.

Zhong, Z., Wu, L.Z. and Chen, W.Q. (eds.), *Mechanics of Functinally Graded Materials and Structures*, Nova Publishers, New York, 2012.

Zhong, Z. and Yu, T., "Analytical solution of a cantilevered functionally graded beam", *Composites Science and Technology*, vol. 67(3–4), pp. 481–488, 2007.

Zhou, D., Cheung, Y.K., Au, F.T.K. and Lo, S.H., "Three dimensional vibration analysis of thick rectangular plates using Chebyshev polynomial and Ritz method", *International Journal of Solids and Structures*, vol. 39, pp. 6339–6353, 2002.

Zhou, D., Cheung, Y.K., Lo, S.H. and Au, F.T.K., "3D vibration analysis of solid and hollow cylinders via Chebyshev–Ritz method", *Computer Methods in Applied Mechanics and Engineering*, vol. 192(13–14), pp. 1575–1589, 2003.

Zhu, H. and Sankar, B.V., "A combined Fourier series–Galerkin method for the analysis of functionally graded beams", *Journal of Applied Mechanics*, vol. 71, pp. 421–424, 2004.

Życzkowski, M., *Strength of Structural Elements*, Elsevier, Amsterdam, p. 298, 1991.

Author Index

Subject Index

admissible function, 220
adverse effects, xi
aerospace structures, xii
all round clamped, 77
all round clamped plates, 51
alternative mode shape, 262
aluminum, 56, 85
aluminum plate, 63
aluminum square plate, 65
appropriate boundary functions, 30
arbitrary load, 74
attendant postulated, 98
axial coordinate, 90
axial load, 107
axial stress, 84
axisymmetric buckling, 215
axisymmetric constraint, 217
axisymmetric vibration, 251

balance equation, 221
bamboo, ix
benchmark case, 153
benchmarks, 249, 252
bending moment, 229
Bernoulli–Euler beam, 97, 98, 102
Bessel equation, 219, 248
Bessel functions, 219, 227
bones, ix
Boobnov-Galerkin method, 216, 217, 225, 227
bottom surface, 78
boundary conditions, xii, 25, 32, 55, 93, 94, 100, 105, 107, 109, 229, 245, 246
buckling, 1

buckling load, 123, 124, 216, 222
buckling mode, 215
bulk modulus, 18

Cartesian coordinate system, 4
cenosphere, 15
central deflection, 77, 79
ceramic material, xi
ceramic plate, 26
ceramic volume fraction, 26
changes in sign, 114
Chebyshev polynomials, 24, 30–32, 71, 75
chemical arrangement, ix
chemical vapor deposition, 15
circular isotropic plates, 241
circular plate, 58, 145, 155, 156, 215, 241
circular simply supported polar-orthotropic plate, 156
circumferential bending moment, 130
circumferential direction, 130
circumferential flexural rigidities, 130
circumferential rigidity, 163
clamped, 37
clamped plate, 32, 42–44, 46, 47, 51, 53, 56, 57, 59, 64, 66, 79, 223, 241
clamped rectangular plates, 24
clamped–clamped, 102
clamped–clamped beam, 92, 94, 105
clamped–clamped inhomogeneous polar orthotropic plates, 177
clamped–clamped orthotropic plate, 177
clamped–clamped polar orthotropic inhomogeneous circular plate, 192